T0260862

Crossing the Zambezi

The Politics of Landscape
on a Central African Frontier

Crossing the Zambezi

The Politics of Landscape on a Central African Frontier

JOANN MCGREGOR
Lecturer in Geography
University College, London

James Currey
Weaver Press

James Currey
www.jamescurrey.co.uk
is an imprint of Boydell & Brewer Ltd
PO Box 9, Woodbridge
Suffolk IP12 3DF, UK
www.boydell.co.uk

and of

Boydell & Brewer Inc.
668 Mt Hope Avenue
Rochester, NY 14620, USA
www.boydellandbrewer.com

Weaver Press
Box A1922
Avondale, Harare

First published 2009

1 2 3 4 5 13 12 11 10 09

British Library Cataloguing in Publication Data
McGregor, JoAnn. Crossing the Zambezi: the politics of a Central African frontier. 1. Zambezi River – History. 2. Zambezi River Valley – History. I. Title
967.9-dc22

ISBN 978-1-84701-402-3 (James Currey cloth)

ISBN 978-1-77922-077-6 (Weaver Press paper)

Typeset in 10.5/11.5 Monotype Ehrhardt
by forzalibro designs, Cape Town

Contents

List of Illustrations

Acknowledgements

This book began its life as a rather different project. Initially, I had planned collaborative research in a vast swathe of the Zambezi hinterland with Jocelyn Alexander and Terence Ranger, stimulated by interest in the layer of pre-Ndebele settlement we had begun to document in *Violence and Memory*. That research would have ended in a very different book, or perhaps series of books. My thanks to both for graciously allowing me to go ahead alone, and for the intellectual support and friendship they have offered throughout.

In financial terms, the research was made possible by a Fellowship from the University of Reading Research Endowment Trust, and a grant from the British Academy. My thanks to colleagues at the University of Reading for giving me a teaching load that allowed me to do the bulk of the writing, and to my more recent colleagues at UCL, where the manuscript has been completed.

Elizabeth Colson has been a source of encouragement and inspiration from the outset, and I owe her a particular debt. It has been wonderful to work in an area with such a rich history of research. The archive she has created through her writing provides an invaluable resource and wealth of insight for later scholars such as myself. My grateful thanks go to her for encouraging me to commission diaries from local people and for commenting on everything I have written. Among the UK scholars who have provided helpful criticism and feedback on chapters, I would like to mention William Beinart, Hugh MacMillan and Lyn Schumaker. Particular thanks to Terence Ranger, who also read the full manuscript.

In Zimbabwe, my thanks must go to the District Administrators and Council Executive Officers in Binga and Hwange who allowed me to do research at a difficult time, gave me access to district records, and upheld the clearance granted by the Research Council of Zimbabwe despite some opposition from 'war veterans'. The research would have been impossible without the help of my research assistants and translators, Andrew Chiumu Mudenda and Alexius Chipembere. I owe particular thanks to the ward councillors who provided introductions and shared their insights, the many other individuals who gave interviews and the Binga fishermen who kept diaries for me. In Hwange, Lawrence Chinyati and Noah Musimanga provided companionship and enthusiasm during memorable visits to the Nambya stone ruins and across the border to Livingstone; they and David Kwidini provided important feedback on drafts. Chiefs Nekatambe and Shana made me welcome and shared their insights into the Nambya past. The staff of the Binga Council Guest House and Hwange Colliery Guest House looked after me, but many others opened their homes, providing friendship, conversation and intellectual engagement, including Synod Mudenda, Herbert Sansole and David Kwidini. In Bulawayo, I had a home from home with Shari and John Eppel.

Regarding archival assistance, I would like to extend thanks to the staff of the Zimbabwe National Archives, the Bulawayo Records Office and to Friday Mufuzi for showing me relevant documents in the Livingstone Museum archives.

For the photographs and other images, I would like to thank David Kwidini (photo 7.1), Colleen Crawford Cousins and the Panos Institute for permission to use Colleen's photograph of a malende shrine (photo 2.3), and Dr Weinrich for permission to use her photograph of chief Siachilaba, the National Archives of Zimbabwe (photos 5.2, 5.4, 5.5, 5.6, 6.1, 6.4), Thomas Cook Archive (photos 5.1 and 5.3), Museum Africa in Johannesburg (for Baines' painting of chief Hwange), the Livingstone Museum (photo 5.7), Emil Schulthess Archive (photo 6.1), BaTonga Museum (photo 8.1), Elizabeth Colson and Manchester University Press (photo 6.5). I endeavoured to gain copyright clearance for permission to reproduce the 1950s dust jackets of *Operation Noah* (photo 6.3a) from William Kimber and Co. and *Kariba* (photo 6.3b) from Methuen who directed me to the source of the image on the original cover (the Federal Information Dept.) and the copyright holders of the first edition, author (Frank Clements) and Northumberland Press Ltd., which I could not locate. I would also like to thank Miles Tendi for helping me with references from the Bodleian at the last minute.

It is testimony to the importance attached to writing history, reversing discrimination and achieving development in both Binga and Hwange, that I have received so much input from local intellectuals, and testimony to the deepening crisis in Zimbabwe that some of them can no longer work towards these ends in Zimbabwe. The rapidity with which the exodus has gathered pace over the course of researching and writing this book means that the usual distinction between those who helped 'in the field' and those who supported 'at home' is inappropriate. Bernard Manyena, one of Binga's energetic activists and a council administrator, who helped me with initial contacts in Binga in 2001, surely did not imagine that he would still be helping seven years later, now as a member of the British academic community, sharing insights into Zimbabwean politics, discussing interview material and providing comments on chapters. His feedback has been invaluable.

Final thanks must go to my husband, Mark Leopold, who has helped in so many ways, and whose unfaltering enthusiasm for this particular project kept me going.

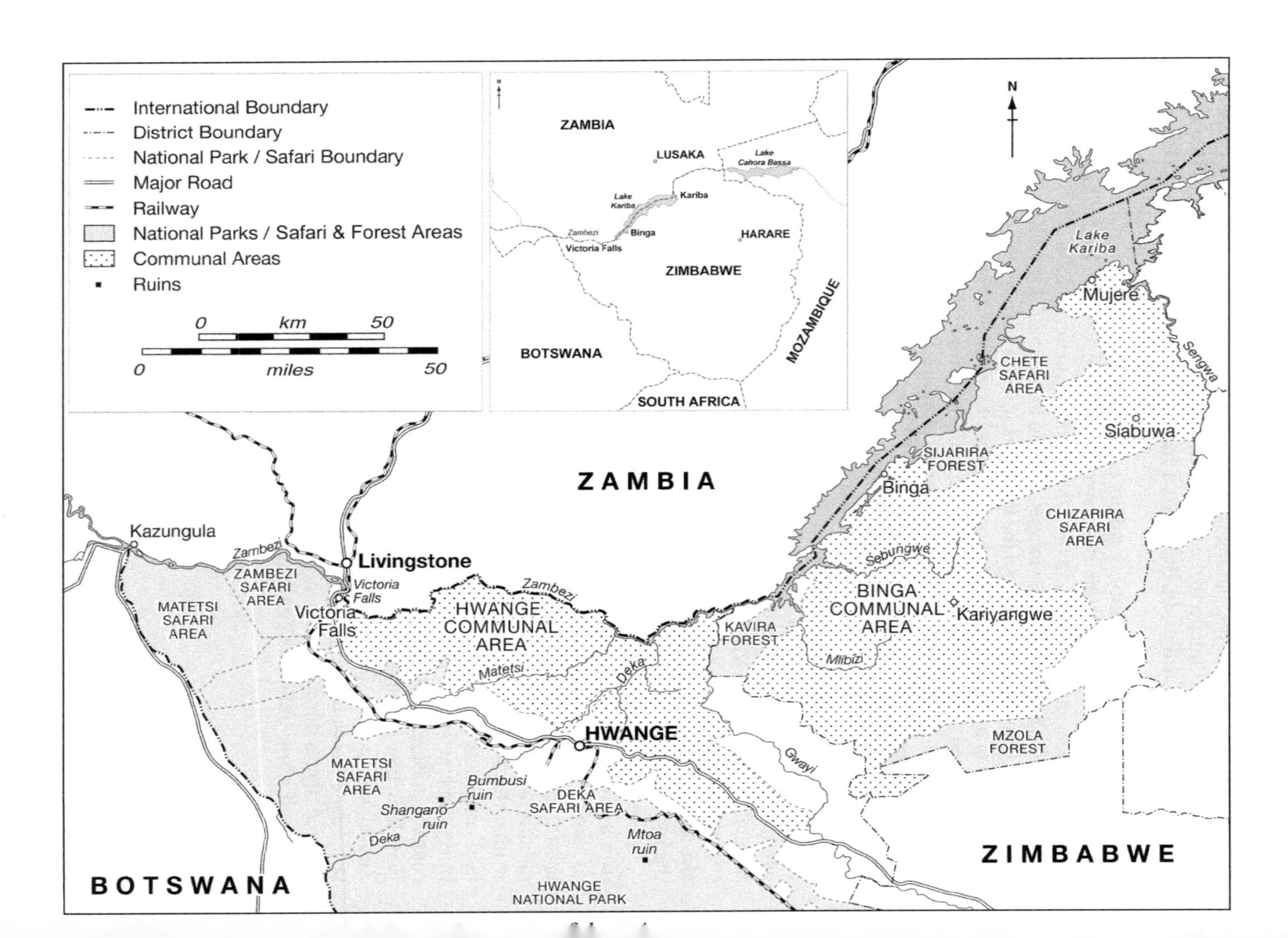
International Boundary
District Boundary
National Park / Safari Boundary
Major Road
Railway
National Parks / Safari & Forest Areas
Communal Areas
Ruins
0 km 50
0 miles 50
ZAMBIA
LUSAKA
Lake Cahora Bassa
Lake Kariba
Kariba
Zambezi
Binga
HARARE
Victoria Falls
ZIMBABWE
MOZAMBIQUE
BOTSWANA
SOUTH AFRICA
N
Lake Kariba
Mujere
Sengwa
CHETE SAFARI AREA
Siabuwa
SIJARIRA FOREST
Binga
ZAMBIA
CHIZARIRA SAFARI AREA
Sebungwe
Kazungula
Zambezi
Livingstone
ZAMBEZI SAFARI AREA
Victoria Falls
MATETSI SAFARI AREA
Victoria Falls
Zambezi
HWANGE COMMUNAL AREA
BINGA COMMUNAL AREA
Kariyangwe
KAVIRA FOREST
Mlibizi
Matetsi
Deka
HWANGE
MZOLA FOREST
Gwayi
MATETSI SAFARI AREA
Bumbusi ruin
DEKA SAFARI AREA
Shangano ruin
Deka
Mtoa ruin
ZIMBABWE
BOTSWANA
HWANGE NATIONAL PARK

1

Introduction
The Politics of Landscape on the Zambezi

In March 2001, a young fisherman on the shores of Lake Kariba in Zimbabwe told me 'I can't say where I learnt to fish, I grew up fishing. We are Tongas, people of the river… I was born on the river bank, I was born into this industry like my fathers before me. What I know is that I found myself fishing … when I started school I was already a fisherman. You don't have to teach a Tonga to fish'.[1] By making this claim, the young man implied long-standing intimacy with the landscape of the Zambezi that stretched back through his family line into the mists of time and evoked a natural connection between Tonga people, the waters and the necessary skills to secure a livelihood from them, as well as an inalienable and privileged right to work the resources of the lake. His claim was echoed by other Zambezi 'river people'. In Livingstone town in Zambia, Chief Mukuni publicized the special relationship between his Leya people and the famous waterfall at Victoria Falls. The chief invited tourists to appreciate the different aspects of the 'thundering mists', to understand that the 'mists of the dead' invoked Leya ancestors and to witness ritual river crossings to the island above the waterfall where Mukuni's ancestors had lived and commanded the Zambezi fords.[2] By so doing, Mukuni also claimed political, economic and cultural rights by invoking natural, ancient, enduring mystical relationships with the landscape for the Leya people through their ancestors.

In both cases, of course, the claims were strategic in their essentialism. They were deceptive in so far as they implied not only unchanging tradition but also relations with a stable landscape. The forefathers of the Tonga fisherman had been displaced by the Kariba dam, an ambitious hydro-electric project that transformed the landscape and ecology of the river beyond recognition, creating a vast man-made lake – the largest in the world at the time. Traditional skills for fishing in the Zambezi, based on knowledge of the river's currents and pools and the annual pattern of inundation and retreat, needed total revision to be of any use in exploiting the wide expanse of the new lake. At Victoria Falls, the landscape might have looked more stable, but it was also marked by the technological interventions that laid the basis for urban and industrial development in central southern Africa which dramatically altered the Zambezi's strategic and economic role. The giant steel bridge that carried the railway across the river at Victoria Falls in 1905, allowing for tourist development, created a new geography of connections, linking colonial mines and towns, funnelling migrant labourers southwards and creating competition for land in its vicinity. It made those who controlled crossing at the old Zambezi fords redundant, and Mukuni's forebears were among those evicted by the new developments around the waterfall.

These colonial infrastructural projects along the river transformed the landscape symbolically as well as materially, as the waterfall and lake became focal points in the developing myths of white settler identity and British colonial rule. The landscapes celebrated in these myths were new and different, and neither Leya ferrymen nor Tonga fishermen had any place in them. These were landscapes of natural beauty and wilderness, which celebrated British imperial heroes, where 'engineering wonders' were juxtaposed with their more famous 'natural' counterparts and provided tangible evidence of the developmental role of colonial occupation and white settlement. Victoria Falls invoked its 'discoverer' David Livingstone, but also Cecil Rhodes, through its location on Rhodes' fantasized 'Cape to Cairo' route, while the dam was later cast by Rhodesian settlers as the ultimate triumph of a long history of European struggle to harness the Zambezi. These white claims to the landscape were radically opposed to those of the Tonga fisherman and Chief Mukuni in that they justified different structures of authority and led to racialized dispossession. But in other ways they were remarkably similar: advocates of British imperial rule and settler politicians rallied landscape and ancestors to their political cause in the same way as the young fisherman and the chief. They too provided evidence for the validity of their claims through genealogies that bound landscape to identity and created a sense of attachment to place; they too used the landscape to naturalize their claims to power, making them appear self-evident and incontestable. But these are just some of the ways the Zambezian landscape has been claimed as it has been transformed, and just some of the ways in which successive claims have differed and converged over time. The forebears of the Tonga fisherman and Chief Mukuni would not have understood the ethnic prism of today's claims, nor would they have understood the equally important discourses of rights, citizenship and development. They had their own idiom for talking about the relationship between landscape and power, their own traditions of using and physically shaping it, and well-honed strategies for dealing with more powerful others who intervened in Zambezian claim-making long before the cast of European actors invoked in the myths of white Rhodesia.

This book is a history of claims to the Zambezi, focussing on the stretch of the river extending from the Victoria Falls downstream into Lake Kariba, which today constitutes the border between Zambia and Zimbabwe. It is a story of 150 years of conflict over the changing landscape of the river, in which the tension between the Zambezi 'river people' and more powerful others has been centrally important. The Zambezi is one of Africa's longest and most important rivers – securing access to its waters and control over its banks, traffic and commerce were crucial political priorities for leaders of pre-colonial states no less than their colonial and post-colonial successors. The book is about the ways in which the course and flow of the Zambezi have shaped history, the river's shifting role as link, barrier or conduit, the political, economic and cultural uses of the technological projects that have transformed the landscape, and their legacies in the conflicts of today. By investigating how the claims made today by Zambezi 'river people' relate to a longer history of claims and appropriations, the book contributes to long-standing debates over the relationship between geography and history, landscape and power. It offers fresh perspectives on the politics of landscape as revealed in the dynamics of a pre-colonial African frontier, the role and uses of colonial science, settler and African nationalist movements, and the politics of recognition in post-colonial contexts. The book provides a new case-study of the making of a colonial state border, and its use in times of peace and war, highlighting the shaping of cultures of colonial and post-colonial state power in borderlands, and the politics of border identities. The political, economic and cultural claims to the Zambezi that accompanied the physical transformation of the landscape highlight changing understandings of the natural world.

I have called the book *Crossing the Zambezi* to reflect the strategic and symbolic importance of knowing the water and how to cross it, for those who regarded themselves as 'river people' and who controlled its fords in the nineteenth century. They depended on the river's floodwaters for a second annual harvest, which made life possible in an otherwise arid and inhospitable environment. But the river was important beyond its role in the human ecology of survival. It could also confer strategic advantage in relations with the succession of centralized African polities based on the plateau to the south or on the open floodplains upstream to the north. Crossing implied many things, for the small, decentralized groups of 'river people' on the margins of these centralized African polities. First and foremost, crossing implied knowledge and skill; indeed the name Zambezi itself is said by Tonga intellectuals to derive from the phrase *kasambavesi*, 'crossing depends on knowledge'. The ability to cross distinguished 'river people' from others who could not do so; it transformed the broad, dangerous, fast-flowing river into a link when for others it was a barrier. Crossing allowed escape from pursuing warriors, who could be stranded on islands and opposite banks. Crossing thus conferred independence from the important regional powers that also made claims to the river, and sought to subordinate those who lived along it.

The interests of more powerful others in the mid-Zambezi mean that this history of claims has been shaped by political and economic interests and networks of ideas that stretch far beyond the immediate locality. The cast of actors in the book includes not only the 'river people' themselves, but the external powers with which they interacted. Thus the book revisits debates about pre-colonial African frontiers in relation to the expansionism of the Ndebele and Kololo/Lozi states towards this part of the Zambezi. It examines the political and ideological dimension of frontier relations, which involved both violent raids on, and stereotypes of, frontier peoples as dangerous and different. The book also sheds fresh light on European explorers' relationships to the societies they moved through, by examining how the Zambezi's most famous explorer – David Livingstone – reproduced local frontier discourses from his African interlocutors and reframed them in ways that led to further imperial interventions along the river. As British agents mapped the boundaries of the Rhodesias, the frontier status of the mid-Zambezi was reproduced through its designation as a 'natural border'. Subsequent interventions along the Zambezian frontier in the course of colonial rule were sporadic, associated with the monumental projects of the Victoria Falls bridge and Kariba dam, which fundamentally transformed the river's strategic role, altered its ecology, redirected its energy, and re-configured access and control. They also had the effect of reproducing old frontier ideas of the river people as different and backward.

The story told here examines these political uses of the projects along the river, investigating their role in legitimating colonial rule and white settlement. As the 'river people' were displaced and lost access to the Zambezi, their claims were not, however, silenced. Rather, the historical grievances generated through this transformation and loss fuelled resistance, contributing to African nationalist mobilization in a strategically important border region, as well as providing grounds for post-colonial mobilization around minority identities and rights, involving globalized networks and challenges to states and international development bodies.

At the time when the field research for this book was conducted, in 2000–2001, the mid-Zambezi river people were making their demands to an increasingly global audience. Their leaders had not only joined national opposition movements and petitioned state leaders, but also sought to influence and seek redress through international agencies. Leaders of the Zimbabwean Tonga, for example, sought redress through the Zambezi River Authority, the World Commission on Dams and the World Bank alongside various local and international NGOs interested in culture and rights. They instituted commemorations of their displacement, developed international exhibits of Tonga heritage with evocative titles such as 'The Tonga used to cross', founded a new Trust for 'the people of the great river' to claim reparations and development, and made common cause with victims of other large dam schemes around the world, drawing attention to their plight through hunger strikes outside the gates of the World Bank in Washington.[3]

These claims, framed in terms of indigeneity and rights, bear the imprint not only of the contemporary politics of belonging, but of the conflicts and claims of the past. They testify to a history of interaction and mutual influence, drawing on modern ideas of heritage and identity, as well as understandings of development, rights and citizenship. Few contemporary protagonists want to undo the controls over the riverine environment that have transformed it, though demands are shaped by new sensitivities to the social and ecological costs of large engineering projects. Their arguments are not anti-science or anti-modernization, but about access to development, accountability and the distribution of benefits from interventions. Given these convergences, the book aims to unravel the ways in which each layer of claim-making relates to its predecessor. Thus it considers European explorers' claims to scientific knowledge in the context of African ideas and political hierarchies in the places they wrote about, highlighting the influence and interests of African informants and intermediaries. Similarly, it discusses the mapping of the colonial state border along the river in relation to the ways in which pre-colonial African states related to their frontier regions. The book also examines how European claims to the landscape influenced African claims, through the development of ethnohistories and notions of heritage, ideas about attachments to the landscape and expectations of progress and development shaped through interactions between missionaries, colonial state agents and African intellectuals. The 'minority' groups themselves are partly a product of the drawing of the border; their strategic essentialisms and the content of their grievances are shaped by cultures of state power, histories of nationalism past and present, and more recent international networks of influence.

Any history of competing claims and ideas about the transformation of the natural world would be ungrounded if separated from the materiality of the river itself and the shifting political economy of its resources and the broader landscape. There are thus two storylines that structure this book. The first concerns a series of 'dramatic episodes' and their legacies, occasions on which claims to the landscape of the mid-Zambezi frontier were elaborated and the potential of the river was debated in centres of power away from the region itself.[4] Each of these episodes was related to changing structures of authority, and shifts in the river's strategic role as conduit, barrier or link; some involved technological interventions that transformed the landscape physically, redirecting the water's energy and altering the hydrology and ecology of the valley. The episodes include Livingstone's 'discovery' of the Falls in the context of his (and others') explorations of the Zambezi in the nineteenth century, the construction of the Victoria Falls bridge in 1905 and the Kariba dam in the late 1950s, and the legacies of these interventions as they shaped post-colonial conflicts over the waters of the Zambezi.

The second storyline is related, and revolves around the process and consequences of defining the river as a national border, the extension of state authority in the border region, and the ways in which border peoples and others have used this divide. The river's status as an international border has been persistently important, but has been particularly charged during periods of major intervention, technological or otherwise, and in contexts of war or economic crisis on one or other side. Although the border cut directly through the dense network of social ties between those who lived along the Zambezi's banks and regarded it as a link, the riverine boundary bore some relation to pre-colonial political and ideological hierarchies, imaginations and uses of the landscape on the part of the region's centralized African states. Despite the history of interchange across the Zambezi, the border is intriguing because it continues to matter, and because cultures of state power differ so markedly across it, despite this human traffic. At the time the research for this book was conducted, cross-border economic activity was rejuvenated and its dynamics were changing rapidly as the Zimbabwean economy was beginning to change gear and starting its downward plunge.

Rivers & History

By writing this book as a history of claims, I have cast the claim-makers as its primary actors. The flow of the river itself has, of course, actively shaped the contours of the landscape and will continue to do so, but it does not articulate claims or disputes. Nor do the animals and fish that live in the river. By relegating such a powerful force as the river and such evidently dynamic occupants to the status of being fought over, while giving elevated consideration to the ideas and political interests of the humans doing the fighting, I am writing against much recent interest in the agency of the non-human world, which has involved various means of trying to equalize human and non-human actors and to blur the categorical boundaries that separate them. Yet, in the period I consider here, the impact of human agency has been disproportionate, and it is important to understand the ideas that drove these interventions. David Blackbourn argues that the metaphor of 'conquest' over nature was central to the making of modern German landscapes, connected to very different political philosophies, and linked to episodes of state expansion, conflict and war.[5] Yet non-representational conceptions of power (used to incorporate non-human actors) can result in depoliticized and ahistorical narratives and are thus inappropriate for this history of claims to the mid-Zambezian frontier.[6] In the context of colonial Central Southern Africa, the notion of a European struggle to achieve mastery over nature was also potent, particularly in relation to frontier regions. Linked to episodes of expansion over people deemed primitive, conquering the environment involved violence and coercion, but was far from the moral equivalent of war.[7] Rather, environmental conquest provided a moral justification for colonial rule, and played a central part in ideas about race and the legitimation of white privilege. Understanding the contemporary ideas driving the changes that so radically altered the landscape does not mean abandoning a concern with the materiality of water and its political effects, or the ways these are altered through interventions.

Some of the human claim-makers who feature in the story have, however, imputed social agency and supernatural powers both to the river and its occupants, particularly crocodiles and a magical river snake. In some parts of the narrative, such as in discussion of fishing in Lake Kariba today or stories of past relations with the river, the reader gets a glimpse of the ways in which daily bodily interactions with the water, wild animals and fish, can create contexts for ideas about the environment that reflect a sense of reciprocity with it, and entangle powerful features of the natural world, ancestors and other spiritual or hidden forces in social life. When discussing ideas about interactions between such 'assemblages' of powerful forces, I try to reflect the perspectives, discourses and actions of the human actors involved in these networks (in so far as language allows), and retain an emphasis on power and politics.[8] In the fishing camps that border Lake Kariba, livelihoods were risky and precarious, and the struggle to make a living increasingly involved 'poaching' in restricted waters, or a stake in cross-border trades, that necessitated daily navigation of a network of adversaries – national park guards, poachers, thieves, crocodiles, hippos, weeds and witches. Though state officials were only a minor component of this assemblage, state power was thoroughly implicated in the structure of risk and relationships on the lake. Indeed, the mythologized figure of the crocodile, which featured so prominently in fishermen's stories of the dangers of the water and difficulties of fishing could stand as a metaphor for the state itself, embodying both state conservationist priorities and fishermen's marginality, enmeshed with older ideas of power. These politicized interpretations of the structures of power governing the lake, which protected life-threatening wildlife at fishermen's expense, have had the effect of heightening demands for development, human rights and the eradication of animal pests, as well as fuelling political opposition movements.[9] But for many of the claim-makers and sources in this book, the discomforts and exposure to harm involved in bodily interactions and dependencies on the environment are memory or romantic myth.

Indeed, some of the most well-known myths of the river today gained their prominence through the pens of those who did not know the Zambezi in this intimate way, from nineteenth-century European travel-writers to more recent popular authors, journalists and tour operators. Stories of river journeys or of the harnessing of the Zambezi through the Kariba dam have commonly involved personifying the river, and investing it with supernatural force, represented by a river god, Nyaminyami. Indeed, the white writers whose river stories pitted European science against the forces of a primeval river god appeared to need myths of a powerful, enchanted natural world more than those who lived with or were displaced from the river. The latter opposed their displacement with a range of pragmatic, economic and African nationalist arguments, and turned to ancestral rather than river spirits for support.

Much of the recent academic and popular interest in rivers is reflected and shaped by an interest in the relationships between different parts of entire catchments, created by the flow of water through the land. Basin-wide perspectives have given important insights into relations up and downstream, and between rivers and their hinterlands as these are shaped and transformed by particular interventions. By focussing on simply one stretch of a huge river, many such interactions and connections are lost from sight in this study, important though they are. There are several reasons why I have chosen not to write about the whole length of the Zambezi. Not only would doing so require an immense scope of research, reading and insight, but constructing a narrative encapsulating the different and shifting connections over such a wide area would excise marginalized groups and voices, such as many of those who feature prominently in this book. Some of the more popular means by which authors have constructed narratives for entire river networks, such as through 'river biographies' or 'river journeys' risk reproducing imperial forms of writing by personifying the river or narrating journeys through different 'peoples' and landscapes passed through from source to mouth, feeding static, essentialized notions of the connection between people and territory, and the exoticized panoramas of indigenous culture so popular today in the tourist and heritage industries. In a southern African context, such books have involved representations of local people juxtaposed with images of animals that go well beyond verging on the offensive.[10] They form a genre I hope to subvert here.

This book also differs from treatments of rivers and history that narrate the development of 'hydraulic societies' through a sequence of irrigation technologies.[11] Such studies have provided illuminating correctives to political histories ungrounded in changing material relations with environment and technology, but, as McKittrick has argued, can be narrow in so far as they imply that societies without irrigation are 'less profoundly shaped by water and its scarcity'.[12] Although the damming of the Zambezi transformed the landscape, large-scale irrigation was not one of the by-products of the hydroelectric project – indeed this is one of the grievances of those displaced by the dam, who continue to suffer acute water scarcity despite living in proximity to a huge man-made lake.[13] This book also tries to avoid the 'optimistic' and 'pessimistic' narrative frames that characterize many water histories. The pessimistic narrative which has perhaps predominated in recent years, has been critical of both the ecological and social effects of the human domination of nature and the political systems that have achieved this. Thus, for Haarden, the Columbia river is 'dead', it 'does not flow, it is operated', and for Worster, the power of the dry American West is appropriately symbolized by the modern irrigation ditch.[14] Yet as Blackbourn argues, both optimists and pessimists may eliminate complexity and obscure the views of contemporary actors: 'both have a tendency, from their different perspectives, to make the process of change seem too straightforward, too smooth and self-evident'.[15] Both types of narrative, but particularly the pessimistic one, have also tended to cast the period before interventions began as one of 'harmony' between nature and society subsequently destroyed. Some of the claim-makers in this book also look back to life along the river through such rose-tinted lenses. But life along this part of the Zambezi before the changes wrought by large-scale technological developments was shaped not only by the human ecology of survival in an arid climate where access to the river's waters was all-important, but also by the external and internalized violence of a contested frontier.

The Politics of Landscape

The two storylines in this book both investigate the politics of landscape ideas. I use the concept of landscape because it facilitates an approach that combines the realm of ideas with material relations, as the term has a double meaning and implies both cultural construction and physical reality. In African studies, scholars of 'land' have long noted that the term can imply an unduly fixed and solid set of properties. [16] Embracing 'landscape' has been one means of destabilizing and politicizing claims to knowledge and control.[17] Although there are many traditions of writing about landscape, much recent work has been produced under the influence of poststructuralist ideas, giving it a rather different orientation and flavour than older 'environmental history' writing.[18]

This is not the place to review the content of landscape studies,[19] but it is worth spelling out some of the criticisms of this work, partly because I hope to respond to them here. Despite the potential of the idea of landscape to combine ideas, physical materialities and political economy, in some studies 'imagined landscapes' drift unconstrained by the world outside the mind of the observer, or appear pre-determined or even transcendental. Donald Moore argues that some recent anthropological work on landscape has 'run the risk of escaping ethnographic and analytical precision', especially when 'radically unmoored from spatial practices'.[20] This criticism of an excessive focus on culture has been levelled in particular at writing on 'landscape and memory'. Kerwin Klein contends that this literature can reinforce dangerous ethno-racial nationalisms by elevating identity, replacing critical analysis with vague invocations of mystical relationships with place and a collective unconscious, and suppressing narrative by concentrating on the power of visual or bodily experience.[21]

Such criticism has been made of Pierre Nora's notion of 'sites of memory', developed in relation to popular nationalism in France.[22] For Nora, these evocative sites were imagined rather than physical places 'where memory crystallizes and secretes itself' (though they were often elaborated in relation to material landscapes, buildings or other artifacts).[23] Building on Halbwachs' emphasis on the social framing of memory, Nora's work usefully highlighted how identity can be rooted and reproduced through such imagined places that 'anchor, condense and express' collective memory.[24] Yet it was also criticized for reinforcing the exclusivity and nostalgia it aimed to deconstruct, for depoliticizing 'memory' and tradition, for suppressing narrative and casting 'sites of memory' as relics of more authentic connections with the past eclipsed by the rise of modern historical consciousness. In so doing it implied an unhelpful dualism between memory (understood as unthinking, embodied and spontaneous) and history (considered as rational knowledge).[25]

These criticisms of the scholarship on landscape are not intrinsic to the use of the term. By considering narratives, images and experiences of place in the contexts in which they are produced, it is possible to avoid contributing to an essentialized phenomenology of landscape.[26] Indeed, scholars have used the idea of landscape precisely to illuminate its 'duplicity', drawing attention to its capacity to naturalize and deceive by concealing both unequal relations of power and the historical investments of human labour that have physically moulded it.[27] The European history of the term itself illustrates the important relationship with changing structures of authority: Cosgrove links the rise of a landscape aesthetic in renaissance Italy to the spread of property relations, and the extension of structures of legal control from urban centres over new rural frontiers.[28] In the British context, too, the development of elitist landscape 'ways of seeing' was associated with new definitions of property rights and acts of enclosure.[29] The idea of landscape has been embraced as a means of historicizing human relations with the environment and, conversely, 'locating' history and deconstructing nationalist narratives, enabling the dynamics of particular localities to be considered in the context of broader networks and flows that have helped shaped them.[30]

In African studies, the rubric of 'landscape' initially fostered an unbalanced corpus of work, which was sophisticated in its treatment of metropolitan, settler and colonial ideas in relation to processes of state and nation-building, but paid too little attention to African ideas.[31] The imbalance was exacerbated by an initial reluctance to use the term 'landscape' in relation to traditional African concepts, on the grounds that the humanized, practical and sacred values they incorporate were embedded in day-to-day life and did not privilege the visual or stem from comparable understandings of property, perspective and aesthetics.[32] This controversy over whether or not Africans traditionally responded to the environment in ways that could be considered comparable to western landscape traditions of seeing became rather circular (and echoed parallel debates over whether or not Africans had a tradition of 'narrative').[33] Much depended on how landscape was being defined, whether loosely as a cultural construction, or by more strictly defined criteria.

Now, however, there is a growing body of work on landscape that has combined European and African politics and perspectives, and cultural and material struggles. In the Zimbabwean context, Ranger's history of the Matopos hills traces contestation over a site that was an important symbolic focus for both white Rhodesian and black African nationalisms, while Fontein analyses the 'power of heritage' at another centre of pre-colonial political and religious power – Great Zimbabwe.[34] But sites of pre-colonial authority have not been the only focus of landscape studies; other scholars have found landscape a useful means of approaching conflicts over margins and borderlands.[35] Frontier contexts have proved fertile ground for research on the politics of landscape because of contestation over material and conceptual boundaries at places where state, nation and local identities are supposed to converge.[36]

This study of claims to the mid-Zambezi frontier extends this body of work. It is concerned with the power relations, the investments of work and technology and succession of disputes that have created the frontier landscape both physically and imaginatively. I have retained Nora's term 'sites of memory' in this book, as it usefully highlights the power and emotional charge of landscape in its capacity to anchor and naturalize collective identity. I hope to have avoided the criticism levelled at Nora, by highlighting contestation, counter-narratives and alternative sites, and by emphasizing the historicity and politics of traditional no less than modern forms of relating to the past. The various aspects of the Zambezi that have been created as 'sites of memory' at different historical junctures by different actors have always been contested and subject to multiple interpretations. The study has a different orientation from accounts of conflict over sites of pre-colonial political authority, whose subsequent importance derives in part from their role as traditional religious centres and their ability to provide a source of legitimacy independent of the colonial or post-colonial state. This is because the importance of the river is not located primarily in ideas about the sacred, but rather in its material qualities and strategic role, and its status as frontier.

The Zambezi was not considered a sacred place by those who occupied its margins, perhaps, as Elizabeth Colson suggests, because it was so important in everyday life to those who lived along it.[37] It did have sacred sites within it: both the river and its margins were made familiar by multiple ancestral 'land shrines' and 'places of power' in Colson's terminology. But these sites were only known and used as such by those living in the immediate vicinity and did not draw in pilgrims from further afield.[38]

Given the violence and instability of life on the margins of expanding pre-colonial states, these sacred places validated decentralized political structures and tended to eclipse the specific content and depth of the past, invoking continuity in vague and inclusive ways to mask impermanence, fragmentation or slave origins. The encroachment of state authority and competition over resources linked to the monumental interventions of the colonial period provided further disruptions. Indeed, although some such sites have come to act as 'sites of memory' for those displaced by the Kariba dam whose old places of power were submerged, the rupture of the displacement itself arguably came to act as the prime memorial site. Nor has the marginality of those along the river been reversed in a post-colonial context. These relations of power are centrally important to the politics of landscape on the Zambezi, as on-going water scarcity, a governance regime that gives privileged place to animals and tourists, and the lack of state accountability to politically insignificant 'minorities' have reproduced long-standing processes of marginalization.

The River as Frontier

The story of the creation of the mid-Zambezi as a frontier does not begin with the drawing of the colonial border. Rather, the mid-Zambezi has a deeper history of being a frontier in the sense implied by Kopytoff in relation to pre-colonial African state formation. For Kopytoff, the term frontier conflated the idea of a region with a boundary, and actual relations of power over space with imaginative constructions of marginal places.[39] As Worby elaborates, a frontier can be seen as a 'limit … just beyond the edge of what is under political and representational control. Alternatively, a frontier can be imagined as a two-dimensional space – a zone – in which the state is undergoing consolidation and, more fundamentally, being "made". The frontier in this sense constitutes both an obstacle to, and an opportunity for, an expanding political centre. It is perceived as having a liminal utility … fraught with temptation as much as with danger.'[40] But frontiers can also be revealing in times of economic decline and state contraction, such as in the context of post-colonial neo-liberal reforms, when bureaucratic authority is characteristically undermined, informalization proceeds apace, and new irregular economic networks cluster round the opportunities created by borders. Indeed, Das and Poole argue that borders can offer particular insight into the changing nature of post-colonial state power.[41]

This book elaborates how each layer of expansionist claim-making over the mid-Zambezi has been shaped by the idea of the frontier as disorderly spatial periphery. Yet it also follows the story of the border through times of state retreat, including the beginnings of the recent period of economic austerity, political crisis and hyperinflation in Zimbabwe. Rather than invoking misleading metaphors of 'state collapse' or romanticizing informal economies as autonomous spaces of resistance, I follow Janet Roitman in analyzing the new 'illegal yet licit' networks clustered around the resources of the lake and Zambezian border as infused with state power, indicative not of its failure but of its recrafting.[42] As Das and Poole argue, such reconstituted forms of the state can be notable for their 'illegibility', yet they are nonetheless enmeshed with the bureaucratic forms of statecraft they act to circumvent – thus border authorities, checkpoints and documents retain their importance, even if checks are made only to extract rents, or the papers themselves falsified.[43] At the time of the field research for this book, the reach of centrally important arms of the state was being extended and brought firmly under central control through a process of politicization and militarization, while the burgeoning unregulated economic networks also involved state agents. Moreover, the legacies of technocratic intervention were still very much present, and the legal-bureaucratic framework governing the lake and border continued to matter and continued to structure the risks of poaching or illegal border-crossing.

Though elaborated in relation to transformations of post-colonial state power, Das and Poole's idea that state authority at the margins can appear notably 'illegible' is also apposite for other historical junctures, because the mobility of the frontier, the contested nature of authority, and the opportunities the border provides for playing off two different state powers, are persistent themes. In an early colonial context, when new state structures of control were being erected, the border region was characterised by deliberate obfuscations on the part of those embodying state authorities (or claiming to do so) and those supposedly disciplined by them. Local state structures at that time also looked weak and partial, but the idea of the state and law, backed up by superior force, gradually took hold and ramified through social relationships, even in the remote Zambezian periphery, long before the Kariba dam.

Given the notable heterogeneity of Africa's state borders, the story told here is that of one particular border, drawn along a river through a frontier region. Yet the prevalence of rivers as borders across the continent hints at the need for broader investigation of their role and relationship to power in pre-colonial contexts.[44] At one level, the story told here reinforces the body of scholarship critical of the view that partition and colonial border-drawing were arbitrary European decisions unrelated to older political realities on the ground. As Nugent argues, 'Colonial boundary formation owed more than is commonly thought to historic patterns of trade and politics'; although the ultimate aim was incorporation, colonial 'border policy itself evolved incrementally' in response to a series of 'practical dilemmas'. [45] European claims to influence commonly hinged on African rulers' own (often wildly exaggerated) claims, emanating from geographical centres of power that had not infrequently seen the rise and fall of a sequence of different states; likewise there could be great historical continuity in the marginality of regions regarded as borderlands, often reflected in what are the border regions of today.[46]

Yet I also hope to show that the imaginative as well as the material dimension of colonial and post-colonial frontiers owed more than is commonly thought to the ideological hierarchies of pre-colonial Africa, as these have been incorporated into subsequent discourses. Colonial images of frontier violence and developmental potential, danger and allure, could gain support from the historical dynamics and discourses of internal African frontiers. The stories of crossing that encapsulate the liminal status of the river peoples and reflect geographical relations of marginality with powerful others developed during historical episodes of expansionism into, and contestation over, the mid-Zambezi frontier. They were reproduced in at least two such episodes: in the 1850s and 1860s when Kololo and Ndebele leaders competed for influence over their mutual frontier (during Livingstone's visits) and in the 1880s and 1890s when Lozi and Ndebele polities did likewise (at a time when European powers also competed to demarcate spheres of influence for African client states and themselves). By highlighting this pre-colonial dynamic, I do not wish to imply that the 'cause' of post-colonial marginality lies in pre-colonial relations of power, but rather to highlight both continuity and change: as Leopold has argued, 'historical "layers"... each contain traces from earlier periods and leave their own traces in the later temporal strata (including those encapsulated in the present).'[47] Unlike the northern Ugandan case discussed by Leopold, the pre-colonial frontier images of the mid-Zambezi river people as violent have not been reproduced over time. Rather, stereotypes of violence were replaced in the course of colonial rule by less threatening notions of the Zambezi frontier as exceptionally primitive.

The fact that the mid-Zambezi had been imagined and used as a frontier by powerful African states in the region does not mean that its designation as a colonial state border 'fits' pre-colonial arrangements (though that was part of its justification), as there was more than simply one way of using the river and 'seeing' the landscape. At the same time as leaders of pre-colonial states treated the mid-Zambezi as a barrier and frontier, those who lived along it saw it as a link between opposite banks. No single territorial limit could capture the complexity of frontier politics and alliances, as pre-colonial states were not delimited by fixed borders. The colonial state border introduced a new concept of territorially bounded power, as the legal basis of two separate state jurisdictions, which brought about far-reaching change.

Once demarcated, state borders took on a life of their own, as the differential rights and authorities they defined were exploited by border people, refugees, traders, criminals and others. In this sense, and as Nugent argues, the borders were 'made' by the actions of those in their vicinity, however marginal to central power, and however irrelevant the border to pre-existing political and environmental contexts.[48] Border peoples also 'made' the border in the sense that they filled the roles of colonial and post-colonial state agents on either side, and were both influenced by and contributed to the development of distinctive cultures of state power in their respective states.[49] In the Zambezian context, this divergence became increasingly acute after the end of British South Africa Company rule in 1923, as Northern Rhodesia (ruled from the Colonial Office) turned to implement indirect rule, while the settler colony of Southern Rhodesia continued to rely on more direct forms of administration.[50]

The very different character of the two states produced contrasting histories and cultures of nationalism, which also contributed in important ways to the making of the border. In the mid-Zambezi, the consequences of the Kariba dam were particularly significant in this regard: early resistance to the dam was mobilized by Northern Rhodesian nationalists in disregard of the state border but, following the displacement, each side of the valley was drawn into the respective territorial nationalist movements of two separate states. This developing national consciousness on the part of the resettled was deepened by the increased difficulties of crossing, as communities were now separated by a huge lake and the border was closed during the break-up of the Central African Federation in 1963, and as Southern Rhodesia moved to the illegal Unilateral Declaration of Independence (UDI) and war. In this context, local politicians south of the border used the history of the displacement to institutionalize the Zimbabwean African People's Union (Zapu) and organized to host guerrillas from their armed wing, the Zimbabwe People's Revolutionary Army (Zipra). As incursions began from rear bases in Zambia, so knowledge of the river and crossing gained a new significance, and a new generation of frontier discourses was created by Rhodesian soldiers, state agents, and African guerrillas.

At the same time as border peoples were becoming more conscious of their respective national identities, ethnic categories and meanings were also being consolidated. This was a process in which the position of the border mattered profoundly. In Zambia, the Tonga bloc was one of the larger ethnicities in the eyes of the colonial state, was politically important in relation to the development of nationalism and remained influential in post-colonial politics. On the south bank, in contrast, the Tonga were marginal to the country's ethnic space, which came to be dominated by the two 'supertribes' of Shona and Ndebele.[51] Their already small numbers were further split between administrative districts, they were discriminated against for speaking a minority language and were the object of derogatory caricatures of backwardness.

The story of the border told here examines the history of activism against this discrimination in colonial and post-colonial contexts. The 'politics of recognition' on the part of minority groups has been the topic of much debate recently, with authors attributing its renewed prominence to the effects of neo-liberal reforms on the cohesion of the post-colonial state and the influence of international NGO networks promoting indigeneity and rights.[52] Englund and Nyamnjoh see the political reforms of the 1990s as ushering in 'a torrent of identity politics in the public sphere'.[53] This literature has a polarized tone, with critics warning of the balkanization of the continent, casting these movements as neo-patrimonial or as a challenge to state command over natural resources,[54] while advocates celebrate their potential to provoke broader national debate over citizenship and rights.[55] My narrative does not assume an intrinsic value to the politics of recognition that is either good or bad, or that can be divorced from particular contexts. Rather, the sentiment of local disadvantage is available, potentially supporting an inward-looking schismatic politics, but equally amenable to an outward-looking, cosmopolitan and inclusive politics.

Though flourishing in a neo-liberal era, these movements have deeper roots. In the case of Tonga and Nambya activism in Zimbabwe, they originated in the 1950s and 1960s, when expanding mission networks began to incorporate north-west Zimbabwe, belatedly producing an educated, Christianised local elite. The notions of culture developed at that time have marked similarities with those promoted today through global discourses on multiculturalism and the tourism and heritage industries. The latter have had a particular impact on the Zambezi borderlands considered here, as the river and lake have continued to be prime national tourist destinations, and the Victoria Falls has international repute as a World Heritage Site. Both industries have encouraged a particular language and conception of culture and the past, as static and bounded, tangible, threatened and as tradable commodities. These global interests in heritage have acted through, and in collaboration with, governments, creating funds and opportunities for marketing national culture, and providing the context for local actors to participate.[56] Notwithstanding the constraints, these industries create opportunities at local level,[57] and the story told here explores how the river's role as tourist destination and heritage site has also shaped the politics of recognition in the borderlands.

No story of the making of the border could be complete without consideration of its role as an economic resource, heightened at different times as the economic trajectories of the two states have diverged, creating new possibilities for wealth at times when money is losing value and opportunities elsewhere contracting, and providing sites where new collusions are elaborated and can give insight into the ways state power is reconfigured and expressed. Initially, following Zambia's economic plunge and the re-opening of the border at the end of Zimbabwe's liberation war, new border networks were developed and embodied by Zambian traders and opportunists who were labelled a threat in Zimbabwean state discourse. This also affected local border identities and stereotypes such that in Zimbabwean villages abutting the border at the time of the research, the label 'Zambian' was often interchangeable with 'poacher', 'thief' or 'violent'. Indeed, as Mugabe was awarding medals to the Zimbabwe Defence Forces leaders who finally finished their clearance of landmines laid along the Zambezi between Victoria Falls and Mlibizi in late 2005, local chiefs complained of an immediate surge in armed cross-border cattle and livestock rustling, which was blamed on Zambians acting in collusion with unspecified individuals said to know the location of old arms caches from the liberation war.[58] But after 2000 any sense of Zimbabwean superiority over Zambians faded rapidly in the context of unfolding political conflict and economic crisis in Zimbabwe, as violent state repression and hyperinflation was combined with punishment, withdrawal of food aid and deliberate neglect of those in the mid-Zambezi borderlands, whose activism for civic rights and alliance with the new political opposition had gained them the reputation of disloyalty to the ruling party. Efforts to reverse a history of marginality produced by a sequence of interventions along the river were silenced and, for many, the struggle for survival and food took over. As elevated transport costs began to disconnect the Zambezi borderlands from Zimbabwean economic centres, so Zimbabweans turned in increasing numbers to the informal cross-border trades they had so recently condemned, heightening old grievances about access to the lake and marginality, and underlining the relevance of old ideas about privileged relations to the river and the importance of skills to cross.

Episodes of Intervention

The three main episodes of outside intervention in the Zambezi frontier that shape the narrative in this book are separated by fifty year periods and provide a useful means of historicizing colonial frontier discourses.[59] There were, of course, continuities in the ways that contemporary colonial actors in each episode used each project to justify colonial rule, white settlement and racialized resource access, through arguments about the developmental role of superior European science and technology, in counterpoint to the backwardness and unruliness of the frontier. Yet, as Stoler points out, 'the quality and intensity of racism vary enormously in different colonial contexts and at different moments', and as Thomas highlights, the justification for successive colonial projects 'reformulated and revaluated prior discourses'.[60]

It is important to understand discourses of race in the Rhodesias partly because the legacies of racism and racialized dispossession in a post-colonial context have arguably not been fully appreciated.[61] But these need to be considered in terms of contemporary controversy. Focussing on the way landscapes and colonial projects 'work' to promote dominant meanings and efface others can obscure debate and criticism, assume rather than investigate reception, and ignore marginalized views. In relation to the Kariba dam, for example, David Hughes has drawn on popular conservationist writing about the transformed landscape, arguing that it 'performed a white cultural agenda' that 'answered to a deep European longing for water'; yet his account ignores the volatile politics of the Central African Federation in which the dam was embroiled, tends to essentialize white settlers and renders the displaced Tonga 'invisible', such that heated controversy among settlers, colonial officers and other commentators, as well as the range of African views, are excluded from his interpretation.[62] Yet, as Beinart, Tilley, Mackenzie, Schumaker and others have argued, it is important not to overlook the 'counternarratives' often developed by scientists and social scientists who interacted closely with Africans, tried to understand African ideas and practices, and whose relationship to colonial states often involved opposition and criticism.[63] Even if these individuals had little influence at the time, over the long term and in post-colonial contexts, their writings and networks have proved influential.

The first project I analyse is Livingstone's investigation of the river. My account contributes to the historiography of exploration by discussing his views and writings on the mid-Zambezi in the context of contemporary African understandings and uses of the river, the role of his African interlocutors, and the impact of his writings on subsequent interventions.[64] From the time he first saw the mid-Zambezi, through to his 'discovery' of the Falls and subsequent movements up and downriver, Livingstone constantly moved with an entourage of Kololo guides and interpreters, with whom he could communicate directly; I argue that he often saw through Kololo eyes. Livingstone's ethnographic insights incorporated aspects of indigenous mid-Zambezian African political and moral hierarchies, reproducing them from a Kololo viewpoint, which involved discourses of difference and notions of superiority over the 'river peoples' of their frontiers, whom they had only recently subordinated, feared, and looked down upon as savage. Livingstone abstracted the insights he had gained from his Kololo guides, separated them from the relations of power they reflected and incorporated them into his own project to make it more credible. The existing scholarship has contributed to our understanding of this process of reframing, by highlighting the representational devices that pointed towards or justified imperial intervention, as Africans were represented through contemporary European ideas about race and class, and the landscape was recast in terms of natural history or aestheticized by drawing on strands of romantic thought that privileged travellers' individual, visual responses.[65] Further understanding of the ways these European representations both incorporated aspects of local discourse and were used to justify specific subsequent Zambezian interventions can enrich the literature on empire, exploration and mapping, which has paid insufficient attention to the history, politics and ideas of the places being described.[66]

Half a century later, the mid-Zambezi landscape was again the object of outside intervention, with the construction of the Victoria Falls bridge in the immediate wake of colonial occupation of the Rhodesias. The bridge's symbolic role was as important as its effects in re-shaping geographical connections. As the highest bridge of its kind in the world, it was the object of a triumphalist discourse, that envisaged 'a new Chicago' on the banks of the Zambezi, and its engineers were lauded for their role in the taming of the central African environment. The new tourist resort at the waterfall popularized the myth of Livingstone more widely than perhaps ever before, as well as providing a site where it could be linked to the developing cult of Rhodes. The landscape was flexible in the ways it could foster different strands of white identity, at once celebrating British imperial expansion and providing a genealogy for white settlement.[67] The triumphal rhetoric surrounding this episode masked the violence of colonial extraction and related upheavals, and totally (if temporarily) effaced local African histories of settlement and uses of the site as a sacred place, though it gave symbolic recognition to Lozi royalty, reflecting their elevated status in the then Protectorate of NW Rhodesia, and the elite ornamentalism of cultures of British imperial authority.[68] My discussion of this episode does not lose sight of local African experiences, as shaped by the violence of regimes of tax and labour extraction and competition for land. Nor does it lose sight of local African discourses, which invoked continuities with the violence of the preceding era through the idiom of slavery.

By the time of the third intervention – the Kariba dam – the two Rhodesias were joined with Nyasaland in the short-lived and controversial Central African Federation, which had been opposed by African nationalists from the outset. As the Federation's flagship project and the basis for industrialization in the region, the dam was also embroiled in controversy. The book revisits the dispute over this episode, looking at the different ways in which the story of the dam was narrated at the time and subsequently, not only by Federal politicians, Southern Rhodesian white supremacists, popular writers and journalists, but also by social scientists concerned about the effects on those displaced by the lake, and by African nationalists and displaced Tonga communities themselves. The dominant discourse of the dam as the triumph of a history of European endeavour to tame the Zambezi, and as evidence of 'white man's genius', is important. It justified turning the Southern Rhodesian shores into a white playground devoted to conservation and tourism, and allowed settlers to monopolize the new multi-million dollar fishing industry that grew up around the lake. At the same time, it turned the Tonga into an icon of the primitive and justified their exclusion. The counter discourses of the dam, narrated through the detailed studies of the displacement by Elizabeth Colson and Thayer Scudder, did not influence the resettlement at the time, but were influential over the long term, as they helped shape international policy towards dams, and 'development-induced displacement' more generally, as well as influencing developments in the Zambezi valley on the Zambian side of the border. The final narrative of the dam and displacement I consider is the version of the story told among the villages of the displaced Tonga themselves, and the ends to which it has been used.

The legacies of these three periods of intervention have shaped the contestation and claims of the post-colonial period discussed in the later sections of the book. Thus Livingstone's legacy has been hotly disputed in the local politics of the Victoria Falls region – with different actors claiming his memory or trying to excise it from their history – and the commoditized landscapes of the tourist industry. The legacies of the dam are also unsettled, and the grievances of the displaced have become louder and gained support over time, from NGOs, international actors and Zambian (though not Zimbabwean) policy makers. By exploring the contemporary and retrospective use of the stories of these different episodes, the book aims to trace the successive layers of claim-making, showing how each layer relates to its predecessor, entangling sedimented meanings in new contexts.

Sources & Methods

This history of claims to the mid-Zambezi draws on oral and archival research. The fieldwork focussed predominantly on the districts of Binga and Hwange in Zimbabwe, though some interviews were also held in the vicinity of Livingstone town in Zambia. Because the book tells the history of a particular feature of the landscape, different groups of people come in and out of my narrative, as they have made claims to and interacted with the river, or ceased to do so.

I conducted my interviews between January 2000 and September 2001, at a time when political violence in Zimbabwe was mounting, and the economy was in free fall. This was a difficult context for undertaking field research and making new contacts, and precluded greater local collaboration. It had a particularly negative impact on the research in Binga, which in the June 2000 parliamentary elections had returned the strongest support for the new political opposition party, the Movement for Democratic Change (MDC) of any rural district in the country. At the time I was trying to make my initial contacts, the district was being punished for its opposition support, and local leaders were being dismissed from their jobs, beaten and harassed by Zanu(PF)-supporting war veterans, receiving death threats from officials of the Central Intelligence Organisation (CIO) and being brought before ruling party dominated 'tribunals'.[69] The politically charged atmosphere made meeting people and interviewing difficult, especially at the administrative centre, where suspected opposition supporters were closely monitored and local institutions trying to promote rights and development were disrupted and forced to close.

In Hwange and Victoria Falls, the research was also subject to surveillance. Although the same process of politicized intervention affected local institutions and ruled out the possibility of investigating some topics, I was able to work with members of the Nambya Cultural Association, as it was seen by the ruling party as marginal and politically unthreatening in a place where opposition politics was dominated by the activities of the union movement in the towns, and the politics of occupying white commercial farms and conservancies. The Nambya Cultural Association will be disappointed with this book, as, by focusing on the river, it distorts Nambya ethnohistories – although Nambya chiefs dominate the Hwange rural areas that abut the river, Nambya constructions of heritage do not look to the river as a symbolic focus, even though the leaders of the Hwange dynasty were among the more prominent of those who lived and traded along the Zambezi in the late nineteenth century. At the end of the nineteenth century, the dynasty's leaders moved away from the river, back to the environs of the stone ruins of the ruling seats of the pre-colonial Hwange dynasty at its apogee. Over the course of the twentieth century, Nambya ethnohistories have become closely intertwined with Hwange town and the ruins inside the Hwange National Park, neither of which are central to the narrative in this book.

In addition to interviews on the recent and remoter past, and biographies of important individuals, Chapter 8 uses the texts of written diaries that I commissioned from fishermen living in Binga's fishing camps. The diaries provided vivid commentary on the precarious nature of livelihoods at a time of increasing economic hardship and highlighted a moment of transition in the fish trade and border economy, as well as revealing changes in the nature of state authority in the borderlands. Diarists' embodied conceptions of the increased effort necessary to secure a living were inseparable from escalating risks, which themselves reflected the ways state power continued to structure the violence and lack of protection intrinsic to illegal activity on the margins of the state. Their stories of the exertions of paddling long distances across the lake, of encounters with crocodiles, thieves and armed Zambian poachers, and of dramatic shifts of fate, help to bring to life not only the shifting material interactions with the water, but also the changing political economy of the borderlands.

The historical sections of the book also draw on a range of archival, contemporary literary and press sources, including the surprising volume of popular writing covering each of the episodes of intervention. Native Commissioners' reports on the huge Sebungwe district of Matabeleland, subsequently divided into the three districts of Hwange, Binga and Gokwe were, of course, basic sources. Two chapters on the early history of the state border draw on district court records, which proved particularly revealing of the tensions and violence surrounding the construction of state authority in the first decades of colonial rule, expressed through cases relating to labour relations, assaults on and by local state agents and a series of murders at Tonga funerals.

Outline

The narrative begins in Chapter 2 with a discussion of river stories, as they are told today by those who claim the status of 'river people', most of whom no longer interact closely with the Zambezi on a daily basis. I analyse these stories at the outset for their historical content – for insight into pre-colonial modes of discourse, the political and ideological hierarchies of the late nineteenth century frontier, and material interactions with the river. The importance to pre-colonial frontier dynamics of the river's strategic role and command of its fords emerges clearly from these stories, and is underlined further in Chapter 3, which turns to imperial discourse about the Zambezi. My aim in examining European narratives is to highlight how they drew on aspects of African mid-Zambezian political idiom, but also to examine their effects in shaping subsequent Zambezian interventions. I am particularly concerned to investigate the association of the river route with violence, and the shift from the idea of the river as a highway to the view that it constituted a 'natural border'.

The consequences of the drawing of the border for those living in its vicinity are the topic of Chapter 4, which examines early state efforts to control the valley, legitimated by casting it as a place of violence and disorder. I argue that the idea of the state took hold despite the rapid withdrawal of the police, as local chiefs and others began to use the law, invoking the power of the state to contradictory ends, both to continue a process of violent extraction and also to contain the internalized violence that was the legacy of the pre-colonial frontier.

Chapters 5 and 6 are about the two major engineering works along the mid-Zambezi associated with the extension of state controls over the borderlands – the Victoria Falls bridge and Kariba dam. Both chapters investigate the political ends to which these projects were used, exploring their material and symbolic significance in laying the basis for urban and industrial growth, altering the landscape and functioning as sites of memory in the myths of imperial expansion and white nation-building. Each episode provided opportunities for invoking and re-assessing past European interventions, giving efforts to tame the river a repetitive quality, even as each set of colonial actors was keen to distinguish itself from its predecessors. The chapters demonstrate how each monumental project reinforced racist ideas about the superiority of European technology, contributed to an expansionist ethos in Southern Rhodesia, acted as propaganda for the maintenance of white rule and entrenched the idea of the frontier and its inhabitants as backward. The later interventions provided the momentum and justification for a vast expansion of the state conservation estate, bringing about displacements and a racialized resource access regime whose legacies persisted into the time of the research.

But these interventions were important beyond the way they were narrated in dominant white discourses. Chapter 7 explores counter-narratives of the dam, as they were deployed in African nationalist and ethnic mobilization as well as being developed in Colson and Scudder's studies of the social effects of displacement and subsequent research. I focus on the local intellectuals on the south bank who shaped both ethnic and national consciousness in the borderlands by reframing understandings of culture and the past focussed on particular sites of memory. They developed a politicized and angry narrative of developmental neglect and broken promises that encouraged local communities to facilitate guerrilla incursions, recharging the strategic importance of the river.

The last three chapters bring the book round to where it started out, by returning to river stories as they are told today, but examining the context and political uses of their telling in relation to the present, highlighting why the idea of being a 'river people' has retained its political charge. To this end, Chapter 8 explores the politics of recognition in the Zambezi borderlands, focussing on the defensive mobilization of cultural difference on the part of Tonga intellectuals after independence, and exploring the subjectivities and strategies of successive generations of local leaders who have struggled to reverse old stereotypes, deliver development and secure inclusion in the Zimbabwean nation. Chapter 9 examines the material relations of everyday life on Lake Kariba, and the attitudes and practices of Tonga gill-net fishermen, who invoke stories of crossing and of being a river people to justify fishing in restricted waters and informal cross-border trade. The concluding Chapter 10 further completes the loop of the narrative by discussing the post-colonial politics of landscape at Victoria Falls, where the idea of being a river people was less about access to the material resources of the river per se, than having a stake in the revenue and international audience generated through the tourist and heritage industries.

By the end of the book, I hope the reader can see in claims to privileged relations with the river made by marginalized groups today, the influence of both past and present configurations of local, state and international interests, and related discourses and conflicts. Even as the landscape conspires to erase history and politics, deriving its power from an implied continuity, stability and permanence, the parties to these debates show this conspiracy for what it is, as they contest interpretations of landscape history, links with collective identities, rights to resources and development, and practices of state and nation-building.

Notes

1 Sanders Mwinde, Mlibizi, 25 March 2001.

2 Interview, Chief Mukuni, Livingstone, 14 September 2001.

3 See The Basilwizi Trust, http://www.basilwizi.org.zw/index.htm [accessed 2 July 2006], R. Jensen 'Hunger Strike Remembers the Victims of World Bank Policies', 23 January 2004. http://www.commondreams.org/cgi-bin/print.cgi?file=/views04/0423-10.htm [accessed 2 July 2006].

4 In many ways, this is a conventional framing for water histories: Blackbourn, 2006, is built around a series of 'dramatic episodes' of transformation. Drawing on Braudel, Tvedt argues that such short term, dramatic episodes are one of the three types of temporality a history of water should consider, including engineering works and water projects, as well as sudden floods or droughts, Tvedt and Jakobsson 2006: x.

5 Blackbourn 2006: 18.

6 I share the doubts others have raised regarding the applicability of actor network theory to situations of conflict, Woods 1997: 321-40, Castree and Macmillan 2001: 222-3. On the difficulty of using 'relational' theories (such as Whatmore 2002), see Castree 2003: 203-11.
7 Blackbourn 2006: 18 makes a strong case for this in the German context.
8 Theories of 'assemblages', in contrast, seek not to understand the perspective of human actors, but to analyse the constraining role of provisional and material arrangements of human and non-human things, to provide an interpretation of the power of non-human actors. For a discussion of the idea of 'assemblages' by Deleuze and Guattari, Rabinow, and Labour's related notion of 'articulations', see Moore 2005: 22-5.
9 On 'pestilence discourses' in comparable situations of marginality, see Knight (ed.) 2000.
10 E.g. Teede and Teede 1991includes picturesque photos of Zambezi valley people displaying genetic deformities (two-toed feet), which are part of the de-humanizing caricatures elevated in the colonial period that minorities are still fighting against today.
11 Debates over Wittfogel are thoughtfully reviewed in Worster 1985.
12 McKittrick 2006: 450.
13 Nor did precolonial systems of managing river water involve irrigation, unlike more centralized polities further upstream and elsewhere, which depended on mobilizing labour for digging canals and erecting mud dams on broader and open floodplains. See Gluckman 1941 and Kreike 2004. See discussion in Scudder 1962.
14 Haarden 1996, Worster 1985.
15 Blackbourn 2006: 12.
16 Berry 1993.
17 Fairhead and Leach 1987, Wolmer 2006.
18 For a review of environmental history, see Beinart 2000: 269-302.
19 See 'Introduction', Beinart and McGregor 2003, McGregor 2005: 205-19.
20 Moore 2005: 22.
21 Klein 2000: 127-50.
22 Nora 1989: 7-25. These criticisms tend to be outlined in relation to this introductory essay, rather than the multi-authored collections Nora and Kritzman (eds) 1996, Nora (ed.) 2002.
23 Nora 1989: 7.
24 Nora 1989: 24.
25 Legg 2005: 481-504, Klein 2000.
26 See Matless 1998, Bender (ed.) 1993, Ranger 1999.
27 Mitchell 1996 and 2003: 233-48.
28 Cosgrove 1998.
29 Cosgrove 1998: 13, drawing on Berger 1977.
30 Stewart and Strathern 2003, Hirsch and O'Hanlon 1995, Samuel 1998.
31 This literature is reviewed in Beinart and McGregor 2003. On travel-writing, see Pratt 1992. Critiques include Guelke and Guelke 2004: 11-31.
32 Luig and Van Oppen 1997: 7-45.
33 The debate over narrative, between J. and J. Comaroff (who use an implicitly restricted definition) and John Peel and Terence Ranger (using a broader conception) is summarized in Comaroff 1997: 43-52. See also Peel 1992: 328-9, Peel 1995: 581-607, Ranger, n.d.
34 Ranger 1999, Fontein 2006.
35 Moore 2005, Hughes 2006a.
36 Wilson and Donnan 1998.
37 Colson 1997: 47-59.
38 Ibid.
39 Kopytoff 1989: 9.
40 Worby 1998: 55.
41 Das and Poole 1994: 9-10.
42 Roitman 2006: 248 and 1994.
43 V. Das and D. Poole 1994: 9.
44 Asiwaju 1985, Nugent and Asiwaju 1996.
45 Nugent 2003: 16.
46 Kopytoff 1989, Donham and James (ed.) 2002, Nugent 2003, Leopold 2005.
47 Leopold 2005: 8. Leopold disrupts the implication of causation by moving backwards in time.
48 This has been the focus of Nugent 2003 and other studies of border regions. See also Kreike 2004 on the Namibia/Angola border.
49 For an interpretation of local agency in the development of such differences across the Mozambique/Zimbabwe border, see Hughes 2006a.
50 Alexander 2006. There was a shift to empower chiefs and elaborate tradition in Southern Rhodesia, but this was much less pronounced than north of the border.

51 The term 'supertribe' is Richard Werbner's; see Werbner 1991.
52 Englund and Nyamnjoh (eds) 2004, Berman, Eyoh and Kymlicka (eds) 2004, Werbner 2004.
53 Englund 2004, in Englund and Nyamnjoh (eds) 2004: 1.
54 See for example, Berman 1998.
55 See for example some of the commentators on Botswana; Werbner 2004, Solway 2004.
56 See McGregor and Schumaker 2006.
57 McGregor and Schumaker 2006: 659, Fairweather 2006, Flint 2006.
58 Chief Shana, Hwange, cited in 'Cross Border Cattle Rustling Rife', The Zimbabwean, 17 September 2006. Shana claimed 420 head of cattle were poached through these networks in a two month period.
59 On historicizing colonial discourses, see Thomas 1994.
60 Stoler 1989: 135-6, Thomas 1994: 17.
61 Raftopoulos 2006.
62 Hughes 2006: 807-822.
63 Beinart et al 2005, Tilley, 2003, McKenzie 2000, Schumaker 2001.
64 On the role of intermediaries, see Hamilton 1988, Johnson 1981.
65 Pratt 1992, Duncan and Gregory 1999, Harraway 1989, McClintock 1995, Jardine and Spary 1996, Miller and Reill 1996.
66 Driver 2001, see particularly chapters one and two; Phillips 1997.
67 McGregor 2003.
68 Cannadine 2001.
69 I have described this process in McGregor 2002.

2

Crossing the Zambezi

Landscape & Pre-colonial Power

Kasambavesi – crossing depends on knowledge
Tonga intellectuals' explanation of the name 'Zambezi'

Crossing is a recurrent theme of stories about past relations with the mid-Zambezi told today by those claiming the status of 'river people'. They tell tales of running to the river to escape powerful pursuers, of ferrying others across or cunningly manipulating enemies' ignorance, such that knowledge of crossing the Zambezi created a link between opposite banks, when for others the river was a barrier. In these stories, both the river itself and extraordinary non-human occupants – fish, large animals, spirits and monsters – are attributed with magical powers and are entangled in social life, often demonstrating a special relationship between the Zambezi and the 'river people' who lived along its banks and regarded the valley as home.

This chapter examines these stories with the aim of providing insights into the politics of landscape on the Zambezi and related modes of discourse before the first Europeans arrived in the nineteenth century, so that European explorers' accounts (discussed in the next chapter) can be situated in the context of the place they were writing about. Of course, oral sources – traditions and histories collected by myself or recorded by others at various times – work against the possibility of historical reconstruction in a variety of ways. Not only do storytellers conflate discrete historical episodes of crossing and telescope historical actors, but the stories play a political and moral role today, linked to modern, ethnicized notions of identity, bids for chieftaincy and reparations, or a romanticized, nostalgic view of the past as heritage in which ritual and 'sacred places' have been revalidated as 'sites of memory', reified as threatened, potentially preservable cultural relics, or developed as tradable commodities. The intervening period has also left its mark, not least in the language and terms in which the stories are told, which have been profoundly shaped by biblical idiom, modern cultural nationalist validation of heroes and resistance, and notions of pre-modern 'authenticity'.[1] Only towards the end of the book will contemporary and intervening influences on the politics and use of these stories become clear to the reader. Here, however, my focus is on traces of the past. Like Carolyn Hamilton, in her exploration of myths of Shaka Zulu, I am interested in examining continuities as well as changes, the constraints and possibilities of historical invention, and the interactions between

African and European ideas which are usually downplayed as the two have tended to be cast as diametrically opposed.[2] Stories about the river provide insights into ways of talking about and understanding the past in dialogue with the present.[3] Even the claims to the extraordinary are important, as Luise White has elaborated, less because they tell us what happened than through the insight they provide into popular idiom in the past.[4]

I focus on two particular types of story – stories of crossing and tales of the enchantment of the river. The stories of crossing and the related figure of the ferryman, highlight the strategic role of the river and its frontier status in the mid-nineteenth century (and perhaps earlier), when relations with more powerful others were an unwelcome political reality for mid-Zambezi river people. One of the functions of stories of intimacy with the river and skill in crossing, when told today, is to give power and status to those with knowledge of how to cross; they are an idiom for talking back from the margins, conveying moral values and differentiating 'us' from 'them'. Although this is now understood primarily in ethnicized terms, the same stories seem to have performed a similar social role in the nineteenth century before contemporary ethnic categories were consolidated, when those along this part of the river lived on the raided and tribute-paying margins of powerful, centralized nineteenth-century African states. Despite being told from the perspective of those for whom the river was a link within a homeland and not a margin, stories of crossing are nonetheless infused with a sense of relationship with powerful others based beyond the locality.

The magical powers of particular features of the riverine landscape, which were related to powers over rain and fertility, mostly validated decentralized, matrilineal political authority, and predated the rise of the centralized states that impinged so much on nineteenth-century mid-Zambezian river life.[5] In the traditions of the more centralized states that encroached on the valley, those attributed with a particular relationship to such places of power are often respected as autochthons. In retelling accounts of ritual at these places, I aim to emphasise historicity and politics, and try to write against Nora's (and others) romanticized view of pre-modern memory as rooted in 'gestures and habits, in skills passed down by unspoken traditions, in studied reflexes and ingrained memories'.[6] Implicit in the bodily ritual at these sites were contested narratives related to who had the right to organize them and decide their content, itself related to competing understandings of descent and kinship, succession, identity, status and political power. Moreover, as we shall see, distinct traditions with a different sense of historical time and different cultural repertoires were perpetuated by different groups living along the river.

River Crossings & Nineteenth-century Political Hierarchies

The section of the river that concerns us here extends from the Victoria Falls downstream through a series of gorges and rapids into what is now lake Kariba, and for convenience I shall call it the mid-Zambezi.[7] The river's course and behaviour over this section helped to shape the politics of landscape in significant ways. The Victoria Falls brings a dramatic end to the expansive Zambezi floodplains, and after

2.1 Mpalira island above Victoria Falls (sketch by Emil Holub, 1881)

2.2 The gorges below Victoria Falls (David and Charles Livingstone, 1855)

plunging down over the lip of the waterfall, the river is suddenly constrained to a tiny fraction of its former width, its force concentrated into a deep channel. Before the damming of the river at Kariba Gorge, the river below the waterfall flowed very rapidly. This was important because the strength and rapidity of the river's movement, and regular interruptions in the form of rapids and waterfalls, inhibited navigation up and downstream and made crossing a highly specialized skill. Another important feature of the river's flow was the fact that it was shaped by the seasonality of rainfall in the upper Zambezi catchment, such that annual floods brought floodwaters to the mid-Zambezi in the hungry dry season allowing for a second harvest. The aridity and dissected nature of the land through which the river flowed in the mid-river limited the sites attractive to settlement, confining them to the narrow patches of alluvium where flood-retreat agriculture was possible. Away from the river valley, the steeply rising terrain up to the plateaux on either side was barren, broken, hilly and scarcely inhabited.[8] The 'river people' who clustered along the river lived in often sizeable villages and depended on the Zambezi both for its fertilizing floods, and its strategic potential. Emphasizing skill in exploiting an ecological niche without considering the river's strategic effects would have the effect of reifying the local, and failing to capture the importance of connections beyond the locality.

These 'river people' (*bamulwizi)* distinguished themselves from the people of the hills,[9] though they shared language and cultural traits with Tonga and Leya communities stretching up onto the plateaux both north and south.[10] They lived in small matrilineal groups and lacked strong traditions of centralized authority.[11] People were often referred to by their leaders' name, though today's ethnic names – Tonga, Toka, Leya – were also current. In the mid-nineteenth century, these 'river people' were part of the unstable, raided and tribute-paying frontier of two major nineteenth-century state systems. To the north was the heartland of Kololo rule and subsequently the Lozi state, centred on the wide, open flood plains of the Zambezi.[12] To the south was the Ndebele state, centred on Bulawayo, which had replaced the authority of older Rozvi states in the early decades of the nineteenth century. In the 1850s, Ndebele warriors destroyed the Hwange dynasty or Nambya state, a north-western outpost of Rozvi authority established by a breakaway group of migrants from the Zimbabwean plateau in the late eighteenth century.[13] In the period that concerns us here, Hwange and the remnants of his people fled across the Zambezi river, where they set themselves up on the north bank under Kololo protection and, like the 'river people' they lived amongst, were embroiled in a complex politics of subordination to and alliance with more powerful African others. The Nambya were also known by the older names of Kalanga and Nyai, and though they lived along the river, their leaders held out against identifying themselves as 'river people' by clinging to the memory of past leaders and past authority, located in the graves and material remains of their former settlements away from the river.

The hierarchical relations with the region's powerful states are apparent in the older meanings of what are now ethnic names. The term Tonga, for example, appears to have originated as a label used by others, applied to subject or chiefless people. As such it had derogatory overtones.[14] The Ndebele used the name Tonga very loosely in ways that overlapped with the 'slave' category, *amahole*. As such 'Tonga' derived from the passive form of the verb *kutonga* (to judge), meaning to be judged or enslaved.[15] The Kololo shared this perspective and had particular contempt for those they labelled Tonga (which becomes clear in the next chapter, as their views are incorporated in

David Livingstone's writing). The Nambya state also had an internal hierarchy. Its elite claimed descent from an original cast of Kalanga or Rozvi migrants led by a certain Sawanga, who had conquered those they found in the north-west and who ruled from a series of stone-walled capitals, the largest and latest of which was called Bumbusi. Nambya myth honoured the Leya leader Nelukoba as an autochthon who 'shared the land', but looked down on the others they found, fought or subsequently incorporated (not only Tonga and Leya, but also San and other hunters, known as Dama and Haka).[16] The Nambya also used the name Tonga in derogatory fashion and called the Tonga and Leya they incorporated Dombe, meaning 'the dirty ones' (a reference to Tonga practices of smearing themselves with ochre).[17]

The 'river people' themselves, however, infused these names with different, more positive meanings. Sometimes they did so by emphasizing a relationship with the river. It has been suggested, for example, that the term 'Tonga' was derived from *mulonga*, river, or *mudonga* the big river.[18] Similarly, some Dombe explained their name in reference to the skills necessary for crossing – interpreting its meaning as 'the strong/brave ones', referring to the strength of the young men who carved the canoes that carried them across the river. Others explained their name as referring to piles of detritus dumped by the river, or to a fish species (*indombe*) which is unresponsive to bait.[19] The label 'chiefless' was also valued in itself without reference to the river, in contempt of centralized political authority, in affirmation of egalitarianism and in pride at independence and judging issues for oneself.[20] In this sense, its meaning is attributed to the active form of the verb *kutonga*.

The names referring to these decentralized groups – Tonga, Toka, Leya and Dombe – were not mutually exclusive: Leya was a Tonga clan name, while Dombe referred to those Tonga and Leya who were at one time under the authority of the Hwange dynasty, and the term Toka originated as the Kololo/Lozi pronunciation of Tonga, and came to be a self-identification in areas close to (and sometimes recognizing) Kololo and later Lozi authority.[21] The centralized states of the nineteenth century were incorporative and multi-ethnic, such that Tonga, Toka, Leya or Dombe could become Nambya, Lozi, Kololo or Ndebele. But cultural difference and an understanding of diverse origins were perpetuated within these states, as, in the context of a recognized overarching authority, such diversity constituted no threat.[22] The small, decentralized groups on the other hand, were fluid and assimilative in a rather different manner, as the absence of overarching authority created great pressure for cultural conformity, thereby erasing memories of other identities and other historical origins. As Elizabeth Colson elaborates, 'In the past, absorption and obliteration of alien status was the common fate for those who entered the Tonga ambit. Slaves, invaders, refugees have all been subject to the pressures of Tonga neighbours who first called them visitors or guests, then introduced them to the idea and responsibilities of clanship and ended by submitting them to the discipline of kinship. In the next generation, it was clan and kin affiliation which was remembered rather than any association with alien group or alien home.'[23] We shall gain further insight into this dynamic in practice, in relation to Nambya leaders' struggle not to forget their home away from the river, and not to succumb to Tonga assimilatory pressures during their period in exile on the Zambezi and subsequently.

People living along the river also used more than one of the names associated with the marginal decentralized groups, switching instrumentally as a means of hiding from more powerful others. As one old man explained:

> We are the real Leya, the Mukuni from around the Falls ... We were hunters of elephant and people of the river. What happened, because of war, after crossing the river, you'd sink the canoes, tying them under water. Then when the enemy comes, you say, 'No, we're not the ones you're looking for, we've no canoes'. But then when the danger is past, someone can dive in and fetch them. Dombe was a type of camouflage – those who followed us, they'd be told, 'No we're not Leyas here, we don't know them, we're the Dombes'. It was a security measure.[24]

Another recounted:

> When there was war on the Lozi side, we'd come here. Then when the Ndebele made war on this side, we'd cross back again. This is our land. When we cross to the north, we are Leya, then when we come back we are Tonga. If anyone pursued us, we'd say, 'No, we're not the ones you're looking for, we're just Tongas'.[25]

These stories of crossing and running are told as moral tales, and convey ethical values – a pragmatism in the face of superior force and danger, an absence of shame in cowardice, running away and evading battle, a pride in the cunning to hide and knowing when not to make a stand.[26] As a Tonga youth told Fr Torrend in the 1880s: 'The Karange submit ... the Shukulumbwe fight, the Tonga neither submit nor fight, but they cross in canoes, and come to live on this [south] bank, returning to their homes, when they no longer fear the Lumbu.'[27] The recurrence of the theme of crossing, and its infusion with moral overtones means that particular historical conflicts and associated episodes of crossing have been lost and tend to merge into one another.

Another way in which stories of crossing can establish a special relationship between 'river people' and the river is through myths of origin. There was no common Tonga or Leya myth of origin in the past (though Tonga intellectuals in Zambia have since invented one), and those along the river often claimed to have originated in the vicinity.[28] However, disparate myths often have a common theme of symbolic and magical crossings of the river by founding figures. Nelukoba's Leya for example (remembered in Nambya dynastic histories as autochthons) claim to originate north of the river around the Victoria Falls but were led to the south bank by a certain Simbalane who made the crossing by walking under the river on the river bed holding a lighted torch that the waters did not extinguish.[29] Elizabeth Colson recounts a myth in which Namansa Mutonga led his people across the river in the opposite direction: when he reached the Zambezi, the waters parted and his people walked across on the river bed. His wife had forgotten her pipe, but when she tried to cross back, the waters flowed together to prevent her doing so and she turned into a crocodile, which haunted the river.[30]

All of these crossing stories, whether told in the context of myths of origin or fleeing, underline the strategic importance of command of crossing in the context of the mid-to late-nineteenth century frontier, and the formidable barrier the river posed to those lacking such knowledge. Immediately above the Falls, the Zambezi was an expansive two kilometres wide and crossing required knowledge of sandbanks and currents, as well as a familiarity with the habits of crocodiles and hippos. An unskilled boatman risked being swept over the edge of the waterfall or capsized by an aggressive animal. Here, in the mid-nineteenth century, crossing points are associated with several different chiefs, including Sekute and Mukuni, who lived on Kalai and Siloka Islands respectively.[31] Though much less wide below the Falls, the river flowed with such a force through narrow gorges and rapids, that crossing in unstable dugout canoes was an equally skilled affair. Moving downstream from the Falls, each neighbourhood and the larger leaders all had their own crossing points, harbours and ferrymen.[32] Most of these leaders also had their own islands and some have traditions of running to the river or moving onto their islands for security. One Dombe group tell of how Monga Mapeta led them to safety across the river to escape wars on the north bank between the Kololo and Lozi, first to the south bank and subsequently to the security of Mapeta island at the Deka confluence.[33] Similarly, Elizabeth Colson recounts the story of Mwemba's flight to Masanga island in the Zambezi,[34] and in general, islands are remembered as 'places you could run to.'[35]

Those who regarded themselves as river people crossed the Zambezi repeatedly and maintained strong links across the river. Indeed, the closest social relations were not up and downstream, but between opposite banks and inland, up the Zambezi's tributaries.[36] The preoccupation with crossing the river was not matched by a parallel interest in where the Zambezi came from or where it went to (though many did possess such knowledge in general terms). The fast-flowing currents and rapids in this section of the river made extended navigation more or less impossible, such that it did not act as a highway of communication for those who lived along it, and Chikunda and Portuguese traders from the East coast tended to follow the course of the river overland by foot rather than by boat.[37]

The landscape itself can provide reminders of the power that command of crossing could confer. The island just above the Falls, known to some as 'Ndebele island', was named after Ndebele raiders whom the Leya had promised to carry across but left stranded on the island to die.[38] Further downstream, another Leya elder recalled, 'Canoes worked in this country! Too much! You see this bend in the river? That was used by the ferrymen to trick the Ndebele, because you couldn't see what was happening around the corner – they would put ten Ndebele men into a big canoe, but they could upset the boat downstream. Coming back to fetch some more, they'd say, "No, I've crossed those others." Repeating like that again and again! It was a good trick! They could enjoy killing! They weren't friendly to Ndebeles!'[39] The treachery of the ferrymen along the river was infamous amongst powerful African groups away from the river who had to depend on them for raiding excursions, and it was a reputation that those along the river found useful to uphold. We shall see in the next chapter how Livingstone recorded stories about the deception and ferocity of the river people from his Kololo guides and hosts.

Some crossing stories can, however, be more readily related to discrete historical events. One such constitutes part of the tale of the destruction of the Nambya state and its final ruling seat at Bumbusi, and tells of the flight of its people into exile across the river. The explorer Chapman recorded a version of this story from the then Hwange himself shortly after Bumbusi's destruction in 1853, and the story is also a well-known aspect of Nambya ethnohistories as they are told today.[40] Myths of the fall of Bumbusi are intensely political, and have a persistent influence in disputes over Nambya chieftaincy, ritual and tradition.[41] They hinge on the politics of succession following the death of the Hwange (Shana) who had built Bumbusi. The cast of possible successors included the deceased Hwange's three sons and his sister's son. But when the sister's son – Lusumbami – triumphed, the sons were disgruntled, and one of them – Chilisa – invited the Ndebele to attack by provoking them with tales of a powerful rival to the north who had 'two hearts'. The Ndebele captured and killed Lusumbami, ripped out his heart and took many captives. Chilisa took over as

Hwange, but his alliance with the Ndebele did not secure his people from raids, so he fled across the river to the Kololo for protection, while his people were scattered and the state was destroyed. Lusumbami's two hearts are said to reflect the Nambya's position between two more powerful states, and the dangers of maintaining relations with both. A leitmotif of one version of this episode (and other aspects of dynastic history) is the moral chaos caused by the patriline's disruption through Lusumbami's succession, sometimes interpreted as evidence of the incorporated matrilineal Tonga peoples' distortion of their rulers' patrilineality (disputed by the house of Lusumbami). A second casts Chilisa and those descended from him as traitors and destroyers of Bumbusi (contested by Chilisa's successors).

According to the myth, it was possible for Hwange and his people to flee across the river, out of Ndebele influence, thanks to the assistance of the Dombe leader Mapeta. In Nambya versions of the story, Mapeta was a lowly Tonga whose status and name reflected Nambya gratitude for his service as ferryman, for which he was rewarded with an honorary title and a Nambya daughter to marry. In Dombe stories, in contrast, Mapeta appears as longstanding independent autochthon in the region of the Deka confluence, with a special relationship to the landscape and river. The following version is from a descendent of Mapeta:

> Now, when the Nambya ran from Bumbusi [their final stone-walled capital], the Ndebele were coming behind. They ran to the river where they ... said 'please can you take us across.' ... So Mapeta sent his man Sinechigani, giving him a canoe to take the Nambya across ... When the Ndebele arrived hot on their heels, Mapeta took his horn and blew it. The Ndebele heard it and all their weapons were thrown into disarray ... The Ndebele could see the Nambya chief on the island [in the Zambezi] on top of a hill, and thought they could cross, as the river is narrow at this point, although its current is fierce and the water deep. They tied big logs together with fibre to make a raft, but the current swept it away and they fell into deep water, some dying in this way. Then they saw some Tonga men from the far bank swimming across, and they thought it must be easy to do so. But the Tonga knew the currents and where to cross. The Ndebele did not know this and more of them died trying.[42]

Named features of the landscape provide a reminder of these stories. The island bearing the name '*Chakona*', 'the preventer', derives its meaning from the well-known phrase '*Chakona, chakakona banyai*', which refers to the role of the Zambezi as barrier, and can be translated as 'the preventer, it prevented those who were sent from crossing'.[43]

Downstream, similar stories of flight and crossing revolve around each of the named islands associated with today's Tonga chiefs, though they are less readily dated. For example, Siansali and Siachilaba's people were neighbours along the river in the nineteenth century. The now submerged Tobwe island is associated with the graves of Siansali and the neighbouring Simwaka island with Siachilaba.[44] Siachilaba's stories of the past tell of how Siansali and his people came running as slaves (*vadzike*) with warriors in pursuit, and were given Tobwe island as a place to hide.

> Siansali's people came running from the warriors, carrying their babies and goats to the river bank. Then chief Siachilaba was called, and said, 'How can we hide these people?' They took them and crossed at a certain island where they found plenty of green maize, even the goats were taken across, everyone was ferried across together. Everybody went to Tobwe. So the chief called the locals, his own people, off the island, 'Hey you locals come off that island, so that we can hide the newcomers here. 'The visitors didn't know how to paddle, so when the harvest was finished,

> they would say, 'Please come and collect us'. The men would be trying to swim, but would be attacked by crocodiles. Some just remained on that island – they were foreigners, slaves, we kept them.[45]

The story underlines the recurrent themes of the river and its islands as places of refuge, the role of knowledge of the river as a marker of difference, and also highlights the fluidity and assimilatory capacity of riverine society, in which slavery was an insult hurled at others and an idiom for talking about hierarchy as well as being a silenced recent past. The details of the story are controversial because Siansali's versions (unlike those of Siachilaba recounted above) do not hinge on slave status and gratitude to Siachilaba's people. Rather, they make claims to being first.[46]

Although command of crossing thus gave those along the river a degree of independence, the power this conferred should not be exaggerated. Frontier status made life along the river precarious and insecure, which is also reflected in the tales told about the landscape. Riverside caves, for example, were important for hiding, whereas circular hollows in the rocky banks are said to have been used as mortars for stamping grain when in hiding, as the grain could be pounded less noisily than in a wooden pestle and mortar. The extraordinary dissected landscape of the Victoria Falls likewise invokes insecurity, through stories of hiding, running and hardship. One of the waterfall's places of refuge was 'at the very end of the basalt spur extending between the third and fourth gorges', accessible via a 'knife edge', the ascent of which was possible only at one, very easily defended, place.[47] An evocative description of a second hiding place was given at the turn of the twentieth century by the first Conservator of the Falls, after exploring the gorges below the waterfall in the company of an old Leya man.

> ...We descended by a zigzag path the towering face of the gorge wall. Huge masses of overhanging rock, seemingly on the verge of toppling over, were passed during the descent. When about three parts of the way down, Namakabwe, the Old Man of the Gorge, pointed out a cave partly hollowed out by nature and partly by hand. During a portion of the year when the water is low, he occupies this place, living on the different varieties of fish to be found in the still pools adjoining the main stream below. Descending into the gorge bed we clambered over huge basalt boulders worn by the action of the water to a glassy smoothness, across patches of fine white sand finally reaching the edge of the river. Here it comes rushing down at a great pace, swerving off at a tangent on meeting an opposing wall, thence, after swirling, eddying and boiling, in a couple of hundred yards it is forced back in an opposite direction to its original course, and finally disappears round the corner of a perpendicular wall towering upwards nearly 500 ft. This locality was a favourite place of refuge in the old days.[48]

The violence of the frontier, however, was not only attributed to warriors of the centralized African states. Rather it was internalized and the 'river people' needed protection from each other as much as from outsiders. Oral histories today tell of interminable fights with neighbours and of obscure local politics, in which some groups were more successful than others, and making your lineage grow – particularly through the accumulation of women and slaves – was the ultimate aim and mark of achievement. Such accumulation was possible by hunting elephant, exchanging ivory for slaves, and by raiding others to obtain slaves directly. Old people have stories of their predecessors raiding women from those deemed 'low', attempting to buy them back again or using women to repay for help and protection given to others; Mapeta, for example had to buy back a sister from a Leya group on the north bank after she had been taken in a raid, at the cost of a large box of beads.[49]

In short, these stories of crossing highlight the mid-Zambezi's use as a defensive barrier on the part of the major African state systems, such that the obstacle it posed to crossing did at times allow for a break between Kololo/Lozi influence on the north bank and Ndebele influence to the south. Those who lived along the river, commanded its crossing points and moved in and out of alliance with these more powerful others, however, regarded the river as a link not a dividing line. Rather, for them, knowledge of crossing was a powerful idiom for distinguishing themselves from others and asserting independence. The places that appear in these stories are landmarks that acted as aids to historical memory, created a sense of belonging and continuity, even if the depth of time evoked was shallow and communal. Many had an additional significance as sacred sites, visited in the course of ritual.

Power, Sacred Sites & the River

My aim in discussing sacred places along the river is partly to counter the idea that people believed in a river god, which has become one of the standard myths of tourism in the valley, and is also repeated in the BaTonga Museum in Binga. Indeed, as we have already seen, the landscape of the river was notable less for being sacred than for its material and strategic role. Colson and van Binsbergen have argued that the importance of rivers in everyday life and their association with movement, flow and crossing, make African ideas about them rather different from ideas about features of the landscape valued as sacred, such as hills, caves, pools and trees, which are associated with permanency, immobility and fixity: the emphasis on crossing in historical tales can make rivers seem 'one dimensional'.[50] Yet the size and perennial flow of the Zambezi distinguished it from other rivers, giving it an exceptional importance, which could imply permanence as well as movement. Moreover, it had dimensions in so far as it was inhabited by fish, crocodiles, hippos, ancestral spirits, and an extraordinary river snake or monster.

The exceptional importance of the Zambezi is perhaps reflected by the fact that it had no name. Those who lived along this section of the river did not use the name 'Zambezi', which is sometimes said to be a foreign word, brought by the Portuguese from the coast. People now explain that the term was derived from the phrase '*kasambavesi*', meaning only those who know can cross, and underlining the symbolic importance of knowledge of the river.[51] The river was (and is) most commonly called 'the big river', 'the big one' or simply 'the river' – in marked contrast to smaller rivers, all of which have names.[52] Old people say this does not cause confusion: 'How could we confuse a river of such size with any other?' One old woman elaborated: 'The river is so big it can never [run] dry: it is the only truth that will never end.'[53] Old expressions confirm the importance and exceptionality of the Zambezi, and its association with power in a pre-Christian era. People swore by the river, saying 'By the river', I never did that '*nolukhulu*', or '*nolukhulu kasa*' in Nambya or '*aleza upamanzi*' (by God who gives water) in Tonga. An elderly court interpreter explained to me that such expressions would be commonly heard in court, and that old people regarded them as more meaningful than an oath on the bible.[54] He elaborated:

> People would swear by the river because it is a great provider to them, which has never disappointed

> them, it is a holy, God-given river. People associated the good spirits with the river because it is a provider. People never lived far from it, they depended on its fish and flood waters... God would come down and give them food through the river. It was a holy gift, something revered. It provided food like a mother.

This explanation is obviously now entangled with Christian idiom and romantic ideas about mother nature. However, it also captures aspects of what seem to be older ideas associating the river with powers that could not be appropriated directly. Stories about God and river spirits convey ideas that are very different from the pagan river god imagined by successive generations of Europeans, which are now peddled through the tourist and heritage industries along the river.[55] Two old Tonga men explained to me that the spirit of Leza (God) had no fixed location in the landscape: '[it] has no special place, it is on the tongues of people who give praises.' But 'People would go to their sacred places by down by the river and say "If you are there, the spirit of our ancestors, please rise now from the water and help us".'[56] Elizabeth Colson argues that in pre-Christian Tonga ideas, Leza was unknowable, did not live in a specific place, and was not personified or gendered (Leza shared grammatical forms with powerful animals and people) and could not be engaged through ritual.[57] Though Leza was understood through effects, most common misfortunes – disease, drought, death – were not attributed to Leza. Leza was invoked when explanations failed.[58]

It is clear that those living along the river thus had a primary engagement with ancestral spirits (*mizimu)* in religious practice (and not river gods), as it was considered possible to enter into dialogue with them. Powerful *mpande* mediums had a particular responsibility for rain and were possessed by ancestors and sometimes also spirits of the wild (*basangu)* or lion spirits *(mondolo)*, usually identified after a period of sickness and initiation by other mediums. Although the area considered here crosses what are now considered ethnic boundaries (between Toka/Leya, Tonga and Nambya), accounts of past ritual practice have much in common, including some shared terminology and idiom, not least because in Nambya areas both Nambya and Tonga languages are spoken, and the original Kalanga immigrants were subject to a powerful assimilatory drive from below over the course of the nineteenth century through inter-marriage, kinship and neighbourhood interactions with the Tonga and Leya among whom they lived. Many important shrines or sacred places are (or were) close to the river simply because people lived there, not because proximity to the river was *per se* important. Thus family shrines *(ntumba* or *numba*) were in homeyards or riverbank gardens, and graves (*magambo)* were located along the river for this reason alone. Neighbourhood shrines (*malende)* could also be along the river for purely contingent reasons, marked by a baobab, small huts and sometimes also a pool.[59]

Although the river itself was not regarded as a shrine or the resting place of God, it evoked respect and awe, and some of its pools and waterfalls *were* seen as sacred places, associated with magical phenomena and distant communities of ancestors, which have become centrally important recently in the politics of reclaiming the riverine landscape as 'cultural' rather than 'natural' heritage. These are places Elizabeth Colson has termed 'places of power' as opposed to 'shrines of the land', though this distinction is not reflected in local terminology, and the two can be linked. For example, rain rituals at a *malende* shrine might involve a trip to the river to collect water from a particular site regarded as sacred, and some of the senior *mpande* mediums with

2.3 Malende rainshrine, Mola (photo by Colleen Crawford Cousins, courtesy of Panos Press)

rainmaking responsibilities had biographical and spiritual connections with particular places in the river, and the powers of the river itself.[60] Such mediums claim to have been initiated (and to have healed and initiated others) by being taken down into the Zambezi.[61] The river did not have a monopoly on such 'places of power', but I shall focus here specifically on those associated with the river. These riverine sacred places had entirely local constituencies; each social unit along the river had such sites, none of which attracted visits from pilgrims from further afield. Rather, mediums along the river sometimes visited more powerful mediums on the better watered plateaux to the north and south.[62]

Sacred places within the river were usually spoken about in terms of vague former communities of unspecified ancestors rather than particular named, individual forebears, as Tonga inheritance and ritual worked to very rapidly eclipse such memory. Most were pools – all of which had an individual name – where people recount how supernatural phenomena could be seen or heard, usually taking the form of the sound of invisible people, of grain being pounded, children playing, drums being beaten or cattle lowing.[63] Chief Saba's *malende* shrine, for example, was near the pool Fufu, located at the junction of the Mlibizi and Zambezi: Fufu 'was a special place ... you would see young girls there or a whole village with women pounding and children crying, and you'd have to clap for permission to pass, and the girls and everything would disappear down into the pool. You couldn't fish there, the fish were fearful and strange.'[64] If fishing in some such pools was forbidden, in others it was possible only at certain times of year, under instruction from those responsible for the *malende* shrines.

Ritual at riverine places of power was not only about rainmaking; some types of fishing in the pools of the Zambezi involved a ceremony, or at least respectful behaviour, to secure protection and success. For example, in Siamupa's neighbourhood, before fishing, the mother and sister of the chiefs went to a pool called Nawutimba where a tributary met the Zambezi:

> Before anybody could go fishing, an old lady from Siamwenda and one from Siamupa went there to ask the spirit ... first those ladies would go and then we would all follow. First the water would be shallow, but as soon as the fishing started, it would fill up and up until it's up to your chin. But a lot of fish would be caught. After those big catches, the two old ladies would come to get their share first, before the others. *Vasenga* is the name for that type of fishing – you couldn't just say you're fishing or you wouldn't get anything. That was a way of respecting. In that pool there would also be crocodiles, but you couldn't mention that. You had to give them a nickname, '*makuba*' – if you say 'crocodile' you would be eaten.[65]

Other important events, such as the launching of a canoe or the killing of an elephant or hippo, could also involve ritual at the river in one of the places of power. This sometimes required an immersion in the river; a deep, still pool in the Zambezi called Chipito, near to Siachilaba's *malende* shrine, for example, was used for the community to immerse themselves after killing an elephant and before trading its ivory which was stored nearby.[66]

The more important spirit mediums, who could draw on a constituency beyond their immediate neighbourhood, had associations with particular sacred pools and elaborate biographical connections with the river. One such Leya medium, called Siawumbe (alias Jelekuja), was associated with sacred sites in the gorges below the waterfall. Siawumbe is said to have immersed himself in the river only to emerge with arms laden with agricultural produce. As some elders remembered: 'Siawumbe the great *mpande's* place was Kanokonoko, in the gorge below the Falls. He would jump into the water and stay under for seven days. After a week he'd emerge carrying all sorts of different kinds of seeds – millet, sorghum and nyemba [beans].'[67] The current medium of this spirit, a woman named Mambaita, explained to me how the life of Siawumbe and that of his descendants was intimately involved with the river, and how the spirit was passed on to his daughter Kasoso during an immersion in its waters:

> Siawumbe's daughter was Kasoso. Siawumbe's *mpande* spirit took her and threw her into the river two days after she was born. She stayed there in the river, until the community became concerned and asked Siawumbe what had happened. Siawumbe told the community she was fine, and that there was no need to worry. Kasoso was being bathed by the *mpande* spirit. After two days, Siawumbe marched to the river ... He raised a stick, and struck the water with it. Immediately the baby was seen in the river and he commanded his wife to take the baby out of the water. This child was Kasoso. Kasoso came out of the river, close to Chekane. She used to go back to that place for rainmaking.[68]

Further down the river, the medium Mawara – one of the more important regional mediums, who attracted visits from a number of different chiefs – was closely associated with the Binga hot springs into which he would immerse himself. But he too had a special relationship with the Zambezi: 'he would sit on the boiling water [of Binga hot spring] with it coming up to his groin and then he'd go into the river, walking on top of the water like Jesus... He'd be standing on the water smoking his pipe.'[69] Others recalled how he would perform miracles, crossing the Zambezi on top of the water without a canoe, fetching water from pools on the north bank and bringing it back to the southern side.[70]

As the reputation of these *mpande* mediums depended on personal charisma and changed over time, so too did pools regarded as sacred through association with them, though the ancestral *malende* shrines appear to have had a more consistent location. Colson argues that society along the river tended to be more stable than Tonga society on the plateau, as riverine gardens and the presence of riverine alluvium constrained movement. Though such sacred places invoked permanence and belonging, a sense of connection with the past did not involve the memory of specific individuals for long periods, and Colson argues that spirits of individual ancestors quickly merged into vague past communities, historical memory tended to telescope everything into three generations at most, and people lacked a fascination with the past.

Some few 'places of power' along the river do, however, appear to have had a persistent location. One such was the Victoria Falls, which, like other sites in the river, was important for those living in the immediate vicinity but did not draw pilgrims from further afield. Accounts of past practice at the waterfall have much in common with the idiom and ritual at other sites along the river (though accounts from Chief Mukuni have been creatively embellished in rituals performed for tourists). Like sacred sites at other riverine locations, the waterfall had particular places for collecting water for ritual, others for immersion and specific pools associated with diffuse forbears, the sounds and sights of which could sometimes be glimpsed. The Leya name for the waterfall, '*Syuungwe na mutitima*', can be translated as 'the heavy mist that resounds', though the term '*Syuungwe*' itself also implies rainbow, or the place of rainbows, and was associated with water, rain, moisture and fertility.[71] The better known vernacular term for the Falls, the Kololo/Lozi phrase 'Mosi-oa-tunya' or 'the smoke that thunders', has different associations and was popularized by Livingstone who had recorded it from his Kololo guides. The Kololo term reflects the perspective of those who did not live closely with the Falls, who knew the waterfall not as a sacred site but as a geographical marker, imagined primarily as seen from afar like the smoke from a fire, and regarded it as useful in locating oneself in a landscape that was unfamiliar.[72]

In stories about the waterfall told by Leya elders, the resounding noise is attributed to the water beating down on Sekute's drum, which had fallen over the edge in battles between Sekute and Mukuni.[73] Mukuni's ritual sites were focussed on the northern aspect of the waterfall, 'Syuungwe mufu' or 'mist of the dead', which was associated with the memory of ancestors and played 'a central part in the life of the people'.[74] They included what is now known as the boiling pot, at the foot of the Falls ('*katolauseka*' or 'make offerings cheerfully'). It is said that a light used to be seen there, or that one could hear the sound of drumming, of children playing, women stamping grain and cattle lowing. Offerings could be hurled into the boiling pot over the lip of the Falls from one of the islands perched on the edge, but people also clambered down to the pit itself. One old man recalled:

> On a special day, food would be prepared and carried down in clay pots. We would creep down and down until we reached the water point. There we would get water and all the food would remain untouched. So the spirits of the parents, if they could hear us, they'd take the food... When you go down the gorge for fishing, if you go at night, you can see lights. Then you would clap, persuading the spirit to be calm. The following morning you would return home with a good catch.[75]

Leya myth also told of an invisible creature (not a God and not Nyaminyami) that lived in the boiling pot.[76] Another important ritual site was on the upper lip of the waterfall where the water swung round and over the edge of the gorge, creating a pool where the water did not move swiftly and it was possible for people to immerse themselves.[77] The diseased and afflicted jumped into this pool in a cleansing ritual in which they allowed their clothing to be washed away over the waterfall, carrying infection and ill-health with it. A final important place was a secret location where water was drawn for rainmaking and other ritual.[78] Those with special ritual roles in relation to the powers of the waterfall were associated with the original matrilineal Leya, held to have predated and married the immigrant chief Mukuni (who subsequently introduced patrilineal chiefly succession). This ritual authority was institutionalised through the title 'Bedyango', held by a woman alongside the male chief.[79] As society around the Falls was not centralized, however, there were other prominent mediums who had close spiritual and biographical connections with the waterfall, and had their own special places, though most also visited the boiling pot. All the chiefs around the falls – not only Mukuni – used different parts of the waterfall.[80] Even the groups of refugees who had fled to the Zambezi, such as Hwange's Nambya or Musokotwane's Toka, came to respect the place as well as the Leya religious mediums who appropriated its powers.[81]

Although rain-making among the Nambya shared this focus on local riverine sacred places, as they sought out local Leya and Tonga mediums with particular connections to the landscape, the Hwange living in exile on the Zambezi also perpetuated the memory of former dynastic power and distinct, non-riverine origins. This was achieved partly through a focus in the most important rainmaking rituals on past Hwanges' graves located in the ruins of ruling seats away from the river, which were transformed into religious sites after their abandonment. As the locus of important ancestral spirits, the ruins were visited by subsequent generations in pilgrimages to ask for rain, wellbeing and protection against external aggression or disease. The fact that named ancestors were associated with particular sites undoubtedly provided an aide mémoire and helped create a greater depth of past than in decentralized riverine society. Traditional ceremonial at the ruins evoked the past in various ways. During ritual performances, the names of the original group of ancestral migrants were recited, along with those of subsequent Hwanges and a large cast of other dynastic characters, including many powerful women. The ritual also perpetuated notions of past hierarchy by remembering the Leya as autochthons, and deriding the Tonga, the latter made memorable through insulting songs and re-enactments of beating lowly people and chasing them away.[82] Partly through such performances, the Nambya retained their own distinctive history, their leaders did not become 'river people', and, as we shall see, subsequently sought to regroup away from the river close to the ruins of the past, when external threats receded.

This discussion of past ritual and 'places of power' within and along the river has underlined how the river was associated with power in positive ways as a source of refuge, fertility, life, food, and truth. Yet power was ambiguous, and so too was the river, which was also seen as dangerous. Floods were unpredictable and could not be controlled, such that there were frequently years of famine, when floodwaters were insufficient or badly timed, and years of disastrous excess when gardens were washed away in raging torrents.[83] The waters themselves were dangerous, especially in unstable dugout canoes, and they harboured life-threatening animals. People remember the river as fearful because it 'took people'.[84] Crocodiles have already featured in stories about the river and will feature later in the book. The animals are much mythologized and feared, as they are widely associated with occult power, seniority and witchcraft: witches are said to move in the form of crocodiles and the powerful are said to have medicines enabling them to send crocodiles in attacks against others. People tried to protect themselves against crocodile attack by building shelters at water collection points and by using charms and medicines, some of the strongest of which were made from crocodile parts.

Aside from the dangers of crocodiles, stories of the Zambezi also include a symbol of the river itself as a place of danger, taking the form of a mysterious river snake or monster.[85] It had various different names – around the Falls it was called Simwaba (and lived in the boiling pot), and further downstream it was named Simusinsi. The now famous term Nyaminyami was unknown along most of the river and appears only to have been used in the immediate vicinity of the Kariba gorge. These fabulous creatures are usually described in very similar terms, as a great fish or snake that lived in the river and could cause accidents on boats, or cause people to be taken and held down in the river's depths or be carried away by the fast currents. It is said that the monster could spread its beard or tentacles out onto the river bank, and sweep people down into the river. If signs of its presence were seen – such as the water turning red or becoming rough or 'heavy' – it was a bad omen, and one had to move away from the river. Nobody now talks about this monster in relation to everyday life, yet some people still remember stories about it, and its actions are sometimes interwoven in tales of the past. Disappearances of women collecting water from the river banks, or incidents of canoes capsizing, or crocodile attacks, could be explained through the monster's action. Children were told stories of the monster to instill fear of the river in them, and were given warnings not to go alone too near to the banks at certain dangerous places. However, the monster could never been seen – one only had signs of its presence. Nor could it be controlled – it was an intrinsic part of the river that had to be respected. It was also connected with rainbows, which are said to calm strong winds and to be a sign than the destructive rains and wind are over; rainbows are sometimes said to reach the ground in deep pools in the river where the monster lived.

Conclusion

I have described a rich body of idiom for talking about the river, focusing on stories of crossing and extraordinary events at sacred places. Though the telling of these tales has been influenced by the subsequent politics of landscape, the stories do capture some aspects of an older mode of discourse and are suggestive of the old strategic and symbolic role of the river. African landscape ideas focus on specific features, some of which Europeans also picked out as worthy of comment (as we shall see in the next chapter) and others of which went unnoticed, such as the different pools, the piles of detritus and the hollows of the river bank. The meanings attached to places in these tales are specific. It is not the remarkableness of pools in general that is noteworthy, but of a specific pool with a name, such as Fufu, where ancestors and magical creatures could be heard and seen, unlike other named pools where they could not; a baobab tree appears in oral histories not as beautiful in itself (though it may well have been seen as such), but as a marker of a particular shrine or grave. Features of the landscape appear in these stories partly because they invoke memories of ancestors named and unnamed, and are important ritual sites. Showing respect to the memory of ancestors at such sites involved a range of bodily practices, in which specifically visual responses and individual historical memory are not privileged. Sometimes 'seeing' was explicitly forbidden; after rain ceremonies at the river, for example, communities leaving the place were not permitted to look back at the river.[86]

Over the course of this book, we shall see how these stories of past relations with the river have provided a resource and have been used in different ways – to give meaning to life in contexts of hardship and displacement, in assertions of cultural difference and the promotion of ethnic heritage, as well as in efforts to hold colonial and post-colonial states to account. We will see how they have been remembered and revalidated to new political ends, despite being marginalized and subjugated in the course of colonial technological interventions along the river. First, however, it is necessary to investigate how the African material and imaginative interactions with the landscape described in this chapter related to early European encounters with the mid-Zambezi. The idiom explored here, with its emphasis on crossing, ferrymen and links, the creation of security and a sense of place, was combined with an almost complete disinterest in where the river came from or went to. The contrast with European imperial preoccupations – tracing the river's source, mapping its features, describing its beauties, denouncing the violence of its trade, assessing its navigability and strategic potential, and with the physical and fantasized link it provided between African interior, the coast and British industrial centres – could not have been more pronounced.

Notes

1 For a discussion of biblical and other influences on oral narrative, see Hofmeyer 1994:175. Maxwell 1999 makes the case that early colonial era pressures produced enhanced territorial emphases in myth and environmental religion.
2 Hamilton 199:28.
3 See discussion in White et al. 2001:1-30.
4 White 2001:281-304.
5 As argued for example, by Ranger 2003:72-86.
6 Nora 1989:13.
7 This term is my own, and comprises the upstream part of geomorphological designations of the' 'middle Zambezi', which extends from Victoria Falls to Cabora Bassa (Wellington 1955:388). Colson's (1960) and Scudder's (1962) studies of the Gwembe Valley also focussed on the upper reaches of the middle Zambezi. They sub-divided the Gwembe Valley into three sections, the upper, middle and lower reaches. The section of the river investigated in detail in this book includes their 'upper river' section, the downstream boundary of which is the Chete gorge, which corresponds with a cultural and linguistic division within the Tonga language, though my narrative at times extends further down the river to Kariba. I differ from Colson and Scudder by extending upstream by including the area around the Falls. On linguistic divisions between 'northern' and 'southern' Tonga, now understood to cross cut the valley in the vicinity of Chete, rather than follow the valley and distinguishing valley from plateau, see Hopgood 1940:xi, Jones and Carter 1967:94, Matthews 1976:41.
8 See Scudder 1962:130-3.
9 People downstream in Gwembe similarly termed themselves bantu bamulonga or bantu balwizi. Colson 1960.
10 Archaeologists have associated 'Tonga' cultural forms with early layers of permanent settlement in the region, dating back to the early part of the previous millenium On 'Tonga disapora' traditions and the archaeology of the region, see Phillipson 1974:14, Fagan 1967.
11 Though compared to riverine society further downstream, social units seem to have been relatively larger,

and ideas of chieftaincy more institutionalized. Colson 1960:187.

12 The Luyana dynasty was conquered between the mid-1830s and 1840 by Kololo invaders from the South, who were in turn overthrown by a Lozi rebellion in the late 1860s. Jalla 1920, Smith 1956, Wills 1967, Roberts 1968, Mainga 1973.

13 On Nambya history, see Ncube 1994, McGregor 2005a:316-37. Colonial versions of Nambya history are detailed in note 41.

14 On the meaning of the term, which is a label used widely throughout central and southern Africa, see Colson 1960, 1962), Beach 1980:158, Weinrich 1977:9, Posselt 1935:135.

15 On Ndebele understandings and stereotypes, see Weinrich 1977:9, Alexander and McGregor 1997:187-203.

16 See Ncube 1986, McGregor 2005a.

17 McGregor ibid.

18 Moreau (1950) suggests that 'Tonga' derives from 'BaLonga', the people of the river, or 'BaDonga', the people of the big river.

19 And which people killed by throwing stones. Interviews, Philip Ncube Ngonzi, Jambezi, 13 March 2000, Mankonga Mapeta, Simangani, 24 March 2000.

20 Weinrich 1977:9, Lancaster (1874) argues that in the Kafue confluence, the name Tonga was claimed to assert independence from low status vassal people. See also Posselt 1935.

21 In the colonial period and since, the Leya (or the Toka-Leya) have been considered part of the Ila-Tonga (sometimes known as Bantu Botatwe) language group. See Smith and Dale 1920, Jaspan 1953.

22 This is the case made, for a different context, by Cunnison 1951.

23 Colson 1970:45.

24 Interview, Peter Siatoma Musaka, Kanywambizi, 2 March 2000.

25 Interview, Josiah Siamwenda Chuma, Deka drum, 21 March 2000.

26 Colson 1971:19-35.

27 Torrend, 1891:286, cited in Colson 1971:29. See Colson's discussion of the absence of shame in cowardice and running away.

28 Colson 1960. Tonga intellectuals in Zambia, such as Enock Syabbalo, have elaborated an idea of unified origins and migrations. See Matthews 1976:49.

29 Interview, Mpala Siwela Dingane, Hwange, 2 April 2000.

30 Colson ibid: 26. Such symbolic crossings are also common in myths of origin elsewhere, see Ranger 2003:72-86.

31 See K. Mubitana 1990, Clark 1952.

32 Hwange crossed at the Deka confluence, Saba at the Mlibizi confluence, Sinamani by the Sebungwe confluence, Siachilaba and Siansali by Tobwe island, etc.

33 Interview, Nelson Nengwa Munzabwa and Sungani George Munkuni, 14 February 2000.

34 E. Colson pers. comm., Matthews ibid.:150-2.

35 Interview, Chief Siansali, Kariangwe, 3 April 2001.

36 Colson ibid.:17., Matthews, ibid.:26.

37 Matthews ibid.

38 Interview, Maxon Musaka Ndlovu, Chidobe, 27 March 2000. A slightly different version has the rowers leaping into the water, capsizing the boat, swimming away and leaving the Ndebele to drown.

39 Interview, Naison Makwara Pegota, Jambezi, 20 March 2000.

40 Chapman first met the Nambya pleading protection from the Tawana shortly after 1853, Livingstone recorded Hwange's subsequent request for asylum from Kololo leaders in 1855, and Chapman met them once again after Hwange had established himself on the north bank in 1862 (when he was with Baines). See Ncube 1994, Chapman 1868, vol. 1:162, 179, vol. 2:73, Baines 1864:475.

41 Versions of Hwange dynastic history can be found in Hemans 1912, Hayes 1977, Henson 1973, Nambya Cultural Association n.d., Ncube 1994, McGregor 2005a.

42 Interview, Mankonga Mapeta, Simangani, 24 March 2000.

43 Interview, Mankonga Mapeta, Simangani, 24 March 2000, pers comm., Nambya Cultural Association, September 2001.

44 Interviews chiefs Siachilaba and Siansali,19 March 2001 and 3 April 2001, Weinrich 1977:13-14. On assimilation, see Colson 1970:35-54.

45 Interview, Chief Siachilaba, 19 March 2001.

46 See the chiefly histories collected by Weinrich (1977:14) on Siachilaba and Siansali.

47 Clark 1952:72. Here Clark found larger numbers of grain and water pots. He describes another hiding place ten miles downstream in the gorge, where a deep fissure opened from the southern wall, which had been visited and described by Livingstone.

48 Unpublished report by F. Sykes, cited in Lamplugh 1908:145.

49 Koporo Chinderendere Mapeta, Simangani, 30 March 2000. The Dombe hunter Ngonze accumulated his

wives through raiding 'low' neighbours and in the form of gifts for killing elephant. Interview Philip Ncube Ngonze, Jambezi, 13 March 2000.

50 Colson 1977:50, van Binsbergen 1992:159.

51 Tremmel 1994 translates this phrase as only those who know can wash, though the implication is the same, as those who do not know the river cannot use it.

52 Lwizi, Mulonga mupati, or Lukhulu in Tonga/Nambya. The latter term is less commonly heard in Binga. There are parallels in Lozi terminology, who called the river 'Leeambye', which also means 'river'.

53 Interview, Alice Nduwo Ngwenya, Simangani, 11 March 2000.

54 Interview, Mathias C. Munzabwa, Hwange, 3 April 2000.

55 References to a river god in the Zambezi are frequent in nineteenth-century explorers' texts and in the popular writing on Kariba, as we shall see, but it is a foreign idea to the section of the river discussed here.

56 Interview, Peter Siatoma Musaka and Josiah Siamwenda Chuma, Kanywambizi, 21 March 2000.

57 Colson 2004, 2006. See also M. Muntemba 1970:28-39.

58 Colson 2004.

59 Accounts of valley Tonga religious practice include Colson 1960, 1969, 1977,1997.

60 See also Weinrich 1977:84-5. This is not the case further downstream: Colson (1997) did not find rain rituals associated with the river, nor was immersion practiced.

61 See Weinrich's (ibid.:84-5) account of chief Sinakoma's initiation as a medium.

62 Such as at Monze, on the Northern Plateau, and Nevana in Gokwe to the South. On the latter, see Alexander and Ranger 1998:3-31.

63 Such sites are not only along the Zambezi, but also along its tributaries. See also descriptions in Symausoonde 1947.

64 Interview, Chief Saba, Sianzyundu, 12 March 2001.

65 Interview, Chief Siamupa, 1 September 2001

66 Interview, Chief Siachilaba, 19 March 2001.

67 Interview, Peter Siatoma Musaka and Josiah Siamwenda Chuma, Kanywambizi, 21 March 2000.

68 Interview, Esinath Mambaita Kasoso, Milonga, 11 April 2000.

69 Interviews Siampiza Munsaka, Timothy Munsaka and chief Sigalenke, Manjolo, 2 April 2001.

70 Interview, Joel Mudimba, Sigalenke, 24 March 2001.

71 Syuungwe was translated as 'mist' in interviews. Clark (1952) translates it as 'rainbow', based on David Livingstone's translation of the term in Livingstone and Livingstone 1865:250. See also Shapera 1963: 326. The Lozi version of the word is 'Shungu'.

72 Sykes (1905:2) describes a 'native song' with the lyrics, 'how should anyone lose his way with such a landmark to guide him?'

73 Interview, chief Mukuni, Mukuni Village, 14 September 2001. A similar story is told in Mubitana 1990:70).

74 Interview, chief Mukuni, Mukuni Village, 12 September 2001.

75 Interview Maxon Musaka Ndlovu, Chidobe, 27 March 2000.

76 Muntemba 1970:28-39.

77 Also described in M. Muntemba ibid. and Mukuni 1957.

78 Interviews, Mukuni Village, September 2001; Maxon Musaka Ndlovu, Chidobe, 27 March 2000. This 'secret' is now public knowledge; its location is described in Chikumbi 1998..

79 The formalized system of chiefs and headmen described by Muntemba, ibid.:29-30 and by Mukuni ibid., almost certainly reflects administrative hierarchies developed as a result of government recognition.

80 See Chikumbi ibid.: 24-6.

81 Some of Hwange's Nambya, for example, visited the Falls and respected the Leya spirit mediums most closely attached to it, despite retaining a central attachment to graves of their former leaders, and the ruins of their former capitals, some 80 kilometres south of the river.

82 For further detail, see McGregor 2005a.

83 Scudder 1962.

84 Colson (pers. comm.) recalls a Gwembe woman saying to her in 1956. 'You think it is good [the river]. It is bad. It takes people'. She notes that in the Mwemba area, there were places in the Zambezi where outcasts were thrown, such as witches, or children who cut the wrong sets of incisors first. This place was a place of fear, to be avoided. See also, M. Tremmel, 1994:22.

85 Called Simusinsi by the Dombe/Nambya in Hwange and Simwaba by the Tonga in Binga. The Leya also have traditions of a monster living at the foot of the Falls.

86 This is commonplace in central and southern Africa, after various rain ceremonies, not just those at the river, both among the Tonga and others.

3

Mapping the Zambezi
Imperial Knowledge & the Zambezi Frontier

When European explorers began to travel along the mid-Zambezi from the 1850s onwards, they imagined it in much the same way as other important African rivers, such as the Nile, the Congo or the Niger – as a route of access between coast and interior. Fifty years later, however, the mid-Zambezi was understood as a barrier and 'natural border', and had been mapped as a boundary between separate colonial states. This chapter charts the process through which this transition occurred. Like the African ways of talking about the river discussed in the previous chapter, colonial discourse about the Zambezi mattered: it was intimately related to power, and to shifting structures of authority and their justification.

Recent interest in nineteenth-century exploration has been concerned primarily with the metropolitan ideas the writers reflected and helped to shape. My interest here, in contrast, is in the relationship between European discourse about the mid-Zambezi and the politics of the place itself – in how European views were influenced by the African idiom, practices and relations of power, how the river route came to be cast as violent and the effects of this caricature, and how imperial discourse developed over time, and in turn, shaped mid-Zambezian history.

By examining explorers' sources of information, and the perspectives of the African intermediaries who moved with them as guides and interpreters, it is possible to suggest ways in which local discourse shaped travellers' writing.[1] I argue that British imperial discourse about the mid-Zambezi incorporated African ideas of hierarchy and difference, and the violence attributed to the river route and riverine society in some texts could be made more credible by drawing on African discourses and stereotypes of African others, and actual conditions of acute insecurity and conflict on a contested frontier.

As European explorers selectively appropriated African ideas, they also reframed and redeployed them, in ways post-colonial critics have elaborated.[2] Travellers entangled local notions of hierarchy and difference with contemporary European ideas about race, class, tribe, degeneration and 'the tropics', recast African landscape ideas in terms of natural history, or aestheticized the places they moved through, drawing on strands of romantic thought that privileged individual, visual responses.[3] Yet further understanding of how European representations incorporated aspects of local discourse is important, partly because the history of the places being described tends to get lost in this work.[4] This has de-emphasized the constraints of African relations of power, masking the marked historical continuities between some pre-colonial internal African frontiers and the borderlands of today.[5]

Just as African ideas could be reframed to support the imperial discourse of the river route as disorderly and violent, so too could they lend weight to the idea that the mid-Zambezi was a 'natural border'. This does not mean that the border is an 'appropriate' designation that fits pre-colonial geographical, political and cultural realities: pre-colonial states were not culturally and linguistically homogenous, and were characterized by contested frontiers and shifting alliances, rather than by fixed, linear borders. The chapter begins by revisiting European ideas about the river as a highway to the interior and a place of violence, as they developed between the 1850s and 1880s, examining their relationship to local Zambezian politics and discourse. It then turns to the period from the mid-1880s, when the process of concession-making and occupation of the Rhodesias demanded new sorts of knowledge about the landscape, and led to the mapping of the river as a border between Ndebele and Lozi spheres of influence.

European Explorers & the Violence of the Zambezi River Route, 1850s-1880s

My discussion of European discourse about the river begins, as might be expected, with the work of David Livingstone, whose status as 'discoverer' of the landmarks of the mid- and upper-Zambezi and author of the first descriptions of those who lived along it, gave his texts a particular impact. Livingstone's legacies have been far-reaching, not only in justifying imperial intervention in general terms, as Dorothy Helly argued,[6] but more specifically in legitimating a series of Zambezian interventions and constructing the myths of white Rhodesia, discussed later in the book. Here, I start by revisiting Livingstone's texts and the secondary literature on the man, to investigate local African influences on his time in the mid-Zambezi, and to highlight how his writings about the mid-Zambezi 'river people' are infused by the perspectives of the Kololo hosts and interpreters on whom his travel depended: my argument is that Livingstone often saw through Kololo eyes.

Livingstone first reached the mid-Zambezi in 1851, at a time of antagonism between the Kololo and Ndebele, when alliances with the 'river people' of their mutual frontier had a particular importance for raiding and defence. The region had not been previously mapped or described by European writers, but it was well connected, particularly with the south (from whence the Kololo had so recently migrated) via routes maintained by Griqua traders, as well as to the west and to a lesser extent, the east.[7] Information as well as goods flowed along these routes, and travellers coming from the south were anticipated when they arrived – David Livingstone himself had been invited to visit. The Europeans who used these routes were part of a highly competitive rush for knowledge and ivory: when Livingstone made his first trip northwards to the Kololo, others were doing likewise, including (according to Tim Jeal) ivory trader H.N Wilson, the naturalist J. Leyland and a son of a missionary, Sam Edwards.[8] There were other explorers investigating routes between west and east coasts, such as the Portuguese trader Serpa Pinto, who reached the Kololo from the west coast with sponsorship from the Angolan government.[9]

Kololo authority in the mid-Zambezi was precarious, thanks to Ndebele raids across the Zambezi facilitated by the mid-Zambezi river people. When Livingstone reached the Kololo, they had just been driven off the Batoka highlands by an Ndebele attack and forced down into the swampy, low-lying, fever-ridden margins of the Zambezi river, upstream from the Victoria Falls. As recent Livingstone biographies have emphasised, the Kololo were keen to host Livingstone for their own reasons: they wanted new sources of arms to fend off raids from their powerful southern neighbours and wanted to use Livingstone's status as son-in-law of the missionary Moffat, (based with the Ndebele), to try to negotiate an end to Ndebele raids.[10] They also needed to shore up their shaky authority over those living along the Zambezi river who repeatedly threatened insubordination and controlled the crossing points, allowing Ndebele access to the north bank. Livingstone depended on the Kololo leaders' collusion with his plans, and represented them as a superior people in Central Africa. In making his case for their superiority more believable, Livingstone could draw on the Kololo leaders' own representations of themselves and the others they deemed low and different, whom they had subjected to raids and only incompletely subordinated.

Livingstone was welcomed by the Kololo leaders as an honoured guest; on his first visit in 1851 (with William Cotton Oswell) he was hosted by Sebituane, and on his return trips (in 1853-5 and again in 1860) by the later incumbent, Sekeletu. When he moved around the Kololo heartland, he travelled with a royal Kololo entourage, and when he ventured further afield he was given Kololo interpreters, guides and teams of porters. Livingstone could communicate directly with these Kololo intermediaries in their own language, as a result of his previous missionary work.[11] As recent migrants from the South, Kololo leaders spoke Sotho/Tswana rather than the languages previously indigenous to the Zambezi valley. Livingstone wrote self-servingly on his first trip, that 'Sebituane had done a good service introducing the Sichuana [language to the Zambezi]... we found people could understand us everywhere'.[12] Livingstone did not learn to speak the other languages of the mid-Zambezi, though he collected some words from them.[13] Although Livingstone's interests were wide-ranging, his primary concerns were with geography, natural history and language. His ethnography, as Shapera has pointed out, was 'brief and sketchy',[14] and took the form of 'relatively scrappy notes, many based on hearsay, not personal observation'.[15] The fact that much of Livingstone's ethnography is based on hearsay is important, as a good deal of what the explorer heard was filtered through Kololo interlocutors. Livingstone's first and most influential book, *Missionary Travels*, is a composite text, assembled from the notes and reflections in his journals, often substantially revised or elaborated. Even the journals, though closer to his Zambezian experiences, were not always recorded immediately.[16]

For the purposes of the discussion here, the differences between these texts are less significant than their similarities; the distinction between the Kololo and the rest, and the particular contempt for those designated Tonga or 'river people' is reproduced in all Livingstone's writings, and emerges particularly clearly in *Missionary Travels*. There are, of course, also frequent contradictions, which is typical of this genre of travel-writing. Although Livingstone (and the Kololo) clearly distinguished between the various mid-Zambezian people, they also used shorthand terms for these other groups, sometimes referring to them collectively as the 'river peoples' or 'black', and sometimes using the various ethnic labels (Tonga, Lozi, Kalanga, Nambya, etc.) loosely to imply subordination and inferior status; Livingstone assimilated the 'subject tribes' to a concept of class hierarchy rather than 'slavery', though he saw their relationship to 'true' Kololo as 'akin to' slavery.[17] Both the distinctions between the Kololo and the rest, and the specific contempt for the Tonga seem to be based on local Zambezian ideological hierarchies, as seen from a Kololo perspective. These categories referring to the bottom of Kololo ideological hierarchies are loose and overlapping, such that there is a conflation and slippage between 'river people' and Tonga, both of which are terms of denigration from a Kololo point of view, even though not all Tonga were river people, indeed most did not live along the river as extensive Tonga settlements stretched up into the hills and onto the northern plateau in particular. The racialized designations in Livingstone's work, in which the Kololo are described as regarding their own lighter brown colour as a mark of status in contrast to the river people they termed 'black', may also have had a basis in Kololo idiom. Livingstone notes the Kololo themselves 'took pride in their lighter complexion', 'as it distinguishes them considerably from the black tribes of the river'.[18]

Some lengthy and glowing passages about the Kololo in *Missionary Travels* appear to be based on notes from what Sebituane and his councillors told Livingstone on one of his first visits of how the Kololo came to be where they were. [19] Livingstone tells the reader that Sebituane's tale 'resembled close the "Commentaries of the Caesar" and the history of the British in India', and represents their narrative of migration and conquest as a civilizing force in the mid-Zambezi.[20] He casts the Kololo as proud warriors led by the intelligent and brave Sebituane, who was 'more frank in his answers than any other chief I ever met. He was the greatest warrior ever heard of beyond the [Cape] colony.' Sebituane was not aggressive by nature but 'was always forced to attack the different tribes, and to this day his men justify every step he took, as perfectly just and right…'[21] Those whom the Kololo fought and subordinated appear in the context of this heroic Kololo narrative as savage and contemptuous, violent, dishonourable and unmanly, as people who prior to their subjugation had exercised power through trickery and ignoble means, and whose defeat was a blessing:

> The Batoka lived on large islands in the Leeambye, or Zambezi, and, feeling perfectly secure in their fastnesses, often allured fugitive or wandering tribes onto uninhabited islets on pretence of ferrying them across, and there left them to perish for the sake of their goods… The whole of Batoka country was then densely peopled, and they had a curious taste for ornamenting their villages with the skulls of strangers. When Sebituane appeared near the great falls, an immense army collected to make trophies of the Makololo skulls, but instead of succeeding in this they gave him a good excuse for conquering them… Sebituane had now not only conquered all the black tribes over an immense tract of country, but had made himself dreaded even by the terrible Mosilikatse [Mzilikazi]. He never could trust this ferocious chief, however; and, as the Batoka on the islands had been guilty of ferrying his enemies across the Zambezi, he made a rapid descent upon them, and swept them all out of their island fastnesses. He thus unwittingly performed a good service to the country, by completely breaking down the old system which prevented trade from penetrating into the great central valley.[22]

This is a particularly stark example of Livingstone portraying the subordinated river people entirely through the narrative of leaders of conquering others. But there are many examples of Livingstone repeating what Kololo interlocutors told him without distancing himself from their views. For example, Livingstone heard from Kololo sources of Hwange's flight across the river in quest of Kololo protection from Ndebele raids and of Kololo plans to disperse them as slaves; in recording the incident, he echoed Kololo moral contempt for people he had never seen or spoken to by noting that 'They [Hwange's people] appear to be thoroughly degraded and vile.'[23] When he personally travelled through those who had once held 'despotic sway' over the islands around the Victoria Falls in 1855, his Kololo guides recounted some of the same stories about the river people. Livingstone's journal account of his visit to Leya chief Sekute's island above the waterfall (Kalai), includes a description of more than 50 mounted skulls and reproduces Kololo stories of how parties from the south were tricked by the Leya and stranded mid-river: 'As the island is surrounded by a strong current, these Batonga felt themselves quite secure, and excelled in pride and cruelty. A party of Bamangwato were, on pretence of ferrying over the river, ferried on to an island and cruelly left to perish...'[24] Livingstone compared such power through deception unfavourably with the Kololo's 'manly frankness'.[25] More stories of skulls mounted 'in the Batoka fashion' are told in the journals and *Missionary Travels* in relation to a visit to Leya headmen Moyara. In this instance, the river people – 'the barbarians of the islands' – are presented as colluding in stereotypes of their own 'fierceness': Livingstone reports being 'assured by other Batoka that few strangers ever returned from a visit to this quarter. If a man wished to curry favour with a Batoka chief, he ascertained when a stranger was about to leave, and waylaid him at a distance from the town and when he brought the head back to the chief, it was mounted as a trophy.'[26]

When Livingstone left royal Kololo company on his first descent of the Zambezi in 1855, he was given Kololo guides and interpreters and a force of 114 men, who were mostly Kololo subordinates – Tonga, Leya, Nambya and others.[27] Of the interpreters, a man named Sekwebu was the most important – he was Livingstone's primary companion and had authority over the other men (which he exerted via headmen for ethnic sections, Monahin for the Tonga, Mosinyane for the Nambya and Luba, and Mokoro for the Subia and Lozi).[28] It is not clear where Sekwebu was born or how he came to assume a position of trust among the Kololo and to be counted among their number, and Livingstone refers to him as Ndebele as well as Kololo. Schapera explains that Sekwebu 'was captured by the Matabele when a little boy and the tribe in which he was a captive had migrated to the country near Tete; he had travelled along both banks of the Zambezi several times, and was intimately acquainted with the dialects spoken there.'[29] Perhaps what is significant is that Sekwebu clearly did not regard himself as one of the Zambezi river peoples, and grew up identifying with the immigrant African powers. After Sekwebu drowned or committed suicide in the Indian Ocean, Livingstone subsequently relied on a 'real' Kololo called Konyata who assumed chiefship of the remaining members of the team.[30]

As Livingstone's party moved from the Kololo heartland eastwards into Tonga country, first passing through villages that recognized Kololo rule, and then proceeding into 'rebel' and independent areas, his views about the people they travelled through were shaped by Sekwebu, and by Kololo attitudes to those living on an unruly, raided and disloyal frontier. The party were on the plateau at this juncture, rather than down in the river valley itself, but the positioning of 'Tonga' at the bottom of Kololo ideological hierarchies is repeated.[31] Livingstone described his own disgust at the Tonga practice of knocking out the top four teeth, but reinforced and justified his reaction with information that the Kololo shared this perspective, and that Sebituane had tried unsuccessfully to ban the practice.[32] Livingstone's criticism of Tonga forms of address and nakedness were also bolstered by his Kololo interlocutors' reaction to cultural practices they regarded as different and inferior. When the troop moved out of areas recognizing Kololo rule, into Tonga villages that had recently experienced Kololo raids, Sekwebu was clearly wary of what kind of reception they might receive, and his unease shaped Livingstone's own reactions. In some instances, Sekwebu managed to use their visit as an opportunity to collect tribute and assert Kololo authority, but in others this was clearly not possible. Livingstone reported how his men were always on guard for signs of hostility and deception, how they mounted a night-time defensive vigil in one village, feared being used by Tonga hosts for their own purposes, were suspicious of being given bad directions, and on several occasions were on the point of resorting to force of arms. His descriptions of Tonga untrustworthiness and use of deception, their propensity to resort to violence, or their lack of moral fibre and cowardliness can be read partly as based on the hierarchy Sekwebu had constructed within their own party, and Sebwebu's reactions and attitudes to passage through those the Kololo regarded as low, and who they feared and distrusted.[33]

Indeed, Livingstone's party clearly reproduced Kololo ideological hierarchies among themselves, distinguishing between the superior Kololo and the rest, and singling out the Tonga as the worst of the lot. With Sekwebu in charge, the others, 'though not slaves', were, 'looked on as an inferior class.'[34] Livingstone himself often echoes Sekwebu's sense of Kololo superiority over these others, and the particular disregard for the Tonga who were 'more difficult to manage than any of the rest of my companions, being much less reasonable and impressible than others'.[35] When Livingstone complained about the behaviour of his men, who stole from, or forced into porterage, those through whom they passed, only Sekwebu appears to be excluded, and the Tonga take the blame. Indeed, his interpreter frequently seems to be included in the designation 'we' in the diaries; for example:

> The Batoka are more degraded than the Barotse. They have less self respect; savage and cruel under success, but easily cowed and devoid of all moral courage. The majority of my company are of this tribe, and a more reckless set one cannot conceive. We have to keep a strict lookout against being involved by their thieving from the inhabitants, in whose power we undoubtedly are. But it is more difficult to manage their tongues. When the people come to make their obeisance, they make wicked remarks in their own language. And some even point to villages, saying 'I killed a man there', 'I broke all the pots in that village', etc. etc., within hearing of the villagers.... When they performed the valiant deeds mentioned, they were in company of the Makololo as a conquering army...[36]

Livingstone thus drew on aspects of local discourse to bolster and make credible his representation of the Kololo as superior to others in the mid-Zambezi By largely suppressing the information he had gained about Kololo raiding and maltreatment of subordinates (which was particularly evident in his journeys upriver with his Kololo entourage, when people fled as they approached), he presented an interpretation of violence in which those subjected to Kololo rule – the Lozi and Tonga – were cast as its source, both through their savage disposition and though their contact with Portuguese slavers and the slave trade, which they were held to have introduced and encouraged to operate in the area, with the Kololo participating only reluctantly through lack of alternative source of goods. Ideas about the violent degraded character of these different subjects, often glossed together as 'river people', gained further support from ideas about the degenerating influence of the unhealthy, swampy, fever-ridden lands in the valley, contrasted to the healthier, drier plains to the south from which the Kololo originated and the healthy Batoka uplands where they had tried to establish themselves and which were the focus of Livingstone's plans for mission, trade and agriculture. Livingstone's views, as presented in *Missionary Travels,* were influential and others repeated his descriptions as ethnographic fact. Thus a volume of Livingstone's lectures edited by the Rev. William Monk includes a summary of Livingstone's contribution to ethnology in which the Kololo are characterized as the only tribe in the mid-Zambezi with 'some European characteristics'. The subordinated others are seen as more degraded, with the Tonga cast as 'probably the most complete savages with whom our traveller has held intercourse in Africa'.[37] The view that the non-Kololo inhabitants of the mid-Zambezi were exceptionally violent and savage was thus taken up in metropolitan Britain among intellectuals who tried to formalize understanding of mid-Zambezian society in terms of contemporary debates about race, degeneration and environment.

Equally if not more influential than his beliefs about the mid- and upper-Zambezi people he encountered, were Livingstone's views on the geography of the Zambezi river, particularly the idea that it was 'God's highway' to the interior, also popularized in *Missionary Travels.* Livingstone's representations of the geography, landscape history and navigability of the Zambezi related to local conceptions in rather different ways than his views of those who lived along it, and were less readily bolstered by local discourse. Moreover, as his Kololo interlocutors were themselves recent immigrants and acquainted quite superficially with only a particular section of the river, and as they had better trade contacts with overland routes to the south (and to a lesser extent the west) than with the east, they could not furnish him with the information he required. Livingstone did, of course, draw much knowledge about the mid- and upper-river system from various (non-Kololo) informants; indeed when he finally explored it in 1853, he commended the 'wonderful accuracy' of a map of interlocking tributaries he had drawn two years earlier based entirely on local knowledge, and described how 'after all the care I could bestow, that the alterations I was able to make in the original native plan were very trifling'.[38] Yet he could find no support in local ideas for his interpretation of landscape history through contemporary catastrophist theories, and his tenacious persistence in the belief that the river was navigable to the sea could only be maintained by ignoring much contradictory information from local sources. The latter is a major theme of the anti-hagiographies of Livingstone, produced from the 1970s, such as Tim Jeal's *Livingstone,* George Martelli's *Livingstone's River: The Story of the Zambezi Expedition 1858-1964* and Timothy Holmes' *Journey to Livingstone: Exploration of an Imperial Myth.*

Livingstone's first sight of the Zambezi, some miles upstream from the Victoria Falls in 1851, was encouraging, however, partly through association with the navigable rivers he knew in Scotland (particularly the Clyde, which is a repeated point of reference), and the commercial and industrial development they had helped foster. As he recorded in his diary:

> All we could say to each other was to express our great pleasure by saying to each other, How glorious! How magnificent! How beautiful! And grand beyond description it really was. Such a body of water, at least 400 yards broad and deep... The scenes of the Firths of Forth and Clyde were brought vividly back to my view and had I been fond of indulging in sentimental suffusions, my Lachrymal apparatus seemed fully charged.[39]

His enthusiasm was not shaken by the stories he heard of a great waterfall not far downriver. When he finally saw it in 1855, he was impressed enough to name it after Queen Victoria, but its size was such that it posed an obvious problem for any plan based on waterborne transport, and perhaps for this reason, Jeal suggests, Livingstone underestimated the dimensions of the waterfall by half.[40] Livingstone once again showed the influence of his Kololo interlocutors in recording the Kololo name '*Mosi oa tunya*' ('the smoke that thunders') as current, describing the Leya name '*Syuungwe*', with its focus on fertility and rainbows, as 'ancient'. Livingstone was fascinated by African riverine gardens, and intrigued by the history of the landscape itself. He tried unsuccessfully to collect oral traditions of a cataclysmic earthquake to explain the dramatic 'rents' of the Falls that could have drained a vast inland lake.[41]

Livingstone felt that waterfalls and cataracts were 'frightful' features that 'spoiled' the river.[42] He did not aestheticize the landscape of the Victoria Falls, as he had done the wide expanse of the river; Jeal notes how his *African Journal* description of it is 'miraculously precise and flat',[43] and that he only inserted into *Missionary Travels* what became famous romantic lines about the beauty of the Victoria Falls at the request of his publisher, John Murray ('scenes so lovely must have been gazed upon by angels in their flight').[44] Reflecting on the obstruction waterfalls posed to civilizing river transport, he noted:

> The sea is after all the great civilizer of nations. If Africa, instead of simple littoral outline, had been broken up by deep indentations of glorious old ocean, how different would have been the fate of its inhabitants. The waterfalls of Mosioatunya [Victoria Falls], Kabompo and others, explain why commercial enterprise never entered the interior of the continent except by foot travellers. I am sorry for it. My dreams of establishing commerce by means of the rivers vanish as I become better acquainted with them. But who can contend against nature? Can these cataracts not be passed by placing boats on frames with wheels? Difficulties are not always insurmountable.[45]

As he moved downstream for the first time, Livingstone initially left the course of the river itself to explore the agricultural possibilities of the high Batoka plateau, and then famously omitted to visit the formidable obstacles to river transport – the Kariba rapids and waterfalls and gorge at Cahora Bassa. He explained away the lack of significant commercial traffic on the river by denouncing the Portuguese as 'effete and syphilitic', 'the lowest of the low', criticizing their boats as unsophisticated compared to British steamships, condemning their inadequacies as a commercial and colonizing power, and supporting his arguments with bad calculations that failed to take account of changes in the river's flow and the vast changes in height over its course.[46]

Back in Britain, Livingstone promoted his vision for cotton-producing, Christianized Kololo communities on the healthy Batoka highlands, linked by steamship to the coast, with British influence and 'legitimate commerce' replacing the evils of the slave trade. The map of interconnecting waterways that accompanied *Missionary Travels* (produced by the geographer John Arrowsmith) was highly influential and helped further his cause.[47] Bridges argues that the map not only reflects Livingstone's growing 'obsession' with lakes and rivers, but also acted as an 'invitation to apply technology in the form of steamboats', given its failure to represent information about flow rates and relative elevations.[48] Livingstone's proposed Zambezi expedition copied both the abolitionist rhetoric of the earlier Niger expedition and its emphasis on technology, science and development.[49] Having secured support from powerful commercial, governmental and scientific patrons and agreeing to recognize Portuguese influence over the Zambezi as far as Zumbo, Livingstone thus left Britain in 1858 as 'Her Majesty's Consul at Quelimane for the Eastern Coast and the Independent Districts of the Interior', accompanied by an influential team of scientists, equipped with a small steamship for the thousand-mile journey upstream and a sugar mill for the Kololo.[50]

The expedition was held up as soon as it tried to enter the shallow Zambezi mouth, as the small steamer first faltered on shifting sandbanks and then struggled against the strong current. But Livingstone was only finally deterred by the Cahora Bassa gorge and waterfall, which ended his ambitions for the mid-Zambezi, and he began to turn his attention to the more accessible Shiri river and its highlands. He deliberately withheld his newfound knowledge of the un-navigability of the Zambezi from his sponsors, however, explaining the delay through the inadequacies of his steamship, and requesting a more powerful vessel.

While he was waiting for a new boat to arrive, Livingstone decided to take home the troop of Kololo men who had accompanied him in his first trip downstream. To do so meant another journey up and down the mid-Zambezi, which he embarked on in April 1860, accompanied by his brother Charles and the botanist John Kirk. The published account of the trip, in *Narrative of an Expedition to the Zambesi,* co-authored with Charles, was less influential than *Missionary Travels*, but is important for our narrative here, as the party actually followed the river's course on this occasion, and there are some notable continuities as well as contradictions with the earlier text in the ways that local mid-Zambezian discourse is represented. Whilst *Narrative* perpetuated many of the views of Livingstone's earlier writing and continued to show the influence of Kololo interlocutors, it also gives a much more critical account of Kololo authority than *Missionary Travels,* and presents information from Kololo informants in a much more distanced manner. The change may reflect Livingstone's own reorientation from the mid- and upper-Zambezi to the Shiri river, and may also have been influenced by Charles' views: Jeal shows from Charles' contemporary letters that he did not share his brother's romanticized view of the Kololo.[51] Moreover, by the time *Narrative* was published in 1865, Kololo authority in the upper-Zambezi had been overthrown.

The Kololo guides who accompanied Livingstone on this occasion were a much diminished troop. Having been left by him in Tete in 1855, the Kololo team had established themselves successfully as traders, raiders and overlords. Many were reluctant to leave. Of the initial party, only 60 set out on the return journey, and half of these deserted before reaching Cahora Bassa. Of the thirty who remained with the Livingstones on the trip upstream to Kololo country, only two were described as 'true' Kololo and the rest subordinates. [52] The Kololo men at the head of the party once again influenced what the travellers heard about the mid-Zambezi people they met along the river, but the Livingstones represent Kololo biases quite explicitly this time, and continually cast doubt on the veracity of the stories their Kololo interpreters told them. Introducing their travel through the 'independent' Tonga chiefs of the river who called themselves Bawe (Mwemba, Sinamani, Sinamakonde and others), the Livingstones begin by casting their Tonga hosts as 'very friendly', generous and hospitable, and on hearing of stories of raids by the Kololo (allied with Batoka from around the Falls), suggest that 'it looked as if the marauders were shutting up the country which they had been trying so much to open'.[53] When the Kololo guides told the Livingstones how violent and savage the Tonga were, the Livingstones recorded their views with scepticism:

In domestic contentions the Bawe [Tonga] are careful not to kill each other, but when one village goes to war with another, they are not so particular. The victorious party are said to quarter one of the bodies of the enemies they may have killed, and to perform certain ceremonies over the fragments. The vanquished call upon their conquerors to give them a portion … and when this request is complied with they too perform the same ceremonies and lament over their dead comrade, after which the late combatants may visit each other in peace. Sometimes the head of the slain is taken and buried in an anthill, till all the flesh is gone and the lower jaw is then worn on as a trophy by the slayer, but this we never saw and the foregoing was obtained only through an interpreter.[54]

Rather, the Tonga emerge in this section as 'men of peace', unlike the Kololo and the 'even more warlike' Ndebele.[55] After passing through sites of ruined riverine Tonga villages destroyed by Ndebele and Kololo raids, the Livingstones reached the Victoria Falls and once again record the heroic story of the Kololo conquest, but this time treat it clearly as self-justificatory narrative rather than as fact: 'Sebetuane always justified his subsequent conquest in that country by alleging that the Batoka had come to fight with a man fleeing for his life, who had never done them any wrong.'[56]

When they met Sekeletu, the Livingstones heard of the fate of the short-lived Kololo mission, sponsored by the London Missionary Society, whose weak and emaciated surviving members had left the Zambezi earlier in the year. The mission had faced continuous hostility from the Kololo and six of the initial Europeans died from fever. Jeal explains Sekeletu's hostility towards the missionaries in terms of Livingstone's absence and the lack of guns, and suggests that Livingstone used Sekeletu's version of events rather than that of the surviving missionaries because it allowed him to absolve himself of responsibility for the team's terrible experiences. Moreover, by blaming the missionaries for their own fate and for destroying Kololo goodwill, Livingstone could justify refocusing on other parts of central Africa without having to admit his own culpability for the deaths or his misjudgements about the river or the Kololo.[57] As the Livingstones left for Tete and moved back downriver, Sekeletu gave them a new Kololo escort, under the leadership of Leshore, whose main task was to shore up Kololo authority over the Tonga chiefs who commanded the Zambezi crossing points downstream, to gain assurances that they would not facilitate an Ndebele attack, and to persuade those they had raided in the past that it would not happen again. The Livingstones described in *Narrative* how Leshore

[b]esides acting as a sort of guard of honour to us was sent on a diplomatic mission to Sinamane. No tribute was extracted by Sekeletu from Sinamane, but… he was expected to act as a guard in case of the Matabele wishing to cross and attack the Makololo … all that Sekeletu asks of him is not to furnish the Matabele with canoes when they wish to cross the Zambezi to attack the Makololo… Leshore's mission, as we here said, was to ratify their vassal-ship, to request Sinamane to furnish us with what canoes he could, and to assure him that [the Toka chief Moshobotwane] had not received, and never would receive, authority from Sekeletu to go on forays among his countrymen.[58]

Despite this more measured presentation of the nature of Kololo rule, and more distanced repetition of information derived from Kololo sources, *Narrative* nonetheless still explicitly casts the Kololo as superior to their subordinates – with the contradictions explained away by the argument that the new generation of Kololo born in the mid-Zambezi had 'been brought up among the subjected tribes', and 'acquired some of the vices peculiar to a menial and degraded race'.[59] The Livingstones added an editors note to inform the reader of the overthrow of the Kololo by the river people in 1864, 'that fate we deplore, for whatever other faults the Makololo might justly be charged with, they did not belong to the class who buy and sell each other, and the tribes who succeed them do.'[60]

Throughout Livingstone's writings on the mid-Zambezi, it is thus possible to see traces of the Kololo's derogatory views of those they raided and subordinated. The Kololo perspectives incorporated in Livingstone's work provide a counterpoint to the views of the river people themselves, discussed in the previous chapter, in which historical tales of river crossings upheld the moral value of independence from centralized authority and the practical value of command over the river, validated trickery to avoid subjection, affirmed the wisdom of knowing when not to make a stand and saw no shame in running away. By reproducing Kololo views, Livingstone could strengthen his anti-slavery narrative and interpretation of violence in the mid- and upper-Zambezi, in which the Kololo riverine subordinates were cast as its source, both through their savage and degenerate disposition, and through their contact with Portuguese slavers and the slave trade. As we shall see, subsequent travellers referred back to Livingstone and an association between violence and the river route could persist despite the contradictions in Livingstone's work, even as this strand of imperial discourse was marginalized.

Though Livingstone's attention was subsequently diverted to other peoples and other places, the idea of the Zambezi as a potentially navigable highway into the interior was not put to rest, and had broad currency among European travellers and imperial strategists. The writers who followed Livingstone were often less detailed in their observations of the mid-Zambezi, give only snippets of information about interpreters and sources, and are generally less amenable to arguments about the incorporation of local discourse (though the growing density of texts can provide some compensation). Many were more preoccupied with the powerful African states – the Ndebele and Lozi – cast as 'warlike' and militaristic (as opposed to the disorderly and chaotic violence of the Tonga chiefs), as well as the Portuguese, who tended to bear the brunt of their condemnation, such that those along the river in its middle reaches were cast as victims rather than sources of violence. In the remainder of this section, I shall discuss these other writers in relation to historians' reconstructions of changing realities of power along the river, highlighting what little insight the texts allow on the perspectives and influence of interlocutors.

In 1862, the trader and explorer James Chapman visited the mid-Zambezi with Thomas Baines, with aims that were remarkably similar to those of Livingstone: to establish a line of trading posts stretching from west coast to east, and to test the potential of the Zambezi as a 'river route'.[61] The commercial side of the venture failed dismally, as Sekeletu refused to trade on terms Chapman regarded as reasonable, and stranded him in the river to force better deals. Chapman did not therefore think highly of the Kololo and described Sekeletu as 'rapacious' and a 'miserable little despot' with a reputation for murdering traders.[62] In contrast to this contempt for the Kololo (and Ndebele), Chapman and Baines wrote more sympathetically than Livingstone of the subordinated and raided Nambya and Tonga along the river, representing them as victims and describing them without the biases of Kololo interlocutors. Chapman had encountered Hwange and his people nine years before, when they had sought protection from Lechulatebe's Tawana, since when they had suffered further Ndebele attacks before crossing the Zambezi river to live as Kololo subordinates. Chapman's guides, porters and informants included Nambya and Kalanga who had experienced these attacks, and now lived under the Kololo. Chapman blamed the Kololo and Ndebele for slavery, arguing that their monopoly of trade in ivory left the subordinated people along the river nothing but slaves to sell, reporting a conversation with a Toka headman: 'if I wished to buy a little boy or girl he could accommodate me, but that the ivory trade was quite monopolized by Sekeletu.... We, said he, who have not ivory, can only sell our slaves if we want to buy cloth.'[63]

Aside from descriptions of the evils of Kololo and Ndebele rule, Baines and Chapman made elaborate romantic descriptions of the landscape and began to establish the necessary components of a visit to the 'natural wonder' of the Victoria Falls.[64] Chapman's photographs of the waterfall were some of the first taken on the continent, and Baines paintings gained a wide audience.[65] Both writers evoked Livingstone, revisiting the sites he had described, such as 'Garden island', where a tree bore his initials from his first visit and inscriptions from his return with his brother in 1860. Through such homage to predecessors in a considerable volume of travel-writing, the landscape of the waterfall began to accrete European names and associations, and British imperial horizons extended north from Cape Town to encompass it.[66]

Although such aestheticized representations of the landscape acted to obscure local structures of authority, the African powers nonetheless continued to matter crucially to travellers. From the mid-1860s, political turbulence within the centralised states on either side of the river brought a halt to traffic up and down the trade routes to the south: to the north of the Zambezi, travellers were kept out after the overthrow of the Kololo due to a period of conflict before Lozi rule was consolidated, while to the south in Matabeleland, the turmoil following Mzilikazi's death was prohibitive until Lobengula's accession in 1870. This turbulence in the two centres of centralized authority was important for those along the river, as it gave scope for independent trade and accumulation.[67] Political authority along what had been a contested frontier briefly ceased to be shaped by alliances with centralized powers who regarded the Zambezi as a separating barrier. This did not, however, bring an end to the realities of violence along the river.

During this period, the Zambezian trade routes and connections with the east coast became much more important: Portuguese and Chikunda trading stations penetrated rapidly upstream and in so doing re-charged associations between disorderly violence, the river route, and the mid-Zambezian region more broadly. The trading station of Zumbo in the lower Zambezi had been reopened in 1862 and over subsequent years several new stations encroached ever further upriver (to include Inhacoe and 'Selous' islands, and Nkalange, just above the Chete gorge/Binga in the mid-Zambezi).[68] In the mid-Zambezi, trade with the Chikunda was on more favourable terms than downriver, yet historians argue that the violence associated with this Chikunda expansionism was more than simply British imperial discourse: Matthews argues that the Chikunda penetration upstream produced a chaotic and violent local politics, in which the more powerful provoked disputes with neighbours in order to extract slaves and accumulate women.[69] The benefits of this trade to the more powerful mid-Zambezian Tonga chiefs were accentuated, according to Matthews, by the British blockade and Portuguese anti-slavery decrees which meant Chikunda traders began moving slaves upriver for exchange with ivory rather than downriver for export.[70]

This Zambezian trade was still expanding as traffic from the south began to pick up over the 1870s and 1880s. The renewed southern trade links were overseen by the

3.1 Chief Hwange trading grain at Baines' house at Logier Hill (Thomas Baines, 1863) (Museum Africa, Johannesburg)

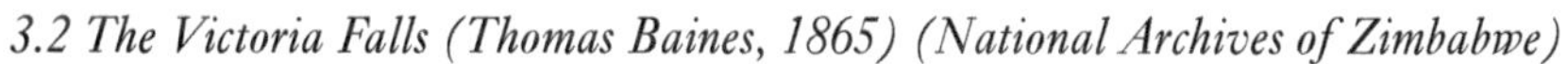

3.2 The Victoria Falls (Thomas Baines, 1865) (National Archives of Zimbabwe)

trader George Westbeech, who was based at Pandamatenga from 1871-88, and ran a network of trading posts on the Zambezi.[71] This new network dealt in ivory, which was exchanged for guns and other goods, and also acted as a hub for labour recruitment – as locals got jobs as porters, and information about opportunities for migrant labour spread, this encouraged the very early flows of migrants to the gold and diamond mines of South Africa, for which the acquisition of guns and escape from subordination were prime incentives.[72] Westbeech also facilitated the dramatic rise in the number of European travellers, particularly from 1876 when he began to act as local agent for hunting excursions, advertised in the London magazine, *The Field*.[73] One of the more influential Europeans Westbeech facilitated was the hunter Frederick Courteney Selous, who made two trips down the river in the 1870s, and whose writings are important in furthering the association between violence and the river, particularly in its lower reaches, through his contempt for Chikunda and Portuguese authority.[74]

Chikunda expansion and associated violence along the river and its hinterland was, however, curtailed and overshadowed by the end of the 1870s, when an old frontier dynamic reasserted itself as the centralized Lozi and Ndebele powers began to compete for the allegiance of those living along the mid-Zambezi and other parts of Tonga country. The small mid-Zambezi chiefs were once again on a contested frontier, sandwiched between two competing African powers. Nambya, Leya and some Tonga along the river thus paid tribute to the Lozi whilst also living in fear of Ndebele; others entered into alliances with the Ndebele.[75] European traders also had to maintain good relations with both major powers, as the Ndebele claimed control over the southern trade route and south bank of the river, while the Lozi claimed the north bank.[76] Westbeech reinforced recognition of the division between their authority, advising Europeans who wanted to cross the river to seek Lozi permission (and used his influence not only to facilitate Selous's next trip,[77] but also to conspire to keep out the Jesuits while letting Protestant missionaries in.[78]

The writing from the mid-Zambezi frontier in this period of Ndebele/Lozi competition showed remarkable similarities with strategic uses of the landscape during the previous era of Ndebele/Kololo antagonism (described by Livingstone from a Kololo point of view). In the 1880s, raiding appears to have begun first from the Lozi side, with raids on Tonga chiefs escalating over the decade, and Lozi rebels seeking refuge among the Tonga, building alliances with some and raiding others. Then Ndebele raids also increasingly crossed the river, likewise facilitated by a complex politics of alliance, in which the role of the mid-Zambezian chiefs was crucial: the raids of 1888 crossed via the Tonga chief Mwemba together with the smaller river chiefs Siachilaba, Siansali, Binga and Sigalenke, who received a share of slaves in return.[79] Tonga elders' accounts of these raids, recorded by Elizabeth Colson, claimed Tonga initiative and maintained that Plateau Tonga chiefs had sent messages inviting the Ndebele to intervene and to oust the predatory Lozi.[80] Colson portrays this use of outside force to resolve local disputes as characteristic of Tonga politics, despite its disastrous consequences when such powerful others were involved, as the Ndebele warriors quickly turned raiders on their Tonga allies.

A particularly revealing account of the strategic role of the river was given by Jesuits who tried to establish a mission among Mwemba's Tonga in the early 1880s on the advice of Selous. Having reached the Zambezi on the south bank opposite Hwange's kraal, they then gave a graphic description of their difficulties in crossing:

> The first question which now presented itself was how to cross the Zambese. This became a serious difficulty which more than once made me shiver... We applied to the old King Wanki [Hwange], who ... answered that he would be put to death by the Barotses [Lozi] if he was bold enough to give us boats so that the caravan could cross... Could we ever cross the river and where could we cross it? It was useless to try at any place within the influence of [the Lozi] King Loboshi. Our only hope was in some tribe of the Batongas which were not vassals...
>
> The great river is a barrier, whose importance is appreciated by the tribes along it, and they keep the barrier carefully in their hands... Notice that all the Batongas on the right bank are vassals of the Matabeles and so it is impossible to settle among them without the permission of Lobengula. Another hindrance!
>
> ... Not being able to cross the Zambese at Wanki's kraal the caravan was forced to go along the right bank and entrust itself to Divine Providence... After having crossed the river Gway... the first Batonga chief whom we met was Tshabi [Saba]. Tshabi received us very cordially but under the influence of Wanki... he also refused to allow us to cross the river...
>
> Two days march after ... we reached the kraal of a young chief Sitcheraba [Siachilaba] ... and we made a bargain with him about crossing of the river. However it was to be an extortionate one. Who could have believed it? When half our baggage was with Fr Teroerde on the other bank our young Sitcheraba had the impudence to have the boats stopped, wanting us to add a certain weight of copper wire to what had at first been stipulated and paid. To enforce this piece of impudence the chief surrounded himself with a troop of young men armed with assegais, who were ready to fall on us.[81]

The Jesuit mission to Mwemba very quickly ended in disaster, as the missionaries succumbed to fever, and survivors fled with tales of poisoning and extortion on the part of their Tonga hosts. This tale of hostility and violence from the mid-river and other Tonga was repeated in the published version of Selous's next Zambezian journey, on which he was accompanied by 'three men who spoke Dutch, Daniel a Hottentot ... Paul, a Natal Zulu [who had married among Hwange's Nambya], and Charley, a lad who had been brought up by one of Westbeech's hunters ... [plus] two of Khama's people ... four Mashunas' and others hired at Pandamatenga.[82] Selous's party tried to move downstream from Hwange on the north bank, at a time when the Tonga along the river were amassed in anticipation of Ndebele raids: they were forced to hand over ever more of their goods, though did not part with the guns that the Tonga men evidently wanted. Encountering similar hostility from the next chief downriver, Selous decided to abandon the Zambezi route 'or the extortion of the Batongas would ruin me...', as in front of him was Mwemba 'the biggest man and the biggest scoundrel of any of them, besides several more of bad repute'.[83] The mid-river Tonga chiefs' reputation for murder and extortion had been built up not only by the fate of the Jesuits and Selous's own experience, but also the death of British trader David Thomas (killed for guns and other property stolen from his Zambezi trading post) and a Portuguese trader.[84] The writings of the 1880s thus had a common narrative strand in the disorderly violence of the Zambezi river route, and the treacherous character of the 'independent' and lawless Tonga people along it and in its hinterland.

For Selous, however, the violence of the 1880s was an 'astonishing change' from the previous decade, which he explained through increased contact between Tonga and Europeans – both through trade with the Portuguese and migrant labour, through which the Tonga were held to have learnt disrespect from the Ndebele.[85] Just as Livingstone had reinforced his derogatory views on the Tonga by repeating his Kololo intermediaries, Selous likewise bolstered his interpretation by reference to the views of his African interlocutors and African discourses of violent others.[86] Selous spoke Ndebele, but did not understand Tonga; he described hearing of the Jesuits' fate in Mwemba first hand from his Zulu worker Paul, who had accompanied the missionaries.[87] He also recounted being warned personally by Tonga leaders of the aggression of their own Tonga neighbours; up on the plateau, chief Monze's councillor, for example, could communicate directly with Selous as he spoke some Ndebele and had told him 'most emphatically not to trust the Batongas, but to hide during the daytime and travel at night ... after walking a mile or so with me they returned home, telling me again not to trust myself in the Batonga villages, or I would certainly be murdered.'[88]

Although these late nineteenth century travellers' accounts thus reflected something of the changing politics of the frontier and occasionally gave insight into local ideological hierarchies, they worked to strengthen stereotypes of the river that were now well established in Britain. Even as the focus of their contempt was increasingly the centralized 'warlike' African powers, their representations of the disorderly and murderous mid-Zambezi tribes and cruel Portuguese and Arab slavers echoed Livingstone and also acted to reinforce an association between the river and violence. Moreover, these texts were now being written and read in an era of high imperialism, which they helped to justify. The river journey was a classic locus for African travel narratives, in which the riverine landscape itself could be used to encapsulate ideas about the continent as 'dark' and provide an imaginative link between metropole and periphery. Conrad's *Heart of Darkness* achieved this particularly memorably (and offensively to post-colonial critics);[89] although the book was about the Congo, it deployed an idiom familiar from a large body of African travel-writing produced over previous decades, including about the Zambezi, and made use of a common nineteenth-century European device of investing rivers with national or racial character.[90]

The most recent biography of Selous reproduces this imperial idiom by using the device of encapsulating violence in the landscape of the river, even as it tries to dispel heroic myths of the man: it describes how Selous first loved the Zambezi, but as he moved downriver he found:

> a chaotic and brutal society in which slaves ... were the common currency, and travellers were in constant danger of extortion... This first encounter with the Portuguese in Africa made a profound impression, convincing Selous that Portuguese colonialism was little short of barbarious [sic] ... anything it seemed, would be preferable to the anarchy here, where individual Portuguese slave-traders conducted their affairs through a network of villainous local cutthroats, either blacks or half-castes. The decay and evil which clung to the Zambezi valley grew more oppressive the more they moved on.[91]

Selous forms a direct link between nineteenth-century exploration and travel-writing, the popular quest and adventure novels that built on them and British imperial manoeuvres on the ground in Zambezia.[92] His conviction of the evils of Portuguese and Ndebele rule in the region did more than justify British expansion to metropolitan reading publics. He had important field knowledge of the river's middle and lower reaches, and of the extent of Ndebele and Portuguese authority, which other imperial strategists lacked, and was able to provide crucial intelligence to Rhodes and others who planned at a distance in ignorance of the landscape and political uses of space. Selous thus spans the transition between the mid-nineteenth-century preoccupation with river routes and trade with the interior, and the late-nineteenth-century concern with occupation and territorial control, which required different sorts of knowledge about the landscape, ushered in a period of concessions and treaties with African rulers, and resulted in the drawing of colonial state borders.

Drawing a 'Natural Border'

Throughout the period considered so far, Europeans had travelled and traded in the mid-Zambezi from a position of weakness: they had negotiated their movements with powerful African leaders whose authority they recognized and who used the visitors for their own purposes. Travellers had observed and documented African political uses of the landscape partly because they needed to know about them to move around, but such descriptions had not been their primary interest. However, the rush to put forces in strategic positions on the ground, and negotiate 'spheres of influence', provided a new context for the collection of knowledge about the landscape and required new sorts of deals with the region's African powers.[93] This became urgent in the mid-Zambezi in the wake of the declaration of the British Bechuanaland Protectorate in 1885, which provoked renewed Portuguese claims to a continuous stretch of land from coast to coast, and stimulated German and Transvaal ambitions.[94]

This is not the place for a full account of the complex African and European politics of the period (and their interactions), which others have already provided.[95] Here, my concern is to revisit this period briefly because it provided the context for a shift in European representations of the river, in which the mid-Zambezi was more consistently cast as a barrier, and came to be mapped as a boundary between Ndebele and Lozi influence, which was then taken as the border between separate colonial states. European strategists' preoccupation with the major African powers in drawing up spheres of influence meant that less powerful tributary or independent peoples were mostly ignored: their political uses of the landscape mattered only insofar as they could be used to claim greater territorial limits for major African powers.

British strategists opposed extending imperial influence beyond the Zambezi for pragmatic and financial reasons, but Rhodes' BSAC agents were ambitious and rushed to secure concessions from Lozi as well as Ndebele leaders, based on exaggerated claims of the extent of their rule. Thus, when Lobengula signed the Rudd Concession in 1888, he controversially claimed Ndebele authority over the territories 'Mashonaland' and 'Manicaland'.[96] In return for signing, Lobengula was offered a significant quantity of arms and ammunition, plus a steamship on the Zambezi.[97] Although Lobengula repudiated the Rudd Concession and looked elsewhere for allies, Rhodes used it to secure the Royal Charter of Incorporation for the British South Africa Company, which was granted in October 1889. The concession recognized the Zambezi as the northern limit to Ndebele authority, but the Charter defined no northern limit to Company operations, describing its 'principal sphere of operations' as 'all of southern Africa West of Portuguese territory, North of Bechuanaland and North and West of the South African Republic'. As Gann notes, 'The Charter thus kept the way open for further territorial extension at a time when the Company's own maps showed Lobengula's effective sphere of influence as going no further than the Zambezi'.[98]

But as we have seen, Lobengula had been raiding across the river, facilitated by the mid-Zambezi Tonga chiefs, on the basis of which he made wild claims to authority over 'a fleet of boats on this river for the passage of my people and impis' as well as control over lands far north of the Zambezi.[99] These claims were not taken seriously, as the north bank was already deemed Lozi territory. Yet Lozi claims to authority over much of Tonga country were little stronger than Ndebele ones,[100] and when a Lozi emissary tried to make deals with the mid-river chiefs to stop the Ndebele from crossing in 1893, the Tonga would neither acknowledge his authority nor receive him, instead they crossed the river.[101] Exaggerated Lozi claims to political authority over Tonga chiefs along the river and up on the plateau (including those so recently allied with the Ndebele) were reflected in the Ware Concession of 1889 and the Locher Treaty of 1890, through which the Lozi agreed to Company protection.[102]

In the light of the expansion of British influence thus achieved, the Portuguese renounced their claims to a continuous band of territory from east to west and recognized rights to free navigation on the Zambezi.[103] However, their influence was still a threat and Rhodes prioritised placing a force on the ground south of the river. Assuming (wrongly) that the Zambezi was passable by heavy military steamboat far into the interior, he gained Foreign Office support to send an expeditionary force to Mashonaland via the Zambezi. This strategy was modified to fit geographical and political realities only through Selous's advice that the Zambezi route was impractical – the river was too low to take ships with 300 soldiers and heavy material, and the Portuguese would oppose the force.[104] Selous had previously criticized the Company for exaggerating Ndebele territorial influence, but when Rhodes offered him work he turned Company agent and propagandist, first heading the pioneer column to occupy Mashonaland (by an overland route), and then signing treaties for the Company with independent chiefs in disputed territories. The presence of British forces on the ground critically altered the balance of power in the region, and the treaty with Portugal was re-negotiated in 1891.

As a result of these manoeuvres and agreements, both sides of the mid-Zambezi were recognized as part of the British sphere of influence. Separate administrations were set up on either side of the river not through design, but through BSAC and British government reactions to unfolding events after 1893 (including war with the Ndebele provoked by the BSAC, abusive Company rule, the fiasco of the Jameson raid and the rebellions of 1896). The Foreign Office had planned to extend BSAC rule north of the river by handing over administrative responsibility for Barotseland, but this was impossible in the context of widespread criticism of the Company in Britain.[105] Separate structures of authority thus persisted, differing both in the degree of imperial control over Native Affairs, and in the way settler interests were (or were not) represented; the future of Northwest Rhodesia was envisaged as a crown dependency, whereas Southern Rhodesia was to be a settler state, with the Zambezi cast as the 'natural northern boundary of what will some day be self governing British South Africa'.[106]

Through all these upheavals and negotiations, the border between Northwest and Southern Rhodesia attracted little controversy. In the eyes of imperial and Company officials it had become a 'natural boundary', in contrast to the 'arbitrary' others which were 'simply marked off by beacon' and attracted further dispute.[107] Successive delegations were sent out over subsequent years to demarcate Rhodesia's borders, to reconsider the boundaries of Lozi rule, particularly to the west, and to reassess the navigability of the Zambezi in its middle and upper reaches.[108] None questioned the border status of this section of the river, which had been thoroughly naturalized through a somewhat confused logic. Most straightforwardly, it was a 'natural border' simply because it was a feature of the landscape, but it was also 'natural' through the circular argument that it was treated as such by African rulers and attracted little controversy. It was thus legitimized through its grounding in the supposed territorial limits of pre-colonial African states. Finally it was seen as the 'natural' limit of white settlement, partly for its reputation for unhealthiness, and partly because of the pragmatic need to limit imperial ambitions somewhere.

Despite the known obstructions to navigating the river, each new expedition that assessed its transport potential between 1895 and 1900 repeated Livingstone's earlier enthusiasm for a grand highway linking interior and east coast. In 1898, a year in which the river was high, a small steamship, *The Constance,* successfully made its way upstream over an 800-mile stretch of the middle river above Cahora Bassa, encouraging the Company and two trading companies to conduct further investigations. Possibilities were mooted (encouraged by surveys of Aswan on the Nile and invocations of Livingstone's foresight) for monumental interventions in the river's course by blasting a cutting or creating a system of dams and locks to avoid transhipment at Cahora Bassa.[109]

As these investigations of the river's transport potential were underway, however, the old east-west axis of the trading economy of the river and Portuguese influence was rapidly undermined by the region's incorporation into the expanding hinterland of the South African economy and the development of north-south axes of trade, transport, industrial development and migrant labour. With the growing importance of communication between South Africa and its hinterland, the river was an obstacle to movement and the key questions became where to route traffic to the north, and how and where to establish official crossing points on the river. The story of the bridging of the river at the Victoria Falls, and the development of new political uses for the landscape of the waterfall is taken up in Chapter 5. But before that, the next chapter turns to the consequences of the new border for those who lived intimately with it.

The idea of the river as a 'natural border' might have had some basis in historical ideas and uses of the mid-Zambezi on the part of the region's African pre-colonial powers, but its designation had ignored the practices of those who lived along the river. Moreover, no state border could ever have been drawn to accurately reflect the territoriality of pre-colonial power, as there was not only one use of the landscape, power did not operate in relation to fixed linear borders encircling contained political and cultural units, and the frontiers between competing pre-colonial polities were shaped by a complex, shifting and violent politics of alliance. The designation of the mid-Zambezi as a border between colonial states was thus the basis of far-reaching local changes, as it introduced a new concept of territorially bounded power, and formed the legal basis of two separate colonial states. As new structures of authority were erected on either side of the border, we shall see in the next chapter how the old violence between the major powers along the river was replaced with the new violence of colonial rule.

Notes

1 On the role of intermediaries, see Hamilton 1988, Johnson 1981.
2 Youngs 1994, Pratt 1992, Duncan and Gregory 1999.

3 On race see Harraway 1989, McClintock 1995. On 'the tropics' see Power 2003. On landscape and natural history, see Pratt ibid., Jardine and Spary 1996, Miller and Reill 1996, Driver 2001.
4 Driver ibid., chapters one and two, Phillips 1997, Edney1990.
5 Nugent 2003, Nugent and Asiwaju 1996, Asiwaju 1984, Kopytoff 1989, Donham and James 2002, Leopold 2005, Stone 1995.
6 Helly 1987.
7 On west coast connections, through Mambari (Ovimbundu) traders, see Chapman 1868 II:139. On Kololo southern links, see Jeal 1973 chapter 8, on links with Griqua traders, see Mainga 1973:82-3. On connections to the east coast, which developed later, see Matthews 1976 and 1981:23-41.
8 Jeal ibid.:99, Holmes 1993.
9 Pinto began this expedition in 1853, but had already visited the Lozi in the 1840s. On the possibility that he or the Hungarian Magyr saw the Zambezi before Livingstone, see Phillipson 1990:76-90.
10 Jeal ibid.:100; Schapera 1960:17.
11 His 'Analysis of the Language of the Bechuana' was published belatedly in 1858. Shapera ibid.:xvii.
12 Schapera 1959:150.
13 Schapera 1960:30-1. Due to Livingstone's influence, later travellers erroneously considered Tswana to be the best language for work among the Tonga, and more generally to be the best linguistic preparation for travels north of the Zambezi in central Africa. The missionary F.S. Arnot, for example, who wanted to work among Mwemba's Tonga, was advised at Potchefstrom to learn Tswana at Shoshong in preparation. Arnot 1969:i.
14 Schapera 1960:xix.
15 Schapera 1963 I:xviii.
16 Schapera 1960: xiii. As late as September 1855, Livingstone wrote to Mrs Moffat, 'If I should try and make a book, it would be compiled out of such and similar everyday affairs, which for the first time I have begun to note down as they occur.' Schapera 1960:xix, Shapera 1959 II: 274. Shapera's italics.
17 The metaphor of class is used repeatedly, e.g. Shapera 1963:318-20, see also Wallis 1956:391. In the first of these citations, Livingstone also draws comparison with with the status of Jews in Europe.
18 Livingstone 1857:155.
19 Livingstone 1987:72-6. The initial, much less embellished, recording of this history of migration is in Shapera 1960:17-26, much of it in note form.
20 Livingstone ibid.:74.
21 Ibid.:72, 74.
22 Ibid.:74, 76.
23 Schapera 1963:289.
24 Livingstone 1857: 454; Schapera ibid.:326. The two accounts differ as to whether Livingstone was still in the company of Sekeletu.
25 Schapera ibid. :329.
26 Livingstone ibid.:454, 457; Schapera ibid.:335.
27 Schapera ibid.:324.
28 Livingstone ibid.:457.
29 Schapera ibid.:322 (note 2), 351-2, Shapera 1959 II:290-1; Livingstone ibid.:682-3.
30 Schapera 1963:322 (note 2), 351-2, Livingstone ibid.:513, 682; Schapera 1959 II:290-1. On possible motivations for Sekwebu's suicide, see Holmes 1993:9, 123.
31 Schapera 1963:349.
32 Livingstone ibid.:456.
33 See for example, Schapera ibid.:347, 349. There are, however, notable contradictions: Livingstone liked Monze and some of the other plateau Tonga hosts.
34 Livingstone ibid.:391.
35 Schapera ibid.:335.
36 Ibid.:349 and a similar passage in Livingstone ibid.:473.
37 Monk 1860:236-7 (on the Kololo) Monk ibid: 243 (on the Tonga).
38 Livingstone ibid.:454.
39 Schapera ibid.:38.
40 Livingstone's point of comparison for assessing the volume of water was Stonebyres on the Clyde. Schapera ibid.:327.
41 Ibid.:334.
42 Ibid.:287.
43 Jeal 1973:148-9.
44 Livingstone ibid.:519.
45 Schapera ibid.:287.
46 Jeal ibid.:154; Holmes 1990 and 1993:97.
47 Stone 1995:50. See also I.C. Cunningham in Larby 1987.

48 Bridges 1994:17, 12; Stone ibid.:50, 112.
49 Helly 1987:228-9.
50 Sponsors were the Royal Geographical Society, Manchester Chamber of Commerce, British Association for the Advancement of Science and the Foreign Office.
51 Jeal 1973:230. Livingstone's own journals for this trip are cursory in comparison to his earlier visits, Wallis 1956. Kirk's records for the journey are missing, as he lost his notes in the Cahora Bassa rapids on their return. Foskett1965.
52 Jeal ibid.:230.
53 Livingstone and Livingstone 1865:181.
54 Ibid:186.
55 Ibid.:188.
56 Ibid.:200-1.
57 Jeal ibid.:chapter 12.
58 Livingstone and Livingstone ibid.:245.
59 Ibid.:230.
60 Ibid.:227-8, endnote 2.
61 Chapman 1868 II:127, Wallis 1941:182.
62 Chapman, ibid.:98-100.
63 Ibid.:139.
64 Ibid.:130.
65 Carruthers and Arnold 1995.
66 Other 1860s visitors included W.C. Baldwin, whose (1863) text did so much to stimulate European interest in African wildlife and hunting. See also Foskett 1965, Glyn 1963. Admiral Washington, for example, talked of extending Cape Colony to the waterfall via Kuruman, see Coupland 1928:261-2.
67 As the value of ivory rose, visitors reported that the large tusks that used to adorn Tonga and Leya chiefly graves along the river were replaced by small valueless milk teeth. Livingstone 1857:534, Chapman ibid.:203, Holub 1881 II:152.
68 Matthews 1981: 28. For an extended discussion of the Chikunda, see Isaacmand and Isaacman 2004.
69 Matthews ibid.:34. Colson argues (1950) that this was characteristic of the expansion of slavery in the region. Traders followed the Zambezi, but tended to move on foot for much of the way.
70 Matthews 1976:392-99 citing P. Terorde 1881:309-310, Selous 1911: 290, 297.
71 Including at Deka, opposite chief Hwange's home and at Leshoma, near Kazangula. Tabler 1963.
72 In 1875, Emil Holub employed a Tonga and a Nambya whom he found at Pandamatenga 'on the lookout for employment'; Selous and other visitors also picked up workers there. Ncube 1994:106-119, Holub 1881:213, Zambezi Mission Record 1902. See also Coillard 1897:154, 186, Thorpe 1951, Matthews 1976:39, 403-5.
73 Phillipson 1990:85. The subsequent flood of travelogues includes: Mohr 1876, Tabler 1967, Oates 1881. Holub (1881) had an ethnographic interest in the upper river, but had little to say about the people along the mid-river that concern us here.
74 Selous described running into a Tonga army trying to sweep Lorenco Monteiro off his island to avenge the rape of their women by Chikunda men, then he passed through deserted villages burnt in Chikunda raids and strewn with decaying human remains, before being hosted lavishly by the Portuguese trader Mendonca, where he described his disgust 'as an Englishman … [at] the sight of ten Batonga women, just captured in the last raid, all chained together.' Selous 1893. On the Falls, Selous 1911:110-11. On movement downstream, ibid.:288-98. At Mendonca's, ibid.:297-8. For a discussion of the insights these texts give into Chikunda society, see Isaacman and Isaacman 2004: chapters 6 and 7.
75 Dawney's Journal entry for 15 July 1873, Hist MSS DA4/1/2, cited in Cobbing 1976:327.
76 In 1877 Ndebele warriors expelled 30 traders from Pandamatenga for hunting without Lobenguls's consent. Cobbing ibid.:203.
77 Taylor 1989, Arnot 1969.
78 Such as Francois Colliard of the Paris Evangelical Mission and Plymouth Brethren Frederick Arnot.
79 Weinrich 1977:14, notes on the history of Siansali and others.
80 Discussions of these wars and shifting alliances include Colson 1950:36-7, Matthews 1976, Cobbing 1976.
81 Gelfand 1968:352.
82 Selous 1893:203.
83 Ibid.:203-4.
84 Ibid.:206-7.
85 Ibid.:205.
86 On his inability to speak the languages of the river, see Selous ibid.:223.
87 Ibid.:203-6, 233.
88 Ibid.:233.
89 Chinua Achebe's critical views are discussed in Kimbrough 1983:1-2. See also Edward Said's (1993)

discussion of James Ngugi's *The River Between* and Tayeb Salih's *Season of Migration to the North*, which counter Conrad's river imagery and reverse Kurtz's journey.

90 Schama 1995:part II.

91 Taylor 1989:71 see also 85, Arnot 1969:13.

92 He provided the inspiration for Alan Quartermain in Rider Haggard's *King Solomon's Mines.*

93 Palley 1966:5. This was the first use of the term.

94 Ibid.:6. Treaties with France and Germany were signed in May and December 1886. Britain refused to acknowledge this 'paper annexation', and criticized Portugal's 'archaeological arguments'.

95 Key texts include: Gann 1958, 1965. For an Ndebele focussed discussion, see Cobbing 1976 and Brown 1966. On the Lozi, see Stokes 1966 and Mainga 1973.

96 Gann 1965:76.

97 Gann 1965:78, citing Hiller 1949:202, Hanna 1965:82. Rhodes included the steamship after hearing about Stanley's deployment of steamboats on the Congo (Rhodes to Rudd, 10 September 1888, cited in Brown 1966:80) and after Louis P. Bowler (an English speaking Transvaaler) mooted a scheme for putting launches on the Zambezi and settling Transvaalers in the wake of securing a concession from Mashonaland chief Mcheza. Rhodes put pressure on the Transvaal, and the scheme was abandoned. Gann 1965:96.

98 Gann 1965:97.

99 Cobbing 1976:334, citing HIST. MSS H01/2/1 Correspondence between H.M. Hole and E.A. Maund, Lobengula to Queen Victoria, 24 Nobember 1888. (The claim did not include the area west of the Gwaai confluence up to the Zambezi which was claimed by Khama – in resurrection of older Ngwato claims – and the Lozi.

100 For an assessment of Lozi authority over Tonga country, see Mainga 1973:154.

101 Ibid.:155-6, citing Baldwin's Journal, 26 June 1893,Coillard 1897:384.

102 Lewanika had envisaged an arrangement comparable to that secured by Khama for Bechuanaland, but, advised by Coillard, he signed the treaty with a BSAC (rather than a Crown) agent.

103 This was important after the discovery of a navigable channel in the lower river providing a direct connection with the Shiri river and British territories in Nyasaland.

104 Rotberg 1988:293.

105 Gann 1958:62-3.

106 Ibid.:63. For a discussion of how local autonomy could persist despite the formal structures of external control in Southern Rhodesia, reflecting both settler interests and the 'government within the government' that was the Native Affairs Department, see Holleman 1969:19-20.

107 Gann ibid.:1, see also Hole 1995:290.

108 Gibbons conducted two expeditions, one that set out in 1895, the second in 1898, on which he was accompanied by James Stevenson-Hamilton. See Gibbons 1904, Wallis 1953.

109 Gibbons ibid.:232.

4

Violence & Law in the Borderlands
Early Colonial Authority & Extraction

Drawing the 'natural border' between Southern and Northwest Rhodesia along the Zambezi had ignored the river people who lived along it. Yet these river people were centrally placed to exploit the new opportunities the boundary provided, and over time, helped to consolidate the idea of the border.[1] Although, initially, the mid-Zambezi river people continued to treat the river as a link in everyday social life and deployed old strategies of crossing as a form of resistance to new demands, colonial state-making nonetheless had an influence. As the violence of being on the contested margins of predatory pre-colonial African states was replaced by the violence of early colonial extraction, the presence of government agents at local level and the idea of the law began to change the fractious internal dynamics of frontier society. This chapter examines the beginnings of a longer process through which the 'river people' of the mid-Zambezi were incorporated into two separate colonial states, with a focus on the Southern Rhodesian side of the border.

Recent scholarship has emphasized the role of the law and courts as 'essential elements in European efforts to establish and maintain political domination'.[2] Yet there were particular problems of establishing such domination in borderlands,[3] and in communities lacking strong traditions of centralized authority. My aim in this chapter is to shed light on the way in which the idea of the state began to take hold in decentralized mid-Zambezian society, as the law was used by government agents to bolster their authority and by others to curb abuses of power, and began to constrain the excesses of violence that had characterized frontier society in the late nineteenth century. The chapter draws on early Native Commissioners' reports and records of criminal cases from the Zambezi valley heard before the district magistrates' courts between 1905 and 1923. It pays particular attention to the charges brought by and against local government agents, and to a series of murder cases relating to fights at Tonga funerals.

Despite the different processes of state-making on either side of the border, the early decades of colonial rule on both banks were shaped by a common process of economic marginalization and increasing isolation along much of the river valley, which was remote from the new centres of colonial political and economic power, inaccessible from the main contours of the developing colonial transport infrastructure, and depopulated through the effects of the expanding tsetse belt. As this process of marginalization and isolation deepened, old stereotypes of the river valley and its people as violent were replaced by new caricatures of the place as primitive.

Early Colonial Extraction in the Zambezi Valley

The first attempt to exert state authority over the border region came from the Southern Rhodesian authorities. We saw in the last chapter how the political economy, and by extension also the society of the river, had a reputation for violence and chaos built up over fifty years of European exploration, trade and travel-writing. Although this reputation had proved a potent justification for British colonial rule in general terms, it was less compelling as a ground for immediate intervention along the border itself, once Portuguese influence had been undermined, and as the small political units along the river did not threaten BSAC authority. The communities living along the river might have been caricatured as 'violent', but they were also seen as cowardly and disorganized, and had never been labelled 'warlike' or 'militaristic' – a designation reserved for the Ndebele in Southern Rhodesian discourse. Decentralized Tonga society more specifically was seen as anarchic and incapable of mounting organized opposition. Perhaps more important, the mid-Zambezi frontier was not involved in the 1893 war with the Ndebele or the 1896 uprisings against Company rule (though retrospective evidence suggested that some of the last deals struck along the dwindling Zambezian trade routes might have provided supply lines and firearms for northern Ndebele rebels, via Nambya and Tonga chiefs, Hwange, Saba and others).[4] The BSAC was initially fully preoccupied with fighting the Ndebele, looting and subordinating central Matabeleland and then putting down the uprisings; it was not considered advisable to establish structures further afield until 'Matabeleland proper' was firmly under administrative control.[5]

When attention finally turned to this frontier in the aftermath of the uprising, however, officials in the Southern Rhodesian Native Administration elaborated old associations between the valley and violence to justify intervention. Val Gielgud, for example, who conducted some of the first patrols in the valley and was regarded as having a degree of 'expertise' on the people who lived there, noted:

> One of the chief characteristics of the Batonga race is its disintegration. They have never been known to act in combination as one race, and the inhabitants of one village, and even of one family, will rarely cooperate to attain some common end. Not even for the purposes of defence, nor to resist or avenge a wrong committed against one of their number, will the Batonga unite; and in my early experience in the Zambezi Valley I have heard of the most atrocious murders being occasionally committed with absolute impunity, owing to the disintegrated state of society and absence of any central authority either in the district or at the kraals.[6]

The Zambezi valley 'needed police', Gielgud argued, because everyone was armed, and those who lived there 'have at present no notion of law or government, and kill and steal from each other with the greatest nonchalance'.[7]

The early official reports of violence in the border regions were accurate in some ways, though they echoed earlier imperial discourse by misrepresenting its source. As we have seen, the violence of frontier society was the product of a context of insecurity and relations with more powerful others and half a decade or more of predatory extraction had left a legacy of local conflict and distrust, in addition to a fear of external raids. Elizabeth Colson's elderly Tonga informants in the 1950s looked back at the years preceding colonial rule as a time when fear was pervasive: neighbours, villages and kinsfolk were set against each other, and the potential for resolving disputes peacefully was pre-empted by rapid recourse to force of arms and vindictive action against those who were weak.[8] The years following the defeat of the 1896 uprising in Matabeleland saw a perpetuation of fear and uncertainty in the Zambezi valley, as the Ndebele regiments were breaking up, former captives were trekking home, and people moved up and down in search of food. Some communities spread out, away from their defensive positions along the river, such as many of Hwange's Nambya, who regrouped around the old hub of their nineteenth-century state,[9] while others felt the need to occupy defensive strongholds was as strong as ever. For those along the river, it was unclear whether the military defeat of the Ndebele would bring an end to raids,[10] while the return of Zambezian men formerly incorporated in the Ndebele regiments was also cause for concern: one such returnee took over the headship of the Hwange dynasty, provoking a scattering of other prominent Nambya men who feared the man's warlike reputation.[11]

It was also unclear how the new colonial authorities would use their superior force, and whether or not they would repeat the violent interventions of their predecessors. Though Native Affairs officials justified colonial intervention along the border through reference to the disorder of African frontier society, their first interventions were provoked by the violence of Europeans. Hunters, traders and prospectors poured into the Zambezi frontier region after the uprising to further explore its potential for minerals, trade, game and coal (the latter was discovered in 1894, but investigations had been interrupted by war). As the main road to the north was routed via the Hwange coalfields and the Victoria Falls, other tracks, such as the one that crossed the Zambezi downstream at Walker's drift (in the vicinity of the Sebungwe confluence), quickly fell into disuse. Yet the marginality of much of the valley to the developing transport infrastructure was a positive attraction for some Europeans, as it made it possible for them to use threats, force and false claims to government authority to pursue a livelihood of violent extraction from those who lived in the borderlands.

The first official patrols to the valley in 1897 and 1898 were sent out in response to Tonga requests for intervention and protection against this influx of white criminals. Assistant Native Commissioner Green was sent out in October 1897, for example, to 'procure witness and evidence against two Europeans named Carsons and Webb who were accused [by Zambezis] of murdering an induna' and stealing stock.[12] In the same year, an important Tonga chief from the plateau north of the river (Monze) sent a delegation to Bulawayo to request protection from traders' violent extortion, and many reports of murder and theft came in from the south bank of the river. While on patrol, officials reported pleas for protection, such as from a 'Zambezi native named Utshali, son of Induna Moio … on the far side of the Zambezi', who 'came to tender in the name of his father, the submission of his tribe, and ask for protection from white men trading in their midst, and … taking their sheep, goats and giving in exchange small presents of beads and limbo, whether the Zambezi are willing to trade or not. He also reported white men as taking ten head of cattle from his father's kraal, without payment of any sort'.[13] When patrols went out the following year to begin to collect tax, a number of chiefs visited by Gielgud, 'expressed themselves ready to pay Hut Tax in return for protection afforded them from unscrupulous white traders, of whom they make many complaints.'[14]

Though some Zambezian leaders thus came forward of their own accord to seek protection, others were hostile, and the reaction along the valley to early patrols and demands for taxation was decidedly mixed. Some Zambezian chiefs decided to flee, others launched assaults on the patrols and an African member of Green's 1897 patrol was 'murdered by some Zambezis'.[15] Many chiefs refused to accept Company authority: Chief Siansali, for example, 'would not be convinced, even after an indaba' of the benefits of white rule, and told Green he thought the 'whitemen were worse than the Ndebele … they had so often been terrorized by whitemen that they were afraid of them', while chiefs Pashu and Sibaba told Green cynically that 'the white chiefs of the government appeared to be remarkable plentiful about his district, and that no doubt, a day or two after I had left another whiteman would come and say he was a chief.'[16]

The fact that the first official patrols were led by officials who spoke Zulu or Ndebele (if they spoke any African language) and were accompanied by Ndebele-speaking African staff must have contributed to this uncertainty and fear in the Zambezi valley. The southern banks of the mid-Zambezi fell under the Sebungwe District of Matabeleland, which encompassed a vast swathe of Southern Rhodesia stretching north and west of 'Matabeleland proper'. The Europeans who led the patrols that set out from Bulawayo or Bubi were joined by Ndebele guides claiming past authority over the area. Thus, Green's patrol was joined by a certain Ndonsa, who was described as 'late chief of the Zambezi by Lobengula's appointment'. Green reported that his patrol 'passed deserted kraals and armed natives', who had hidden stock in fear, after hearing rumours that 'Ndonsa was coming up with an impi to avenge the death of Sinegoma, the late Sinegoma having been killed by Ndonsa's principal induna'.[17] Nor did being led by Ndebele guides make for easy movement, as they obviously were not as intimately acquainted with the landscape as they claimed. Gielgud complained:

> It would be hard to overestimate the difficulties of this patrol… Following the course of the Zambezi and crossing the headwaters of its tributaries we had often to cross seven and eight rivers in one day. These same rivers being swollen by the rain, and with steep muddy banks. Our donkies, we had literally, sometimes to throw into the water on one bank, and seize and carry up the other. For this work Matabele are quite useless if the water is deep, and I had to obtain the help of the Batonka. On leaving the Zambezi on December 25th, our guides lost their way owing to the whole country being totally under water, and for two days we wandered at the headwaters of the Uluzi and Lukulu river among mountains, jungles and swamps, and through which jungles we had to cut our way with axes … at night sleeping where we stood, sometimes ankle deep in mud and water.[18]

The Southern Rhodesian Native Affairs Department (NAD) was restructured after the uprising into a more professional force than beforehand: there were greater imperial controls on its activity, and its European staff included a body of experienced cadres recruited from Natal and the Eastern Cape, whose outlook was at once authoritarian and paternal.[19] Though the Matabeleland NAD was Ndebele-oriented and used Ndebele as its African vernacular, its officers understood their role on the margins of what had been Ndebele influence as freeing tribes which had been 'to all intents and purposes slaves' of the Ndebele, and placing them directly under the NAD.[20] Gielgud used his first patrol to the mid-Zambezi in 1898 to identify and appoint Nambya and Tonga chiefs and collection of tax began in the same year.[21] Some chiefs came forward themselves or were easy to identify, such as the leader of the Hwange dynasty, who had long been a key point of contact for Europeans and whose people now lived partly within the newly pegged Hwange coal concessions. Native Commissioners considered the Nambya a 'composite tribe' and Hwange was seen as having powers extending over a huge area and diverse peoples, including Dombe and Leya people living along the south bank of the river around the Victoria Falls.[22]

Further downstream, twenty-two independent Tonga chiefs were recognized in Southern Rhodesia in 1898, most of whom were along the river. Given the decentralized character of Tonga society and mixed reaction to Company patrols, the process of appointing chiefs was more complex. The big men of Tonga society held a degree of political and religious authority through their responsibility for neighbourhood *malende* shrines and an ability to command a force,[23] but some powerful men with sizeable followings pushed forward slaves or others to represent them (and subsequently regretted it),[24] while others ran away and were brought under the authority of neighbours. Disputes over chieftaincy and authority at *malende* shrines then (as now) frequently refer back to the time when 'slaves took over' in these first years of colonial rule.[25]

Old strategies of crossing the river retained their utility as a means of avoiding the new authorities' extractive demands, especially as there were initially no comparable dues on the north bank. Tax collection in Northwest Rhodesia was held up until 1904 by negotiations with Lewanika over how the revenue was to be divided between the administration and Lozi authorities and, related to this, what powers Lozi rulers would have over others they claimed had paid tribute in the past. In this context, Gielgud complained bitterly of the problems of exerting authority and collecting tax, when people could simply run across the river. On his first patrol, he described:[26]

> The Abatonga mostly live at the river front... These people never seem to move from their kraals unless fully armed ... All the natives inhabiting the river front have boats, and on the slightest alarm, they betake themselves to the islands or the far bank of the river ... These people are all most suspicious, and at some kraals on our approach all the people fled and we had the greatest difficulty in inducing them to come back and talk with us. They are extremely reluctant to show their boats, and keep them cunningly concealed in the river ... The river is the main road of the district, and I consider it absolutely imperative that for the proper control of the District, that the authorities should control the river. Without this every man wishing to avoid arrest for offences committed or to avoid taxes etc. will fly immediately to an island or cross the river in his boat.[27]

The Southern Rhodesian authorities initially made considerable effort to control the river through military manoeuvres to physically retrieve those who had fled across it. Chief Mola, for example, had threatened to cross the river with all his people when presented with demands for tax in 1898 and subsequently did so, running 64 miles north of the river into rugged hills, where he was tracked down by a Southern Rhodesian force and brought back south of the river in 1899.[28] NC Carbutt reported on another much less effective effort in August 1900:

> My patrol along the Zambezi was in conjunction with Major Harding [of NW Rhodesia], and had for its object the returning of certain natives who, to evade the payment of their taxes, had crossed to the North bank of the Zambezi from the South bank. Owing to my having had no intimation of Major Harding's intention to carry out this patrol until he was actually at the river, and my having only been a fortnight in the district when it took place, it was not as effective as it might otherwise have been, for I did not know who the natives that had left the district were,

> and could find no record of their names in this office, so that I had to pick up what information I could as we travelled down the river... Owing to the secretiveness and objection the Batonka have to giving out any information… I had not been told about many kraals at all… As a result of the recent patrol 5 kraals have returned… There remain to be sent bank some 15 kraals, with a population of about 200 men, their wives and children. On my return from Kariba, I collected hut tax. From the Kariba to Somnyanka's island there are practically no natives, this being the portion of the river from which the natives have crossed to the North bank. From Somnyanka's island to Tshete, I found that the natives were quite unprepared to pay their tax; moreover, on inspect of their receipts showed that there was not one from whom tax is not due for 1898 and 1899 and in a few cases they had not been registered and had paid no tax at all.[29]

These cross-border operations were part of a significant BSAP and NAD deployment in 1899. As a result of Gielgud's appeal for a force to control the Zambezi and its independent minded people, efforts were made to establish a permanent government presence: eleven BSAP stations were established on the north bank including at Walker's Drift (though most were in the hills away from the river itself, including the headquarters at Monze) and on the south bank, and a permanent NAD administrative post for Sebungwe was set up at Lubu, together with a network of BSAP posts (including Lubu, Tshete and Omay). Yet this official presence was an extreme challenge to maintain: by April 1900, at the end of the first wet season, the north bank BSAP stations had been reduced by half 'from various causes, but principally sickness'. On the south bank at Lubu, three European officials and one African messenger died of malaria and two police troopers were killed by lions at Tshete, a further BSAP trooper died of fever and the Omay post was abandoned for its unhealthiness. The entire staff – African and European alike – had been incapacitated for long periods, food supplies had been precarious, medicines insufficient and other medical aid was, of course, totally unavailable.[30]

The ill-fated Sebungwe administration abandoned Lubu after this disastrous first wet season and moved its headquarters to a station at Tshete gorge overlooking the river, from where it continued to operate for a few years. A new district of Hwange was carved out in 1903, centred on the growing population at the coalmines, and new administrative buildings were constructed for the Sebungwe administration at Kariangwe in 1909, though these were rapidly abandoned as the station was engulfed by the advancing tsetse belt. Thereafter, the Sebungwe administration was based in Gokwe, some 200 miles or so away. Regular contact with the mid-Zambezi from Gokwe was undermined by the expansion of tsetse infested areas to within 15 miles of the river itself, which 'formed almost a complete barrier', militating against effective administration and creating 'undue hardship' for those needing to make the journey.[31] European administrators' and police officers' retreat to Gokwe in 1911 marked the end of a decade of relatively close contact between the new authorities and the people of the river.

One of the aims of intervention in this early period was, of course, to fill the acute labour shortages of the mines and farms, and labour recruiters had been part of the initial influx into the valley.[32] Men from the Zambezi valley did enter the labour market on a large scale – indeed, as we have seen, they began to do so decades before colonial rule was established or taxes demanded. By the time the Native Department was beginning to establish a presence, those labelled 'Barotse boys' or 'Zambezi boys' in the labour markets of Southern Rhodesia and South Africa were 'almost entirely' Tonga.[33] Their prominence by 1900 was such that Robert Coryndon considered Tonga men (from valley and plateau) made up 'the bulk' of labourers on the Matabeleland mines.[34] Tonga men also briefly worked for employers closer to home, such as the Hwange colliery and railways, though were quick to abandon them when conditions proved abysmal.[35] Having started to migrate to work early, the valley Tonga used their own social networks to get jobs and avoided official labour recruitment agents, such as the Rhodesia Native Labour Bureau (RNLB) agent stationed at Walker's Drift,[36] who withdrew after failing to find willing recruits for the colliery.[37]

The particular jobs Tonga men came to be associated with in the new Southern Rhodesian economy were surface work on the mines and night-soil collection in Bulawayo town, where they worked as municipal employees. Men from the valley also exploited opportunities for making money by trading tobacco from riverine gardens, and in years of famine in the valley sought piecework with African families on the plateau in exchange for grain.[38] Though men from the valley thus maintained a degree of independence in securing their livelihoods, the specifically 'Tonga' or 'Zambezi' niches in the labour market acted to reinforce a view of the Tonga as 'low' and different in the eyes of other Africans. On the mines, surface work was less dangerous than underground work but it was also low status and poorly paid; in the rural areas, doing piecework for grain in years of hardship gave the Tonga a reputation as beggars who lived perpetually on the brink of starvation. Perhaps most damaging of all was the position of Tonga men in the rapidly expanding town of Bulawayo, where night-soil collection was stigmatized as 'dirty work', and the municipal authorities segregated the Tonga from other Africans by housing them in a fenced, municipal compound, constraining interactions and underlining their difference.[39]

As Tonga men sought to enter the labour markets on their own terms and evade other extractive demands in the valley, NCs were quick to blame the local chiefs for obstruction, describing them as 'quite indifferent to the interests of the government'.[40] NCs had no hesitation in replacing obstructive chiefs with those they considered more pliable and, as Eric Worby has argued, had scant concern for 'traditional' credentials in the first decades of colonial rule.[41] The first Hwange, for example, became notorious for forbidding all Nambya men from working in the colliery; when he died in 1903 (in unclear circumstances in a police prison), he was replaced by a new chief, Nemananga, the former chief's policeman, who was chosen because he had been a useful informer on Ndebele movements in the 1896 uprisings and had acted as labour recruiter for the colliery. Several Tonga chiefs were deposed and replaced for non-cooperation, particularly after salaries for chiefs were introduced in 1907 (Jobolo and Sibaba were both ousted in 1907, and Siachilaba was removed three years later in 1910).[42] The deposed Sibaba was replaced by a certain Siamate, whom Gielgud favoured as he '... has worked a great deal in Bulawayo and will no doubt be more enlightened than men who have never been out of the Zambezi Valley and less dominated by superstitious fear than they are'.[43] Such impositions often proved counterproductive and were later amended: the new chief Siachilaba, for example – a man named Simariampongo – had been chosen by officials to replace a man who was leprous, but he was rejected by his people as he could not host the spirit of Siachilaba or take on the name, and thus Sianbubi was appointed in 1911.[44]

These interventions to create an effective network of government agents along the border had little impact on border crossing and tax evasion, which were persistent complaints, year after year, along the length of the river, and not only in the more remote parts of the valley. Native Commissioners felt there was a 'floating population, which dodges between this district and NW Rhodesia'.[45] The imposition of a tax regime on the north bank in 1904 did not stop the exodus from south to north, as the amounts demanded in Northwest Rhodesia were lower. In Hwange district in 1903 (when tax registers were being prepared in NW Rhodesia), rumours circulated that the 'whole of the inhabitants of the northern part of the district were anxious to return to NW Rhodesia'. At subsequent indabas with the people of the Falls area, Native Commissioners heard how the Leya on the south bank were reviving old associations with the Lozi in the interest of crossing the border. Some people had responded to summons from the Lozi paramount Lewanika and had given their names as his subjects, whilst others had their names added to the list of his subjects without their knowledge.[46] In 1912, the reasons given for the 'considerable exodus of Baleya' were the advantages of living north of the river, where 'guns can be used, no rent is collected and the tax is only half what it is here'.[47] It is difficult to judge the scale of the exodus from the south bank, but it appears to have been significant, and may help to explain the rapid advance of the tsetse belt after 1901.[48]

The distance of the river valley from a permanent official presence after 1911, the use of crossing as a strategy of resistance, and close social links between opposite banks created a whole series of administrative difficulties for NCs. The fact that some people still lived on islands in the river – now an international waterway – made it unclear where they belonged for purposes of registration and tax.[49] Moreover, chiefs' people frequently straddled the river, as did individual households' homes, fields and livestock. Cross-river marriages, bride service and child-pledging disrupted registers: Gielgud railed against the practice of men doing bride service (*galila* in Tonga, or *garidzela* in other parts of Sebungwe) on the grounds that it emasculated young men living under the 'domination of old women' and men used it to evade paying tax. He complained: 'the custom undoubtedly makes the collection of Native tax more difficult and puts obstacles in the way of the Native Marriages Ordinance and the Native Tax Ordinance'; moreover in the valley 'it allowed a NW Rhodesian man to reside in southern Rhodesia but not pay tax'. [50]

The movement of stock and migrant labourers across the river was also supposed to be regulated in the interests of disease control, which was considered particularly important in the wake of the rinderpest epidemic, in the context of the rapid expansion of the tsetse fly, and outbreaks of measles, smallpox and human trypanosomiasis. Yet people on the south bank rebuilt their herds by restocking from the north bank and moved animals across the river without permission. The spread of the tsetse belt south of the river was blamed partly on people from the valley and NW Rhodesia breaching the pass ordinances and walking to work through the expanding fly zone: those crossing at Binga and moving on footpaths to Bulawayo, Gweru or Kadoma traversed what officials regarded as 'the most dangerous part of the belt', from which homes were evicted in 1913 as tsetse control measures. Native Commissioner Hemans ordered river chiefs to instruct labour migrants not to move through tsetse areas, but doubted that his order would have any effect in the absence of local government agents prepared to enforce it.[51]

From the perspective of those who tried to administer the mid-Zambezi borderlands, the river thus seemed more of a natural link than a natural boundary. Gielgud was the first of many administrators responsible for the practicalities of collecting tax and enforcing the law among those along the river to advocate incorporating the whole area into the administration of one or other side of the river.[52] He also used the on-going close connections between the Tonga on both banks to support arguments for amalgamating Southern and Northern Rhodesia, as the fate of the Rhodesias continued to be debated throughout the period of Company rule.[53]

Although the separate structures of authority erected on either side of the river and the differential extractive demands shaped responses to early colonial rule along the border, the geography of economic and industrial development and the new transport infrastructure created a common process of marginalization and increasing isolation along much of the valley. As the early European interest in the place faded, state officials, police, labour recruiters and traders withdrew from the valley to the healthier, more accessible plateau.[54] The abandoned valley was deemed unfit for white settlement, and there was, therefore, little competition for land. In 1911, the Native Reserves Commission noted that three reserves had been defined within Sebungwe, on advice from a Native Commissioner, in the form of 'arbitrary circles' drawn with a compass around central points on the map. But there was no attempt to move people into the reserves, rather those people who already lived within them were moved out in 1913, as they were tsetse infested. The Commissioners felt it was not worth clarifying the situation until the land was free of tsetse and hence habitable.[55] After this withdrawal of officials to the plateau, the state intervened little in the valley. Early efforts to introduce Egyptian cotton to riverside gardens were abandoned as a failure, and a scheme to develop irrigation from a dam at Kariba gorge, developed by Native Commissioner Keigwin in 1912 resulted only in a new file in the office of the Director of Land Settlement rather than any concrete changes on the ground.[56] Keigwin's Native Development Department, created in the 1920s, scarcely tried to operate in Sebungwe District, let alone in the inaccessible areas along the river.

Despite its marginality and isolation from the competition for land and resources, and the developments that began to transform other parts of rural Southern Rhodesia, and despite the ineffectiveness of efforts to control crossing, the idea of the state nonetheless had begun to take a hold in the first decade of intervention and it persisted after European officials had withdrawn, embodied in the network of local chiefs and other African government agents stationed in the valley. The courts, which sat in the administrative centres of Gokwe and Hwange, extended their reach into the remote parts of the river valley because those living along the river chose to resort to them. Below, I shall explore in what ways and to what ends they did so.

State Authority & the Law

Maintaining law and order and administering justice were key roles for the Native Affairs Department. In remote areas, Native Commissioners had often doubled as Magistrates even before their judicial role as Assistant Magistrates was formalized in 1910 through the Southern Rhodesian Native Regulations Proclamation. This Proclamation empowered them to try civil cases between Africans as well as criminal cases in which the accused was an African, and set up a division between the NCs courts in the rural areas and the Magistrates' courts in the urban areas, both of which were subordinate to the High Court.[57] Gann argues that from 1910, NCs 'began to play a major part in the country's judicial system, as Africans dissatisfied with local decisions in their villages or angry at being accused of witchcraft by their fellow-tribesmen, began to flock to the NCs courts to seek the white man's justice'.[58] In the remote border areas that concern us here, it can hardly be said that people 'flocked' to the new courts in this way; indeed, Sebungwe Native Commissioners commented frequently on the very low level of reported crime from Nambya and Tonga areas, which they interpreted not as reflecting actual crime levels but a disinclination to bring cases before the court. They claimed a Tonga man would 'endure great hardship before making any complaint, for he has a much greater fear of the magic which he believes the accused or his relatives can direct against him than he has faith in the NCs powers to protect and help him'.[59] Moreover, injured parties were 'always agreeable to compensation in the form of a few goats or perhaps a child', making the detection of crime next to impossible. Thus the cases that did come to light tended to be 'those in which an attempt is made on a man's life and he manages to effect his escape'.[60] The huge distances to the administrative centres must have contributed to the constraints on reporting anything but the most serious cases, as to do so necessitated an exhausting expedition of several days. In the discussion below, I shall focus on two characteristics of the criminal cases from Tonga areas heard in the Sebungwe NCs court between 1905 and 1923; first, the significant number of cases involving local level government agents (as victim and accused), and second, the significant number of murder cases resulting from assaults at Tonga funerals.

The cases involving local government agents reveal the character of the early colonial state at local level: how power was exercised by chiefs and others, and how it was understood and challenged by subjects. There were, of course, tensions throughout Southern Rhodesia over the appointment of chiefs as government agents, both regarding the legitimacy of particular individuals, and over their new roles. However, the challenges faced by government agents in Tonga communities appear to have been extreme; there was a strong tradition of disrespect for centralized authority in Tonga society, and, as we have seen, the insecurities of the late nineteenth century had produced a situation where men resorted rapidly to violence when presented with an opponent who was weak. Writing about Tonga communities on the Northern side of the river, Colson argued that government agents were frequently exposed to violence from their own people, and that those who survived in their posts for any length of time were assumed to possess strong medicines and to be a witch.[61] The Sebungwe District court records appear to lend support to these judgements, giving an insight into some of the life-threatening acts of violence chiefs and other government agents experienced from their own people, particularly when they tried to exert authority over others. They show how some chiefs turned to the courts in self-protection, and how subjects also began to use the courts against their chiefs, to curb abuses of governmental power and the violent extractive demands with which it was associated.

The nature of the violent assaults brought to the courts from Tonga villages cannot be understood through the explanations conventionally used to explain the structure of antagonism in 'stateless' societies. As Colson has argued, the cross-cutting ties of kinship and locality did not allow for the development of protracted blood-feuds, though some revenge killings did occur.[62] Nor did Tonga society stage duels or other forms of killing as spectacle, which might be theorized through a notion of ritualized sacrifice, though as we shall see, particular forms of competitive dance commonly resulted in fights, in which a person was occasionally killed.[63] The five instances of assault on government agents heard before the Sebungwe NCs court between 1906 and 1913 involve two murders (of a chief and an NCs messenger), and three cases of attempted murder. All were opportunitistic, with individuals taking advantage of moments when victims were unguarded. For example, in 1906 Simwinza, a subject of chief Pashu, was charged with intent to do grievous bodily harm to his chief. Pashu had been attempting to investigate a dispute between Simwinza and another man, but had been insulted, cursed and attacked. Pashu testified:[64]

> I went to Simwinza's village [to investigate a quarrel between the accused and another man]. I carried my rifle and when I set down I laid it down by my side. I asked the prisoner about the matter and he got very excited. 'What are you conceited about? You are puffed up by the whiteman! You want the people to break your neck in the same was as they broke the neck of your elder brother [the previous chief Pashu, whose death after falling from a tree was attributed to witchcraft]'.

Simwinza called Pashu 'msunakagogo' [mother's genitals] and attacked him with an assegai of the type used 'in war or in pursuit of very dangerous animals'.[65] Two years later, chief Pashu was involved in another case in which violence ensued from his efforts to investigate an assault.[66] This time, Pashu had (perhaps wisely) gone in the company of an 'impi'; his force had arrived at the man's kraal when a beer drink was underway and a general fight broke out, in which two of Pashu's men killed another man. Pashu claimed his own attempts to investigate were 'the start of the trouble', and that his men had retaliated when attacked, though they clearly took advantage of their attacker's weakness, striking him on the head from behind as he was running away with a 'stick shaped like a pick' and a knobkerrie.[67]

It was not only chiefs who resorted to the courts in self-protection or to bolster their authority. A former NCs carrier – Siabalagwe – did likewise on chief Sigalenke's advice.[68] Siabalagwe was widely reputed to be a witch by his neighbours, and had been accused of keeping crocodiles and sending them to murder others.[69] He went to the court after four armed men attacked him to avenge the death of a young boy killed by a crocodile. As Siabalagwe testified:

> About the middle of December, after I had been discharged as a carrier by the NC, I found that a son of Manhunya, the female accused, had been killed by a crocodile and it was said that the crocodile was mine, which I had kept in my grain bin, and that I had caused it to kill Maluma. When I got home, Tshogwa and Manhunya came to my kraal and shouted out 'Now the evil-doer who has killed our child with his crocodile has come home, who keeps crocodiles in the gain bin, let us kill him.' [Having gone off to fetch three men] the four men came back bearing the woman behind and surrounding my kraal with their assegais. They shouted out 'Let us kill him now as he hasn't got anyone to help him'. I evaded them and ran to the chief [Sigalenke]. I told him about it and he said the best thing you can do is to go to the NC.[70]

The violence from their own subjects that government agents in Tonga areas were exposed to in the first decades of colonial rule often occurred in the context of a funeral wake. Tonga funerals were occasions for the performance of neighbourhood drumming and dancing teams, and were one of the few times when neighbourhood groups came together.[71] In the course of the dancing, accusations or old grievances were sometimes voiced, but even where there was no explicit accusation, the dancing itself could provoke vicious fights. The funeral drumming performances were highly competitive, involving special tall drums, a chorus of women and troops of male dancers who enacted choreographed mock assaults, made theatrical displays of aggression, and hurled spears into the ground at the feet of opposing drum teams in the audience. On display during the team's performance were the collective pride, status and identity of the neighbourhood and the dancers' masculinity. The performances were a structured rivalry, in which violence was anticipated; indeed Colson has provided a vivid description of how teams could come to a funeral sparring for a fight and being deliberately provocative, shouting insults or using bodily repertoires of symbolic demasculinization and denigration.[72] When the teams arrived at a funeral, they made separate camps at the hosting homestead, each choosing a pitch from which they could beat a hasty retreat. Although it is tempting to invoke Girard's ideas on the role of the scapegoat in collective violence to help explain the deaths that could occur during these funeral fights – and certainly there was generally little relation between the victim and the initial provocation – my aim here is less to theorize collective violence, than to historicize it, and to understand the contexts that exacerbated or contained the violence associated with a particular cultural practice.[73]

Six murder cases relating to Tonga funerals came before the Sebungwe courts between 1905 and 1923, two of which involved the killing of government agents. These cases stand out not only because so few criminal cases from Tonga areas reached the courts, but also because the collective nature of the violence made it difficult to apportion individual blame, resulting in extended hearings and lengthy testimonies, with a large numbers of witnesses called to give evidence. The extended witness statements are revealing, as they include detailed accounts of the course of events and attempts to quell excessive violence. In one of the funeral murders brought before the Sebungwe district court in 1913, Chief Siansali was called upon to explain to the NC acting as magistrate why the proceedings at Tonga funerals commonly ended in violence. Siansali believed such violence was less frequent than in the past:

> When a person dies, say early in the morning, he is buried almost at once. Then in the middle of the day the drums are brought out and those men at hand come and dance, the method being for the women to do the applauding while the men forming line in their various sections come dancing past the grave leaping about armed with their spears and sticks and displaying prowess generally. On the second day more people come from further away and on the third day those from still further... It is, however, in the dancing that the affrays always start. One section will come up leaping and dancing then the other section, being perhaps worked up to it by the women or a long standing grievance will dash in amongst them in order to disperse them and there ensues a regular battle in which in former days many got killed...[74]

The Sebungwe funeral cases I discuss below help to show how the idea of the NCs courts had begun to impinge on this specifically Tonga cultural arena for violence. The fact that such cases reached the courts is significant; the testimonies show that reporting deaths or excessive violence came to be seen as a potential course of action on the part of aggrieved parties, and that containing and reporting violence was the responsibility of government chiefs. Recourse to the courts became part of local argument, and acted to restrain both intentional acts of vengeance and collective acts of violence.

Magistrates hearing these funeral cases sought to uncover specific grievances that could explain the violence. In some of the cases, unresolved conflicts did come to light, and were treated as causal explanations; they included both accusations against individuals for the death of the person being mourned and quarrels unrelated to the funeral itself. The first funeral case heard before the Sebungwe court was one such, with a clearly articulated accusation and revenge attack. The man who reported the case was a certain Siadenta, who had narrowly escaped with his life at the funeral of his wife, when the woman's male relatives accused him of killing her. When called to give evidence, Siadenta recounted:

> Last autumn when the crops were ripening my wife died. Chief Jobolo came to the mourning ceremony. The next morning, three of Jobolo's young men also came. Their names are Kwata, Tshilopa and Dimangwagwa. When they arrived there was a small crowd of people assembled... Kwata at once commenced bombarding us with stones and shouting 'Where is Siadenta the wizard, the wife killer?' I sprang out from amongst the crowd, when the three at once rushed at me, throwing assegais at me, shouting, 'Stab him! Stab him!' I dodged the assegais except one which went through my left hand. They then rushed me with their sticks, but I warded off their blows with my assegai and battle axe which were broken... Jobolo then interfered and stopped the attack on me.[75]

Although chief Jobolo together with 'the women of the kraal' had stopped the fight, Siadenta reported the incident as a case against the chief. Siadenta explained: 'Jobolo said to me that I was not to report the matter, I would only get him into trouble; he said he would give me a sheep if I did not report it. He has never given me a sheep and that is why I am now reporting the case.'[76]

Another case illustrates how the fight following funeral dancing could provide the occasion for men to act opportunistically upon old grievances, in this instance against a chief's messenger. The messenger was killed after he and another elder had helped the head of the funeral-hosting homestead disperse two embattled visiting drum teams, but on his way home, the messenger had been found isolated and unarmed, and was killed. Those giving evidence explained the course of events that had led up to the man's death as follows:[77]

> Evidence of chief Siansali: Siansambo, the deceased, was my nephew and my official messenger. On the day when the mourning and subsequent fight took place at Mangwato's I was not at home, having gone over to Northern Rhodesia on business. When I got back to the river some women were drawing water on our side shouted out to me that 'people are killing each other over here'. I got into my canoe to go over as quickly as possible, but was prevented by the hippopotami... When I got to the kraal ... I found the dead body of the deceased, Siansambo ... I asked my mother what had happened and went to their kraal to arrest the men... The accused had a grudge against the deceased on account of their being ordered up by me through him to cut grass to thatch the NCs huts. On that occasion he reported to me that the accused [and others] on receipt of my message had called him 'Msunakangogo (genital organs of your mother) wherever there is any work to do for the NC why do you always come and turn us out for it.' The occasion for the gathering on the day of Siansambo's death was mourning for the death of the child of Mangwato.

> Evidence of Mangwato: ... A child of mine died about this time last month and the usual mourning custom ensued. Accused were all present. They came with Mwanamwato's section. When the leaping and dancing commenced they were dancing with Mwanamwato's section and another section of Simankan's dashed in amongst them and a fight ensued. I parted the accused, Siamiliyo and Burinyangu, and eventually with the help of Mwanamwato and the deceased I managed to disperse them to their various kraals. Many people had, however, in the meantime got hurt. I in the meantime had hurt myself by running my foot against a log and went home ... and know nothing of what occurred afterwards...

> Evidence of Buriyangu: [after the fight was dispersed] I went home by the path which passed the accused's kraal... I was seen by Siamiliyo, he went to his hut and then he rushed out with his

> sticks and assegais and came at me and hit me on the head with his stick … then the deceased came … he said 'why didn't you leave him alone, we've already separated you once' then they turned on him…
>
> Mother of deceased: I do not know why they murdered my son, but I think it was because he was the chief's messenger and on one occasion when he turned them out to work they had cursed him and called him 'msunakangogo'. I saw them kill him and I am certain they killed him on purpose, it was no accident, they caught him with no weapons to defend himself, carrying only a tail and took advantage of it.

The final funeral case where a specific unresolved grievance came to light and was upheld as the underlying cause for a funeral murder involved a long-standing conflict between two men, over the killing of an ox, without the owner's agreement, that had been injured by a crocodile.[78] Unusually for funeral fights, the two men had been dancing in the same drum team but a quarrel over an impala horn sparked a fight between them, and the death of one.

In the remaining funeral murder cases, however, there was no underlying cause for fighting. The insults that provoked the violence seem to relate purely to the performance of the dancing itself: one team kicked sand in the face of another team, or began dancing themselves before the previous team had finished, or threw 'ashes about and sliced an assegai into the ground' in a manner regarded as a challenge.[79] On such occasions, the deaths cannot be explained by old grievances and were unrelated to the incident that triggered the fight; the person killed could be anyone who did not manage to defend himself, such as a man vulnerable because he was cut off from others in his own team or because he was temporarily unarmed, or a woman trying to break up the fight accidentally injured or killed. The records from one initial hearing provide a particularly detailed (if disputed) account of how such insults to the pride and identity of a drumming team could lead to a fight and the death, in this instance, of a north bank man – who had not been dancing but got left behind at a funeral on the south bank when the rest of his party retreated to the river after a fight:[80]

> Evidence of Chalichusia (northern party): I remember the death of Sinakaimbi, a headman, on the S. side of the river... Sinakaimbi's heir was a man named Chonga who lives at my kraal... A number of other people from the north side crossed over... We took with us our drums and horns... The people from the south side were assembled when we arrived [at Siankaimbi's kraal]. As we came up to the place of the mourning we beat our drums and blew our trumpets and the noise we made clamoured against that made by the drums and trumpets of the southern people so that Siamupa the chief, told his people to desist for a while till we had drunk the beer he brought us. Siamupa's people also drank beer at the ceremony, the two pots I have mentioned were provide to assuage the first thirst, the thirst of arrival. We and the southern people drank our pots of beer in silence.
>
> The usual custom was followed: first Siamupa's people danced to the beating of their drums advancing close to the other parties, including ours, seated on the ground. The dancers stabbed at the people on the ground with their assegais in dumb show, and then retired to their place and sat down, whereupon Siagoloba's party beat their drums and repeated the same as performance and when they had finished, we from the northern side, also danced and acted our part in the same manner. It was during our turn of dancing, in the second round, that Siamupa's people arose and began the new turn while we were still dancing and acting. They were jealous, I think, of our superior style of performance and this made them break in before our turn was properly ended. As Siamupa's people advanced towards us they kicked up the dust

from the ground in our face in a rude and aggressive manner. Their chief, Siamupa, seeing this, remonstrated with his people for being rude to us, the visitors, and at the same time enjoined us to sit down. Chonga who was, so to speak, the master of our part of the ceremony, then shouted to us to take our drums and to cross over back to our own community seeing that the southern people had thrown sand on us. We then retired along the path leading to the river, beating our drums and blowing our horns to show how we were going away at our own accord and were not being driven off like dogs.

As we walked away Siamupa's people followed us with their drums and noise, and some of them threw stones at us. Then there followed a sort of running fight between the two parties, and a man was struck down on both sides. Siamboti the deceased was not with our party as it retired, we had left him sitting near the grave and did not see what happened to him as we all ran away towards the river… Siamboti was the only one of our party who remained behind, he had been sitting with the women near the grave… Siamupa's people had their assegais in their left hands as they drove us away, but they struck with knobkerries…

Evidence of Tshalitshosia (northern party): Siamboti [the deceased] was not with us, he did not take part in the fight, as the accused says, because Siamboti was not with us, he did not dance with us, he preferred to sit and look on, he was not of a playful disposition, he never took part in our own dances at home…

Evidence of Gdanga (northern party): Siamupa's party came towards us, they put down their drums and took up dust and sand and threw it on us… we ran away, we did not beat the drums as we ran away… [during the fight] I escaped into the grass and hid myself because I was afraid of being killed by Siamupa's men who were more than in the northern party…. [when the body was brought to us] we carried the body down to the river and crossed over to our side… before we crossed [Manganiya came up and handed over the spear of the deceased to the deceased's son, admitting he was one of those to strike Siamboti, but saying Siakaganyana was the first to strike] I interpreted Manganiya's action in handing the assegai to the deceased's son as an expression of regret and a confession of guilt.

Chalisongola (son in law of the deceased): we know of no quarrel between the accused and the deceased. We Zambezis go armed, always.

Wife of deceased: When I saw there was going to be a fight I ran away and hid in the grass. Most of the other women did likewise. I ran away when I saw Siamupa's people kicking up the dust, that was the beginning of the row.

Evidence of Chief Siamupa: the three parties from the southern side and one from the northern side began to dance in turns, as our custom is. After some time, two of the parties, that of Sineusangwe and that from the northern side got mixed up in their turns and each sided tried to drum down and dance down the other. Seeing that trouble might follow I shouted to them all to behave themselves but they would not listen. Soon afterwards, a stone was thrown in amongst my people by one of the northern party, which one of my men picked up and threw back at the other party. This started a general fight. I tried hard to stop this but could not do so and had to give up. As the northern party was retreating… I saw the accused Siankaganyana strike Siamboti on the head with his knobkerrie – they were striking at one another and fencing with their assegais... Siamboti fell … as he fell, Manganiya came up and struck him also with his kerry. When Siankaganyana struck the deceased the northern party had retreated and Siamboti was cut off, he was 'among the enemy'… Both Manganiya and Siankaganyana only struck Siamboti once each. Then the other two accused came up and then also struck Siamboti as he was lying on the ground. I asked them what they would do and say when the northern NC asked for the life of his man and they all said, each in turn, 'what can we say, we have killed him.'

Although these cases that came before the courts were murder cases, where violence had not been successfully contained, they nonetheless help to throw some light on the way in which the idea of the law and of reporting to NCs had begun to be part of local debate, how chiefs were called upon to constrain the violence and saw it as part of their duty to do so, often in conjunction with other elders and the women, and how they intervened at considerable risk to themselves. Not only were chiefs' powers strengthened by recourse to the courts, but the threat to the chief's authority of not reporting a murder case or not trying to control violence, but of having such an incident reported by one of his subjects, provided further incentive still for chiefs to intervene.

The court records show that there was considerable abuse of authority on the parts of chiefs and other government representatives in these early years, and that this was only gradually controlled through the courts. A flurry of six cases brought against government agents between 1917 and 1921 demonstrates the way in which the state at local level remained associated with violent extraction long after the initial abuses that came to the NC's attention at the turn of the century. In this five-year period, two cases were brought against Tonga chiefs for homicide, two cases were brought against Native Department messengers for murder and rape, and a further two cases of extortion were brought against a BSAP official and a man posing as one.[81] The hearing against the BSAP official is particularly revealing. The accused was a constable on the north bank of the river who had obtained sheep and goats from 36 people on the south bank, paying less than their value on grounds that he had orders to secure rations for the Gokwe station; he had assaulted an old man, Simunda, for complaining about the price he offered, handcuffed him to a post, struck him with a sjambok, cut off his beard and instructed him to smoke the hair through a reed pipe; he had extorted stock from chiefs Sinatshembu, Sinesinanga, Siabuwa and Sinempande claiming it was government orders, and the chiefs had complied with the order, which they treated as legitimate and sent out instructions to their people to bring their stock. NC Green, who heard the case, reflected: 'What is interesting [about this case] … is that they could get away with this, saying it is a government order, ie it is expected of the government, such behaviour?'[82]

The question was a pertinent one. Had Green been acting as magistrate in relation to the other cases noted above, he might not have been so surprised, as they show chiefs and other local government agents exerting their authority abusively, using violence and making extractive demands of their subjects. In a case against chief Sampakaluma, five years earlier, for example, the Sebungwe NC heard how Sampakaluma had imposed a 'fine' of eleven sheep, eleven goats and three hoes on a man he accused of causing the suicide of the previous chief, and had ordered a gruesome punishment on a man who had killed another's dog (he was left tied to a tree with hands, legs and penis bound so tightly that his limbs and abdomen swelled up, his penis fell off, and he died when untied the following day).[83]

Partly because such incidents of violence were sporadic, and only occasionally heard before the courts, they did not reinforce old European stereotypes of the Tonga and the Zambezi valley as 'violent'. Indeed, they pale to insignificance in comparison to the flood of murders and violent assaults coming into the courts in the new towns and mines, often associated with abusive labour regimes. The Zambezi valley murder cases were heard by a large number of different NCs, none of whom developed significant expertise regarding the people of the river; indeed official knowledge of 'the Tonga' remained scant in Southern Rhodesia, as they lived in an inaccessible part of the colony, and were marginal to its political space.

Over the first two decades of colonial rule, nineteenth-century ideas about the Zambezi valley as a place of violence and lawlessness were replaced by new understandings of the place as primitive. Native Commissioner Hemans, NC for Sebungwe over the decade of the 1920s, summed up what became the dominant stereotype of the Zambezi valley when he described it as a wilderness attractive only to those who loved hunting and big game; it was a place 'at the back of beyond' where nature was 'still in the raw' and the Tonga people were 'untouched and unspoiled by civilization'.[84] In subsequent chapters we shall see how this reputation for backwardness persisted over ensuing decades and was reinforced when the Zambezi valley was belatedly subject to modernist state intervention through the construction of the Kariba dam.

We have seen in this chapter how the early decades of Company rule marginalized the people of the river, as the old Zambezian trading economy collapsed and the border region was remote and increasingly cut off from the new colonial infrastructure and the political and economic centres of Southern Rhodesia. Related to this process of marginalization, old imperial ideas about the river as a transport highway and a place of violence faded and the valley had come to be seen as a primitive wilderness. Yet the European authors of nineteenth-century stereotypes of the Zambezi were not forgotten; rather they were remembered for other ideas and their texts were popularized to new political ends. Before leaving the early years of colonial rule along the river, it is important to consider the impact of early developments where they were concentrated – around the Victoria Falls – where these new political uses of the landscape were elaborated. Those who lived around the waterfall and in close proximity to the new transport axis had a very different experience of the first decades of colonial rule than those who were increasingly marginalized and isolated downstream.

Notes

1 A similar case is made by Paul Nugent for the Ghana/Togo border, see Nugent 2003:9.
2 Mann and Roberts 1991:9.
3 See Asiwaju 1991.
4 The first Company patrols reported Portuguese trading posts at Saba on the Matetsi and elsewhere, and noted the prevalence of guns in villages along the river. William Reid to administrator, Bulawayo, 23 February 1897, LO 5/6/8. Cited in Matthews 1976:421.
5 CNC, Annual Report, Matabeleland Province, 1898, NB 6/1/1. Cited in Ncube 1994:138.
6 Gielgud, cited in Hole 1905:62-67. See discussion in Colson 1962.
7 Gielgud, Report on Sebungwe District, 8 January 1898 NB 6/5/2/2.
8 Colson 1953: 3.
9 For further detail on the process of Nambya regrouping and controversy over the different Hwanges, see Ncube 1994:120-33. Some Nambya remained on the north bank of the Zambezi in the vicinity of Gambo, where they fell under Leya chiefs.
10 ANC Green, Report on Patrol to Zambezi valley, 14 December 1897, NB 6/5/2/5
11 Ncube 1994:133; interview, chief Nekatambe, Hwange 15 February, 2000.
12 Monthly Report, October 1897, NB 6/4/1. ANC Green, Report on Patrol to Zambezi Valley, 14 December 1897, NB 6/5/2/5.
13 NC Lanning, Monthly Report, August 1897; Monthly Report, Bulawayo, September 1897 NB 6/4/1; Gielgud, Report on Patrol to Zambezi Valley, 8 Jan 1898 NB 6/5/2/2. On Monze, who was also probably motivated from a concern to maintain independence from the Lozi, see Vickery 1986.
14 Monthly Report on Patrol to Zambezi Valley, January 1898, Gielgud, NB 6/5/2/2.
15 Monthly Report, September 1897, Bulawayo, NB 6/4/1.
16 ANC Green, Report on Patrol to Zambezi Valley, 14 December 1897, NB 6/5/2/5.
17 Ibid.

18 Gielgud, Report on Patrol to Zambezi Valley, 8 January 1898, NB 6/5/2/2.
19 On the NAD, which was restructured in 1898, and formally constituted in 1902, see Holleman 1969, Steele 1972). Native Affairs in Northwest Rhodesia was set up with a comparable structure, outlook and staff. Gann 1958.
20 CNC Taylor, 'A Short History of the Tribes of the Province of Matabeleland', Bulawayo, 11 January 1904, A3 18/28.
21 Ibid.; Worby 1994.
22 CNC Taylor ibid.
23 Sebungwe, Annual Report, 1902, NB 6/1/3-4.
24 Interviews with chief Saba 12 March 2001 and Francis Munkombwe 15 March 2001; Ncube 1986:10.
25 On a similar process (slightly later) in Northern Rhodesia, see Colson, 'Comment on the Tonga Report' in Allan and Gluckman 1968:187; Vickery 1986:64-5.
26 NC Sebungwe, Annual Report, 1899, NB 6/1/2.
27 Gielgud, Report on Sebungwe District, 8 January 1898, NB 6/5/2/2.
28 NC Carbutt, Monthly Report, Sebungwe, August 1900, NB 6/4/4.
29 NC Carbutt, Monthly Report, Sebungwe, August 1900, NB 6/4/4.
30 The monthly reports over this period were increasingly desperate, see August, September, October, November, December 1899, through to May 1900 (Microfilm) NB 6/4/4. Ncube 1994:139-40.
31 NC Sebungwe, Annual Report, 1917. See discussion in Ncube 'ibid.
32 On the violence of turn-of-the-century labour recruiters in the valley, see Harding 1905:287-8.
33 Vickery 1986:54.
34 Cited in Vickery, ibid.:53. Vickery explains the initiation of this movement partly in relation to a series of natural disasters – locust attack in 1894-5, rinderpest the following year , drought in 1898-9, the desire for manufactured goods, such as blankets and clothing, (ibid.:54).
35 NC Sebungwe, Annual Report, 1907 NB 6/1/5-7.
36 NC Sebungwe, Annual Report, 1904 NB 6/1/5-7.
37 On RNLB agent at Walker's drift trying to recruit for the Colliery, see Sebungwe-Mafungabusi Annual Report 1904, NB 6/1/5-7. On failures of RNLB see also Sebungwe Monthly Report, December 1906, NB 6/4/7. On withdrawal, see Sebungwe Annual Report, 1908, NB 6/1/8-10.
38 The trade was undermined by the growing taste for manufactured tobacco from the 1920s onwards. On the dynamics of the (larger) Shangwe tobacco trade in this period, see Kosmin 1997.
39 On the association between the Tonga and night soil collection, see Thornton 1978.
40 Ncube 1994:147. NC Sebungwe, Annual Report, 1907, NB 6/1/5-7.
41 Worby 1994:371-92.
42 Cited in Ncube 1994:148.
43 Cited in Ncube 1986:10-11.
44 Ibid.
45 Sebungwe, Monthly Report, June 1906, NB 6/4/7.
46 Wankie, Monthly Report, February 1903, NB 6/4/4.
47 Wankie, Annual Report, 1903, NBH 1/1/1.
48 Explanations usually hinge on the recovery of game populations, due to restrictions on hunting through the Game Ordinances and the rise of migrant labour. See Ford 1971.
49 See correspondence between S and NW Rhodesian authorities over Siambale's island. NC Gwembi District to NC Sebungwe 12 September, 1914, NGB 2/7/1.
50 Gielgud, 5 January 1906, 14 July 1905 and other correspondence relating to garidzela, N23 1-3. Such bride service was cast by officials as originally a Shona custom, that had spread to Tonga areas. Some held it to be practiced particularly by Pashu, Dobola and Saba. See ANC Gokwe 9 April 1918, 'Native marriage ordinance', NGB 2/2-5.
51 ANC Hemans to Supt. Natives 26 May 1916, B3 13/5-7.
52 Sebungwe, Annual Report, 1907, NB 6/1/5-7.
53 On early debates over amalgamation, see Wood 1983:37.
54 There was a parallel process of administrative withdrawal in NW Rhodesia when two valley stations were closed and offices moved to the plateau. The primitive Methodists, who had established a mission briefly on the north bank also withdrew. Colson 1960:30.
55 Surveyor General to Supt. Natives, 4 July 1917, NGB 2/7/1.
56 On the former, see Native Commissioner's reports, Wankie, 1903-7; on Keigwin's plan, see Clements 1959:24-5.
57 Steele 1972:110-111; Gann 1965:147-8.
58 Gann ibid.
59 Sebungwe, Annual Report, 1903, NB 6/1/3-4.
60 Sebungwe, Annual Report, 1902, NB 6/1/3-4.

61 Colson 1971:27.

62 Colson argues (1953) that the *mukova* (matrilineal kinship group) is the most important source of support for any individual, but these ties are complicated by the importance of the father's matrilineal kin, and by the ties created through locality. For a discussion of 'revenge' that does not revolve exclusively around the idea of a blood-feud, see Stewart and Strathern 2002: chapter six.

63 Girard (1997) has provided some of the most extended recent theorization of sacrifice and ritual in acts of collective violence.

64 *R.* v. *Simwinza*, 1906, Sebungwe court records, criminal cases, D3/8/1.

65 Evidence of chief Pashu, *R.* v. *Simwinza*, 1906, Sebungwe court records, criminal cases D3/8/1.

66 *R.* v. *Simagwali and Siamkabi*, 1908, Sebungwe court records, criminal cases D3/8/1.

67 Evidence of chief Pashu and others, *R.* v. *Simagwali and Siamkabi*, Sebungwe courts records, criminal cases D 3/8/1.

68 *R.* v. *Tshogwa*, 1913, Sebungwe court records, criminal cases, D 3/8/1.

69 *R.* v. *Tshogwa*, 1913, Sebungwe court records, criminal cases D 3/8/1.

70 Evidence of Siabalagwe, *R.* v. *Tshogwa*, 1913, Sebungwe court records, criminal cases D 3/1.

71 See the description of funeral drum performances in Colson 1960:181-86.

72 For a discussion of sport and other contexts for collective violence, see Blok 2000.

73 The rootedness of Girard's work in Christian ideas and European psychoanalytic theory, particular traditions of sacrifice and notions of heroism, also undermines the applicability of his ideas. Tonga ritual did not involve elaborate sacrificial rituals, and Tonga ethics had a strong pragmatic element, in which 'fierceness' was regarded with ambiguity, and notions of masculine strength and courage did not hinge upon concepts of heroism and martyrdom. See Colson's (1971) exposition of Tonga ethics.

74 Evidence of Chief Sisansali, in *R.* v. *Gasonda, Siamilenga and Tshalisanga*, 12 December 1913, D 3/8/1.

75 Evidence of Siadenta, in *R.* v. *Kwata, Tshilopa and Dimangwagwa*, Sebungwe, 1907, D 3/8/1.

76 Evidence of Siadenta, in *R.* v. *Kwata, Tshilopa and Dimangwagwa*, Sebungwe, 1907, D 3/8/1.

77 Evidence given in *R.* v. *Gasonda, Siamilengo and Tshalisanga*, Sebungwe, 1913, D 3/8/1.

78 *R.* v. *Siagasungwa*, Hwange, 1923, criminal cases Wankie, 1891-1923.

79 *R.* v. *Samalambo*, Sebungwe, 1905, criminal cases Sebungwe, D 3/8/1.

80 *R.* v. *Siankaganyana, Manganiya, Tshalitshosiya and Siamateme*, Sebungwe, 1919, D 3/8/2-3.

81 *R.* v. *chief Sampakaluma*, 1917; *R.* v. *chief Sinakoma*, 1919; *R.* v. *(Native Department Messenger) Mandaba*, 1919; *R.* v. *(Native Department Messenger Mataya)*; *R.* v. *Manowa and Mipata* 1920; *R.* v. *Tshemonde*, 1921, criminal cases, Sebungwe, D 3/8/2-3.

82 *R.* v. *Manowa an Mipata*, 28 January 1920, criminal cases, Sebungwe. D 3/8/3.

83 *R.* v. *Savuna*; *R.* v. *Sapakaluma and others*, both June 1917. Chief Siakoma was also charged with murder, criminal cases, Sebungwe, D 3/8/2.

84 Hemans 1935:7, 13, 64-6.

5

Bridging the Zambezi at Victoria Falls
Science
& Early Colonial Expansion

The one part of the mid-Zambezi border that did not become rapidly isolated and marginalized in the first decades of colonial rule was the Victoria Falls, where the main road to 'the North' crossed the river. The tourist resort created around the waterfall after 1898 was very much a by-product of this developing transport infrastructure, which linked the industrial centres of South Africa via the Hwange coalfields to the mines of the Copperbelt and Katanga.[1] The building of the bridge over the Zambezi was the occasion for triumphal celebration of European technology and imperial expansion, and the new resort at the Falls popularized new understandings and uses of the landscape of the river, in which the waterfall's position along a 'natural border' was less important than its status as a 'natural wonder' and its location on an imagined transcontinental highway from Cape to Cairo.[2] As such, it became a focal point – a 'site of memory' – in the naturalization and legitimation of British imperial expansion and rule over the Rhodesias and of white settlement. The new political uses of the landscape at the Victoria Falls popularized a genealogy for Europeans in central Africa, which looked back to Livingstone and other explorers discussed in Chapter 3, and promoted a romanticized myth of their activities.

This chapter explores these new political uses of landscape and their consequences for those who lived in the vicinity of the waterfall. For the Leya and Toka people who had commanded the river's crossing points above the Falls, cultivated its banks and islands and propitiated their ancestors at the waterfall, the insecurities of the late nineteenth century were replaced not by a growing isolation, but by their close proximity to the new colonial infrastructure, competition for land and engagement in the new labour markets of the railway and railway towns. The new uses of the landscape competed directly with, and subordinated their use of, the river. As local people's access to the waterfall was undermined, cultures of colonial authority developed at the Victoria Falls initially incorporated Lozi royalty and celebrated their command over the river, reflecting the elevated place of Lozi rulers in NW Rhodesian legal and administrative structures, and the elite 'ornamentalism' characteristic of British imperial practice.[3]

Much has been written in criticism of European traditions of viewing landscape, particularly in imperial contexts. Mary Louise Pratt argues that such ways of seeing, rooted in European romantic and natural history traditions, are inherently colonial and masculine.[4] Husbands writes of the development of tourism at Victoria Falls in terms of the 'industrialization of the gaze'.[5] As we shall see, there are some notable continuities in European representations of the landscape of the waterfall – from the on-going popularity of reproductions of Baines' paintings, to aspects of the tourist itinerary and its emphasis on views and panoramas. However, a focus on continuities in a singular landscape way of seeing is narrow and decontextualized. It downplays social conflicts and is abstracted from changing political, economic and intellectual contexts. Moreover, it ignores shifts and that are particularly significant if one wishes to explore the political uses of landscape in relation to the formation of colonial and subject identities, and in relation to practices of state-making.

The resort's commemoration of nineteenth-century explorers was significant in establishing a lineage stretching backwards, but it also reflected a new sense of distance from, and progress since, the age of the Victorian pioneers. The early twentieth century resort celebrated modernity, the achievements of colonial science and command over nature, epitomized by the railway, the bridge over the Zambezi, and the new conventions of seeing introduced by camera and film. The landscape was experienced not only visually, but also bodily though various forms of sport and leisure. It was centrally important to an expanding tourist industry (as well as colonial settlement) that the environment was tame and safe (even if it was marketed as remote and wild), and suitable for women as well as men. The experience the resort offered to tourists gave no insight into non-elite African life in the vicinity, and the way it was shaped by the violence of early colonial extraction, conditions of work on the railway and elsewhere, evictions, and the far-reaching social, religious and political changes of early colonial rule.

The chapter continues to trace the divergence of state-making practices on opposite sides of a colonial border and their significance in terms of identity formation for those who lived in its vicinity. The fact that the waterfall was split between two separate states mattered in some ways from the outset, but the structure and practice of the two states diverged more dramatically as British South Africa Company rule gave way to 'responsible government' by settlers in Southern Rhodesia in 1923 and as the Colonial Office took over Northern Rhodesia and began to implement indirect rule.

The New Political Uses of Landscape at the Victoria Falls

By the time the decision had been made to route the railway to the north via the Hwange coalfields, the tourist potential of the Victoria Falls had already begun to be realized. The new road from Bulawayo to the waterfall had a regular wagon service from 1898,[6] and saw an increasing volume of traffic as miners and prospectors, administrators, traders and missionaries crossed into Northwest Rhodesia by way of the Falls, and as visitors came up the road specifically to see the waterfall itself. Colonial settlement initially took the form of a handful of pioneer traders and others grouped at Sekute's old crossing point on the river above the waterfall, now known as the Old Drift, where new ferry services were run by Europeans. Sekute and his people, like the other Leya and Toka communities that had clustered around the waterfall and lived on its islands for security began to spread out away from the river as the threat of raiding ceased. Though this initial process of spreading out provoked movements both north and south away from the river, we began to see in the last chapter how there were subsequently large movements northwards across the river in evasion of tax. At the same time, competition for the land in the vicinity of the waterfall and transport route escalated, as speculative interest in the region soared.

The prospect of the arrival of the railway provoked fantasies of dramatic economic growth, and global comparisons. It was confidently assumed that urban and industrial developments would be on the scale of Niagara City and Buffalo, which had grown up on the basis of power generated from the Niagara Falls. The African Concessions Syndicate Ltd. was not unusual in considering it 'a fairly safe prophecy' that such cities would develop, and 'an obvious fact that the industrial future of a large portion of Rhodesia and even of South Africa' would depend on electricity from the Falls.[7] The *South Africa Handbook* of 1903 noted that thanks to the proximity of coal, minerals and water power, the site possessed 'all the factors for the creation of a great manufacturing centre. A new Chicago, let us call it *CECILTON*, will spring up near the banks of the Zambezi.'[8]

Such expectations of growth following the railway spurred the conservation of the landscape, and a body of scientific research that went beyond the immediate practical concerns of engineers and prospectors. The BSAC decided to reserve the immediate Victoria Falls area in 1894, in response to a rumour 'that some enterprising individual was going to "peg out" the land around the Falls and charge gate admittance', an initiative the Company was keen to 'forestall'.[9] The Company wanted 'immediate action' to protect the Falls – and particularly its timber resources – from 'disfigurement at the hands of transport riders, traders and others.'[10] Niagara's precedent was invoked again in debates over conservation, this time offering a negative example. Lord Curzon was not alone in feeling that the Victoria Falls were more sublime than Niagara on the grounds of the 'lack of signs of civilization', and it was widely believed that new industrial prosperity in Niagara had spoilt its aesthetic appeal.[11] A park was designated around the waterfall itself (on both sides of the river), and a Conservator was appointed in 1900, with cross-border responsibilities.

Frank Sykes, who filled this post, was also Civil Commissioner for the Livingstone area.[12] His vision of the Falls was influential, as he designed the modifications of the landscape that made it more amenable to visitors and wrote the first 'official' guide, published in 1905. Sykes thought the landscape needed to be manipulated to 'excite the wonder of the onlooker' and to maintain its 'primitive charm'.[13] He felt it necessary to 'open up views of the river by judiciously cutting down trees', 'to fill up gaps by plantations' and to enlarge hippopotami tracks which were 'the only means of approach to some of the best points of view'.[14] He also wanted to charge admission, a proposal that was dismissed by the Company as impractical and 'undignified'.[15] Sykes was assisted by a Curator, Mr Allen (Forester to the Rhodesia Railways and former employee of Kew Gardens) who was appointed in 1904.

The rush of scientific investigation that accompanied the preparations for the bridge over the river, helped create new understandings of the landscape. The networks of collaboration and dissemination that were used and developed in these investigations underlined the role science could play both in strengthening imperial ties and in contributing to emergent settler identities and related processes of state-building in the Rhodesias and South Africa.[16] Geological and archaeological research was particularly significant. It challenged previous interpretations of the Falls landscape by explaining the formation of the chasm through fashionable gradualist ideas of the daily power of erosion, and rejecting the outdated catastrophist (often biblical) notions held by Livingstone and others, who had attributed it to a sudden mighty crack resulting from 'awful convulsions of the earth's crust'.[17] Geologists provided potent evidence of the antiquity of human settlement at the Falls, when newly discovered archaeological remains were shown to predate the physical retreat of the waterfall, which had cut back through the gravels in which the relics were situated.[18] These new scientific understandings infused visual appreciations of the landscape. Thus the President of the Rhodesian Scientific Association considered that 'the mighty spectacle of the Victoria Falls becomes the more sublime when we think that the forty miles of deep and narrow chasm are the result of the incessant action of water for at least 100,000 years.'[19]

The scientific networks developed in these endeavours were of mutual benefit to regional and metropolitan scientific communities, and important in building both settler and imperial identities and institutions. Botanical 'discoveries' provided new means of appreciating tropical splendour *in situ*, as well as new garden plants for the 'horticulturalists at home'.[20] Ornithologists collected dead birds for the South African Museum at Cape Town as well as British museums.[21] Local settler enthusiasts and Company employees gained the status of membership in metropolitan scientific associations for their assistance to eminent visiting scientists. Thus the photographer and local entrepreneur Percy Clark was made Fellow of the Royal Geological Society for helping Henry Balfour find stone-age axe heads close to the Falls, exhibited his photographs of the views of the waterfalls that Baines had painted at the Royal Geographical Society, and was elected Associate of the Royal Photographic Society in 1925.[22] The new technologies of camera and later film allowed representations of the landscape to reach an increasingly global public – postcards were mass-produced and a cinematograph produced in 1906 was said to have reached an audience of six million within six months of the filming.[23]

These new scientific understandings of the Falls were one way in which, by being better known, the landscape was tamed. But the growth of white settlement and tourism demanded taming of a different sort. It required the recasting of a place known for the discoveries of intrepid male pioneers, as a site of genteel leisured activity, where women were not out of place. Tourism achieved this in a range of ways, even as it commemorated the mythicized achievements of predecessors. All guides emphasized the comforts and amenities of the Victoria Falls Hotel, completed in June 1904. Though much less grand than it later became, this first hotel helped cast the Falls as a modern, luxurious resort, deploying symbols of a global tourist industry and of empire. The hotel's logo was the Lion of Africa and the Sphinx of Egypt, the first owner was Pierre Gavuzzi, who had worked at the Carlton and Savoy hotels in London; the chef was French, the barman an American from Chicago and the first waiters were Arabs.[24] This first wooden structure had been modified from the original railway employees' quarters and was equipped with electric lights and fans, and hot and cold water. The British travel agents, Thomas Cook and Sons, as official passenger agents for the Cape Government and Rhodesia Railways, followed the progress of the railway and the construction of the hotel in their magazine, *The Excursionist* (renamed the *Travellers' Gazette* in 1903), and began offering excursions from Cape Town in the same year. They anticipated a rapid expansion of business at 'Nature's greatest spectacle' when the traveller could 'enjoy European luxury even here in the heart of Africa'.[25] The Falls was a suitable destination for royalty, and others who arrived on the 'trains de luxe' from Cape Town.

The amenities and activities offered by the hotel and other local entrepreneurs, and the itineraries promoted in the guides, represent a profound transformation of the way the landscape could be experienced, and highlight its new political uses. The abandoned islands which ten years earlier had provided places of refuge and were associated with Leya chiefs who controlled crossing, were now renamed after European discoverers, visiting royalty and others, and were considered in terms of their suitability for taking tea and picnics, or for sunset views of the river, accessible by Canadian canoe or motor-powered launch. Visitors were invited to visit 'Livingstone Island', to see views of the 'Leaping Water' (so named by Thomas Baines), to walk through the 'Rain Forest' (so named by Thomas Mohr). They could imagine themselves in a lineage of pioneers, in which the first *white* woman to visit the Falls (in 1882, Mrs Francis, born in Grahamstown), and the first *English-born* woman (in 1884, Mrs Ralph Williams) were also remembered.[26] There were opportunities for sport fishing and shooting (for men), and details of walks of varying degrees of effort for men and women. Later rickshaws, pulled by Africans, were introduced to take tourists from their hotel to the waterside. Guides offered clothing tips to women as well as men, in addition to listing local suppliers.

This modern leisured experience of the landscape offered by the resort evoked the myths of empire in many ways. David Livingstone was remembered as missionary, scientist and servant of empire. His texts were reproduced and publicized perhaps more widely than ever before; his photograph appeared on the cover of early guide books and he was honoured in the naming of NW Rhodesia's capital. The place was made accessible by potent emblems of modernity in the form of the railway and the 'engineering wonder' of the bridge over the Zambezi, and the route also provided reminders of Cecil Rhodes. The bridge was planned, not without some controversy, to pass in full view of the Falls, as Rhodes had wanted. It was designed by the British engineer Sir Douglas Fox and constructed by the Cleveland Bridge Company of Darlington.[27] The striking combination of symbols of the power of nature and of triumphal British science along an important imperial axis gave an excessively jingoistic flavour to understandings of the Falls landscape. When the first train pulled into the Victoria Falls in April 1904, driven by the daughter of engineer Harold Pauling, it flew a Union Jack and bore a board below its headlamp reading 'We've got a long way to go'.[28]

The opening of the bridge the following year, in July 1905, reinforced this imperial triumphalism. *The Financier* declared the bridge 'one of the greatest engineering marvels of modern times and a most important link in the Cape to Cairo Railway';[29] *South Africa*'s journalist could see 'the mosques of Cairo ... already rising on the mental horizon'.[30] *The Globe*'s journalist celebrated this 'interesting event in the heart of Central Africa', important because the bridge could 'claim the distinction of being the highest in the world, has been erected in the heart of the Dark Continent and furthermore, represents the forging of another link ... in the great scheme proposed and started by Cecil John Rhodes.'[31] Guide books promoted the bridge as the place where the best views could be gained, and some entertained the visitor with technical details of its construction, including measurements (of south and north end spans, the centre arch and height in different places) and long lists of the names of consulting and contracting engineers.[32]

The festivities organized at the Falls for the opening of the bridge were significant for their jingoism, their celebrations of the railway, European science and new symbolic appropriations of the river. Most prominent among the eminent people brought to the Falls in the many special trains arranged for the occasion were a delegation from the British Scientific Association, whose president, the astronomer Professor George Darwin (son of Charles), gave the opening address. As the BSAC noted in a telegram, it was 'Very fitting that foremost representative of science should be associated with inauguration of modern engineering. Regret the founder of country is not alive to witness realization of part of his great ideal.'[33] For Thomas Cook and Sons the opening was 'an event second only in importance to the completion of the [Cape to Cairo] line itself', and their magazine described 'the memorable scene' of the five special trains carrying Professor Darwin 'through trackless, uninhabited tropical bush to the renowned falls'.[34] Cook chose the Falls as their emblem for tourism in the southern African region: the image they used showed the waterfall superimposed with images of 'newstyle' transport in the form of a train and counterposed to the old-fashioned discomforts of the wagon (see photo 5.3). Audiences in Britain knew of the event not only through the press and tourist industry: a new set of postage stamps was issued in Britain displaying the Victoria Falls.[35]

Although the propaganda surrounding the opening of the bridge celebrated empire and the taming of an African wilderness by European science, the festivities surrounding the event were also revealing of the way in which colonial culture selectively accommodated aspects of African tradition. The one-day regatta that was part of the opening celebrations offered a glimpse of such incorporations. The regatta set Rhodesian against South African crews, using 'clinker built fours' which had been imported from Oxford especially for the event, and also featured 'native races'.[36] There were many reminders of boating events and leisure back home. The Administrator at Livingstone town had a punt imported from England, which had been launched on the Thames and which he used for the first time at the regatta.[37] Local entrepreneur Percy Clark recalled how the rowing course was lined with 'the usual "joints" – poker tables, "Under and over", "Crown and Anchor" canteens and side shows.'[38] *The Evening News* declared the Zambezi, 'Our New Henley. The finest rowing waterway in the world.'[39]

But Britain and the Thames were not the only reference points in these celebrations on the river. It was of course significant for colonial identities constructed in terms of race, that European and African crews raced separately rather than competing against each other. The 'native' crews were mostly made up of the employees of European-owned launches on the river (one was comprised of BSAC employees, the other of workers of local trader and self appointed 'mayor' of the Old Drift, Mopane Clarke). But notions of African subjecthood were not undifferentiated. The regatta was attended by the Lozi paramount (Lewanika) and included a race of Lozi royal barges, which was won by the crew of Letia (Lewanika's son).[40] Tonga and Leya people, in contrast, though they had a more intimate connection with the regatta course, which traversed their old fords, had no public role in these displays on the river. Their invisibility in relation to Lozi aristocracy and Lozi traditions of display on the river is revealing.

There was a mutual admiration among settlers/administrators and Lozi royalty, each of whom was fascinated by the other.[41] Such admiration had been both reflected and reinforced following Lewanika's visit to England for Edward VII's coronation and his audience with the King in 1902. On his return, Lewanika often appeared at public events wearing the British Ambassador's uniform he had been given in London, which was highly ornate and heavily decorated with gold braid.[42] Tours at the Falls that were organized for visiting dignitaries always included a reception by Lewanika,

5.1 The opening of the Victoria Falls bridge, 1905 (Thomas Cook Archive)

5.2 Boating on the Zambezi, ca. 1905 (National Archives of Zimbabwe)

5.3 Old versus new means of transport to the Falls from Cape Town (Thomas Cook Archive)

5.4 Percy Clark, photographer, Victoria Falls (National Archives of Zimbabwe)

and he played the role of aristocratic host remarkably well by all accounts. The Lozi royal barges themselves, particularly the huge Nalikwanda or 'national' barge, were spectacular, especially as they moved with accompanying fleets and pomp instantly recognizable to European audiences. As Jalla described in 1928: the Nalikwanda had

> a handsome awning of mats surmounted by a [large] carved wooden animal – elephant, giraffe, etc. – occupied the centre of the barge, over a white awning where the king was installed with all his equipage and served from his kitchen. The cook could even make a fire on board. This barge was propelled by fifty paddlers chosen from among the Barotse princes and councillors. They were tricked out for the occasion in special costume, with animal skins on their backs and lion manes on their heads.[43]

The barges were described and photographed repeatedly in the early twentieth century.[44] The Kuomboka crossing received particular attention, and continued to do so as its ceremonial aspects were elaborated over the course of the colonial period. It was the most important event in the Lozi calender, when the Lozi aristocracy led people out of the rising floodwaters surrounding their homes in the wide Zambezi flood plain, onto higher ground.[45] This display, and the crowds accompanying the Lozi aristocracy's movement on the river more generally, were incorporated into events at the Victoria Falls and Livingstone, even though this was some 300 miles away from the Lozi heartland (though other royal centres were nearer).

At the 1910 regatta, for example, when the resort hosted the finals of the world sculling championship (pitting Richard Anst from New Zealand against Ernest Barry from England),[46] the Lozi aristocracy attended as usual. Letia appeared dressed in a black frock-coated suit and a cocked hat with white and red feathers.[47] Lewanika arrived in a royal barge accompanied by 60-80 large canoes and 600 African paddlers with paddles raised in salute.[48] Later the same year, during the Royal Tour of Victoria Falls and Livingstone, Lozi royalty came in full display to the town once again: it was reported that Lewanika and Letia had an entourage of 2,000 Africans.[49] The tour was testimony to the fusing of local and British symbols of status in cultures of colonial authority. Just as administrators' receptions incorporated the Lozi aristocracy and celebrated their royal traditions, so Lozi receptions of visiting royalty came to incorporate British costume and other material artefacts.

But the exchanges of riverine symbols were particularly interesting. Lewanika had a model 'Nalikwanda' barge sent to the King, which was displayed in an imperial exhibition in 1907,[50] whilst British royalty sent Lewanika a Putney barge at around the same time (though being made of soft Norwegian pine, it did not survive the journey up the Zambezi rapids intact).[51] As the Lozi royal barges incorporated hulls of European manufacture, the Livingstone administrators developed their own tradition of barges, with some aspects of their design borrowing from those of the Lozi aristocracy.[52]

We have seen how the Lozi aristocracy received tribute from the small subordinate Leya and Toka groups around the Falls in the nineteenth century. As Livingstone was the capital of Northwest Rhodesia, whose boundaries supposedly reflected the limits of Lozi rule, it was to be expected that the Lozi aristocracy should have a presence in the town. Thus, when Livingstone town was laid out in 1905, away from the swampy malarial margins of the Zambezi, Lozi authority was reflected in the naming of the open park at the centre of the new town's grid plan as the 'Barotse Centre'.[53] The Lozi presence in the capital persisted even as the colonial administration gradually undercut Lozi powers and greatly restricted their territorial jurisdiction to the confines of the Lozi reserve (which did not include Livingstone and its surrounds). After 1904, for example, the tribute the Lozi had collected from the Leya and others was replaced by taxation to the Administration, though for the first three years the Lozi continued to station an induna at the administrative centre who helped collect dues (a proportion of which still went to the Lozi aristocracy), and Lozi indunas initially continued to sit on courts in the Falls district (and other places outside the Lozi reserve).[54] In both Livingstone and Victoria Falls, Lozi was the most widely spoken (and highest status) vernacular, and aside from the English term for the waterfall, the Lozi name Mosi-oa-Tunya (which derived from the Kololo) came to be the most widely used. The Lozi were an 'obvious elite' amongst African workers in Livingstone, and filled most posts requiring education, quite aside from the aristocracy's ceremonial presence.[55] The Lozi also dominated the early curio trade and carving business, as they had a more elaborate carving tradition than the Leya, which was more quickly adapted to tourist demands.[56] Lewanika himself owned one of the first curio shops in Livingstone, and was reported to make £200 a year in 1910.[57]

The Lozi aristocratic presence in Livingstone and the inclusion of displays of Lozi command over the river in public events show the importance of symbolic exchanges with local African royalty in notions of colonial authority in metropolitan Britain. But these inclusions were limited, and were also compatible with ideas about white superiority characteristic of southern African settler culture and promoted through the tourist experience at the resort. Percy Clark's guide to the Falls, for example, includes not only the usual itineraries and homage to Livingstone and explorers, but also appends stories of his own pioneering experience. In them he adopts the manner of an old Africa hand informing the visitor not only of his meetings with individual Lozi royalty, but also drawing on stereotypes of the native as 'instinctively dirty' and 'a curious animal, but very useful to a superior race, after living in the country a few years they know all his little idiosyncrasies....'[58] Furthermore, an elite Lozi cultural presence was not established on the south bank. The south bank managed to maintain its dominance over the tourist trade – much to Livingstone settlers' chagrin – partly because Livingstone town had been moved away from the immediate vicinity of the river and the construction of hotels along its banks had been restricted, giving the Victoria Falls Hotel a distinct advantage. Thus it was the representations of the waterfall and resort on the south bank that came to be most important.[59] The divisions between settler communities and the character of the resort on the two banks became more pronounced as Company rule ceased and the two sides of the river were drawn into the respective colonial administrations of Northern and Southern Rhodesia.[60]

On the south bank, settlers' interests prevailed, and the vote in favour of 'responsible government' in 1923 led the way to policies of territorial racial segregation and 'separate development'. Though chiefs received some new powers, there were limits on the use of custom and tradition as modes of administration.[61] On the north bank, chiefs had more far-reaching powers and tradition played an elevated role, particularly after the implementation of indirect rule in 1924, and after the Colonial Office limited settlers' powers, constraining the alienation of African land and the exploitation of labour. Although Livingstone town did not manage to attract tourists on the scale it hoped and also lost its status as administrative centre (as the capital was moved to Lusaka in 1935), these losses were rectified to some extent by its emerging position as a cultural capital.[62] Governor Hubert Young had an interest in anthropology and archaeology, and wanted to build a museum to encourage the collection and display of artefacts representing a disappearing African social world.[63] Native Commissioners began to collect artefacts in 1930 and the David Livingstone Memorial Museum was constructed in the town in 1934; it became a focus first for archaeological research, and later for socio-cultural anthropology as the home of the Rhodes Livingstone Institute in 1938. Settlers supported these developments, as they hoped they would attract tourists and lure them away from the south bank.[64] The displays certainly reflected now outdated anthropological perspectives and reified the idea of discrete tribes, each with an associated territory, costume and material culture, but at least they acknowledged an African presence of some sort.

The resort on the south bank remained dominated by its natural scenery, its position on the Cape to Cairo axis and the reputation of the Victoria Falls Hotel, which was rebuilt in the 1920s to its current scale and grandeur. The resort retained its associations with empire and luxury, and was marketed as an exclusively white playground amidst wilderness, where Africans were present as an undifferentiated servile force. David Livingstone was commemorated on the south bank, not with an urban museum to preserve and display African archaeology and material culture, but with a huge bronze statue, funded by the South African Caledonian Society, in the park overlooking the waterfall (see photo 5.5).[65] It is significant that Southern Rhodesia's new National Museum, established in 1936 in Bulawayo, was devoted exclusively to natural history, and the Museum service persisted in discouraging the collection and display of African material culture.[66] The tourist industry kept alive the imagined geography of Cape to Cairo long after it had lost its significance in the world of realpolitik. Several travellers attempted the journey in the first decade of the twentieth century, including the novelist Charlotte Mansfield.[67] But the first commercial Cape to Cairo tour was organized by Thomas Cook and Sons in January 1922, for which the appeal was that 'the spirit of wild places can only be grasped on trek', and in which the Victoria Falls was a notable stopover.[68] But southern Africa was increasingly being marketed as a tame environment with a climate suited to healthy winter breaks, and as a fashionable destination. Alongside the natural wonder, Thomas Cook emphasized in 1930, 'There is a splendid and comfortable hotel at the Falls and during the season the fashionable throngs in the grounds and on the verandas are more reminiscent of a European spa than of a retreat in the interior of Africa.'[69] Tourist marketing continued to build on the image of Niagara and successfully created the Victoria Falls as a place of romance for lovers and honeymoons, as well as a venue for international scientific and governmental conferences. One south-bank guide appended a list of suicides and accidental deaths at the Falls (of white people), playing on the relationship between danger and the erotic so successfully exploited at Niagara.[70] Unlike the increasingly detailed historical, archaeological and ethnographic content of the north-bank guides, most Southern Rhodesian guidebooks continued to concentrate exclusively on natural history and European explorers.

One of those who visited the place at this time was Agatha Christie, whose archaeologist husband attended one of the scientific conferences at the resort in the early 1920s. Her mystery, *The Man in the Brown Suit,* was first published in 1924 and is partly set in the Victoria Falls.[71] It is perhaps the ultimate representation of a white playground. The intrepid heroine, Anne Bedingfield, sets out for southern Africa in search of adventure, and ends up heading for the Falls on a train from Cape Town caught up in a plot encompassing two murders, revolt on the Rand, diamond smugglers and a global criminal network. Having discovered who is at the apex of the

5.5 Unveiling the David Livingstone statue, 1935
(National Archives of Zimbabwe)

5.6 Tourists at Victoria Falls 'Big Tree', a former Leya rainshrine, n.d. [ca. 1950s]
(National Archives of Zimbabwe)

network, Anne is tricked into slipping over the edge of the Falls, only to be rescued from the cliff face by the man she loves who takes her to one of the islands above the waterfall to recover. The couple reject his wealthy inheritance and society life in London, and escape back to their tropical island wilderness. The only African who features (briefly) as part of the background in this white playground, is an old Lozi woman servant on the island.

We have seen how the resort created at the Victoria Falls promoted landscape ideas that could celebrate and reinforce Edwardian metropolitan notions of imperial authority in the periphery, but was also flexible enough to support sometimes competing local settler identities. We have also seen how Lozi aristocracy and representations of disappearing tribal cultures could have a place, particularly on the north bank, even as Africans were also cast as a racialized, inferior other and the landscape of the waterfall itself was thoroughly naturalized on the south bank. However, we have lost sight of the Leya and other local Africans with whom the waterfall was historically associated, who had no place in the landscape ideas initially promoted through tourism. We have also lost sight of the effects of the expansion of the railway and developments along it, which had made possible the creation of a tourist resort and affected local African communities profoundly. The next section of the chapter turns to the experiences of the Africans living around the Victoria Falls, and explores the effects of the new political uses of the landscape along the line of rail on their own relationship to, and ideas about, it.

The Political Economy of Expansion: What the Tourist Did Not See

These new political uses of the landscape of the Victoria Falls, which celebrated British imperial expansion and white settlement, sit uncomfortably with the violence of early colonial rule and processes of extraction and subordination, aspects of which were discussed in the previous chapter. Heroic accounts of the arrival of the railway and boasting that the construction of the Victoria Falls bridge involved the deaths of only one European and one African, for example, ignore the great number of deaths that accompanied the extension of the railway up to the Falls, and the appalling conditions of work, particularly in the first decade of colonial rule. Here I turn to examine what close contact with the railway and associated developments meant for those who had lived near to the Falls.

The railways were important employers, particularly during their construction, but also afterwards, and the fact that Livingstone and Victoria Falls were railway towns had a crucial impact on their development. In 1903, when the railway track first approached the Falls, the District of Wankie had some 7000 railway employees, and although the numbers were subsequently reduced, railway work and the railway compounds were permanent features.[72] The experience of working on the railway was shaped by some particular features of the industry – such as the practice of contracting and subcontracting labour recruitment and supervision – which had important effects on pay and conditions.[73] In December 1902, the Sebungwe Assistant Native Commissioner visited railway construction compounds to enquire about the cause of the 'wholesale desertion from work' of local Africans. At one compound he found 'that of about 300 natives about 200 had left work' as food had run out (supplies having been interrupted by cattle disease), and that among those who remained, there was a 'misunderstanding' over wages.[74] NCs reported deaths month after month: in May 1903 for example, the NC for Wankie reported eleven deaths among railway workers, mostly of dysentery and scurvy, and noted disaffection 'exacerbated by the inhumanity of some freight contractors'.[75] Conditions on the railways partly reflected the general violence of labour conditions at the time, but also the specific structure of railway subcontracting. As the NC explained:

> I could not help observing that a most unsatisfactory system seems to prevail there. The work is in charge of the railway engineers, who are there on behalf of the contractors to check the work of the subcontractors. The engineer in charge is, I understand, responsible to his employers for the general satisfactory progress of the working hand and wields a certain amount of authority over the subcontractors. He is, in fact the central authority on the spot. I was therefore surprised to find that he refused to make himself responsible for seeing that the subcontractors carry out their part of the agreement with native workmen, so that if, for instance, a contractor absconded without paying his men or refuses them proper treatment whilst working for him, these have nowhere to seek redress in the first instance, except the nearest government officials for whom it is very difficult to get at all the facts.[76]

The Hwange courts saw a steady stream of cases from the railway camps – against railway gangers and 'truck busters' for assaulting and murdering workers, and against workers for assaulting their gangers or for deaths resulting from fights in the compounds.[77] In 1909, an outbreak of cerebrospinal meningitis on the Bulawayo to Victoria Falls section of the railway saw the death rates soar from 10 to 48 per 1,000, and the Northern Rhodesian government threatened to disallow labour recruitment from the territory. An official investigation by the Southern Rhodesian Medical Director described overcrowding in the compounds, inadequate sanitation and rations and endemic scurvy, and recommended the railways be compelled to provide rations comparable to those on the mines.[78] Yet the railways were able to continue to breach regulations, including new legislation against 'acts of inhumanity' laid out in the Native Labour Regulators Ordinance of 1911, as the structure of employment made it difficult for the police to make charges against employers on technical grounds as foremen and others subcontracted to oversee workers often did not meet the legal definition of 'an employer'.[79]

Although conditions on the railways compared unfavourably with those on the mines in some ways, they were nonetheless able to attract labourers as railway work was perceived as less dangerous. Moreover, the early reputation of some mines for maltreating labourers also spread quickly. At the Wankie colliery, for example, outbreaks of scurvy, dysentery and pneumonia instantly got the mine 'a very bad name' among locals; by 1906, Sebungwe's NC reported mass desertions for 'a chain of ill-treatment ... which is widely talked of in the district...' Discontent at the colliery culminated in 1912 in a 160-worker strike, which brought the colliery further notoriety though it also led to some improvement in conditions.[80] A strike on the railways by black workers in the same year similarly ushered in some improvements in the railways.[81]

The settler farms, pegged out along the line of rail up to the Falls between 1903 and 1910 also competed for local labour, though they were unable to match salaries paid by the mines or railway contractors. Labour conditions on the farms can be illustrated by the example of the relations of a prominent member of the early settler community with his labourers – Albert Giese. The young man Giese was famous for having discovered coal in 1894 and in the years before Native Department officials arrived in Wankie in 1903, he had acted as unofficial 'father of the district', helping new settlers and posing as a source of knowledge about local Africans. Giese owned a store at the Victoria Falls and his farm was one of the first to be pegged out along the railway, in the vicinity of the colliery. Having rapidly got a bad reputation with local Nambya and Leya for violence and sleeping with his labourer's wives, Giese (like other white settlers) resorted to armed raids over the Bechuanaland border to capture San workers. These cross-border operations were justified on spurious grounds and were understood by those who took part in them in the terms of a previous era – the language of slaving and raiding is pervasive. In one such raid (undertaken on the pretext of apprehending a tax defaulter), Giese 'procured a patrol ... of one European policeman and five native constables' to round up labourers. One of the native constables involved testified in court:

> We natives proceeded to carry out Mr Giese's instructions accompanied by his 2 guides, since it was his impi. We ... struck west into Bechuanaland ... visited a dozen or more waterholes and captured as many men, women and children with their belongings as we could. The journey back eastward ... occupied four days. One party had trouble with a certain man found. Either he resisted capture or else it was a row about his relative, the guide's wife. At any rate he got a thrashing, and on our way back we had to leave him near Pandamatenga disabled; and afterwards we heard that he died of his injuries. The guide captured a sister of his wife's to make her his wife, but she escaped on the way back. All told we brought to Mr Giese's some 20 or 30 men, women and children with their belongings.[82]

Settlers also tried to poach each other's workers. Giese became embroiled in a protracted dispute with another white farmer – Mr Cummings – whose farm was just South of the Victoria Falls and who had built up his own labour force partly by offering refuge to those harassed by Giese.[83] The violence of early means of procuring workers on the farms was undermined after 1910 by more formal structures of recruitment, and cases against individual settlers were also brought before the courts. In 1912, Giese appeared before the Wankie magistrate's court, charged with assault, murder, brutality and neglect of workers, and was the object of a lengthy investigation by Native Department officials based in Wankie, who felt his activities undermined the department's role and image. Thereafter, he ceased to have a quasi-official role in the district, the administration of which was in the hands of the professionals in the NAD.

In the light of these conditions of employment, it is hardly surprising that some locals chose to work further south and some avoided formal work altogether and raised tax dues through trading in dried fish to the labourers at the colliery and other mines, selling curios at the Falls, and cultivating vegetable gardens and accumulating stock if they could. Some chose to work in the Victoria Falls resort and in the town of Livingstone, as it grew, where conditions were perceived as better than on the mines and railway, and were more compatible with rural production. Former workers at the Victoria Falls Hotel remember that salaries were reasonable and food plentiful in comparison to other employers, though they felt it was degrading being expected to eat leftovers from the table, being insulted or hit by young white female guests, and having to keep off the few pavements that later existed in the resort.[84]

Migrant labourers from a broad swathe of central Africa were drawn into the two railway towns themselves, and also moved through, as recruitment centres were established at the border posts on both sides of the river to funnel migrants to the labour markets to the south. The Rhodesia Native Labour Bureau (RNLB), which aimed to divert streams of labour to low-wage employers, largely failed to recruit among local Nambya and Leya (just as they failed among the valley Tonga downstream): being border peoples with an early history of migrant labour, they had knowledge of how to cross independently, which they could use to their own advantage. From 1908 the RNLB organized a new form of 'recruiting' at the river's crossing points, such as Victoria Falls and Kazungula, which involved preying on more vulnerable migrants from further afield whom they waylaid with offers of 'free ferries'. The railway's own recruits did likewise: those stationed on the Livingstone side of the border offered free train tickets southwards, and provided free food and accommodation in Victoria Falls before workers had been registered and gained their passes on the Southern Rhodesian side of the border.[85] The association of the ferries with 'chibaro' (forced labour, or slavery) became so well-known that, when new free ferries were started up in 1928, 'for the first few months these facilities were very little used as natives were suspicious that there was some catch and that if they used the free ferries and accepted the free food they would find themselves bound to the Bureau or some similar organization'.[86]

Though they were initially in a position to pick and choose the work they wanted, pressure on the local Leya who lived around the Victoria Falls mounted as they began to be evicted from the lands they had occupied around the waterfall. The evictions were most far-reaching on the south bank, and the voluntary movements across the river in evasion of tax in the first decade of colonial rule were replaced from the 1920s by forced expulsion across the border. In Southern Rhodesia (where no Leya chiefs were appointed), the Leya were labelled 'foreigners' who belonged over the river, and were told to 'go back' to their chiefs on the north bank. As the Leya were moved out, conserved land in the immediate vicinity of the Falls expanded to form a mosaic of game reserves, safari and forest areas, creating the desired image of a wilderness around the resort. Oral testimony describes several rounds of evictions, beginning in the 1920s, with the last vestiges of Leya settlement on the south bank being almost totally cleared in the 1930s; those who resisted are said to have had their homes burned. This seems plausible, though the archival record is scant, even at the height of the evictions in the 1930s. In 1935 the NC Wankie noted: 'Most of the natives who resided in the Victoria Falls Game Reserve were removed in July to Northern Rhodesia'; in 1936 removing people 'from whence they came' is reported as almost completed; in 1937, a further 41 people were 'sent back' to Northern Rhodesia.[87]

The former Leya inhabitants are still remembered in some place names in the Victoria Falls National Park. Jafuta, for example, was an important Leya man who lived in what became the Park, and his name still appears on signposts to the luxury 'Jafuta Lodge'. He farmed a fertile vlei just south of the Falls at Chamawondo and owned a canoe and a sizeable herd of cattle. The village he headed was split by the evictions, some members went to Northwest Rhodesia as the Wankie NC instructed, others moved onto the infertile lands of the increasingly crowded Wankie native reserve to the east and south of the Falls. Jafuta's brother's son remembered how the state broke up this Leya community: 'The white people came to tell us to go across the river to Chief Mukuni, they didn't want us here...'.[88]

Africans remaining in the vicinity of the Falls on the south bank in the wake of the 1930s evictions – most of whom were working at the resort itself and who included many Leya and Lozi – were 'removed to a central village away from the river' in 1940.[89] This second round of removals into the village was justified in terms of malaria control, which involved protecting white visitors by 'reducing the human pool of infection by removing the large majority of natives from that area'.[90]

After the evictions of the Leya population on the south bank, land close to the resort, which had been designated as part of the Hwange reserve, was filled up with other evictees from elsewhere, particularly Nambya and later Ndebele. The Hwange reserve had been created by the Native Reserves Commission of 1914–5 and comprised arid and rocky land. As the Native Commissioner commented, 'out of any block of ground available at present, nine tenths would be practically useless, owing to the want of water and the rocky nature of the soil'; new lands added to the reserve in the 1920s were worse, indeed the NC had not seen 'poorer land in the colony'.[91] Nambya people were forced into the area first by evictions from the coal concessions, then through the imposition of land rent on European land, though movement was delayed as threats of evacuation over the borders contributed to the suspension of the rent until 1919. Most movement into the reserve occurred over the subsequent decade, as rent began to be collected and European farmers exerted pressure on the Native Commissioners through the Rhodesian Agricultural Union. The scale of movement increased as evictions occurred from the huge Hwange game reserve after 1928. The population of the part of the native reserve closest to the Victoria Falls was increased dramatically again in the 1950s, when Ndebele evictees from the white farms around Bulawayo were controversially transported up to the Falls and dumped there, hundreds of miles from their former homes. Although the area was settled by this mix of people, and was administered through Nambya chiefs (and later also an Ndebele chief for the Ndebele immigrants), a prior layer of Leya settlement was recognized, and Leya mediums were sometimes sought out for their privileged relationship with the land, rain and river (even as they were looked down on in other ways).

The expulsion of south bank Leya 'foreigners' did not end cross-border movements: there was close contact between the Leya (now mostly in Northern Rhodesia) and the Nambya living in the reserve south of the river, due to old kinship ties dating to the time of the mid-nineteenth century Nambya state (which had incorporated many Leya, Tonga and others under the label Dombe), which were further strengthened during the time the Nambya had fled to the river to escape Ndebele raids. When the Nambya had crossed back south and regrouped around their former ruins, now within coal concessions and a Game Reserve, they brought with them many Leya people, and Nambya families, including those of the chief, incorporated many women of Leya (as well as Tonga) origin, and the Tonga language and Tonga cultural forms flourished in Nambya homes, alongside Nambya. When Ndebele evictees were being resettled, some Leya took up the opportunity to re-establish homes in the vicinity of the waterfall.

On the north bank, land alienation was also severe – particularly so by Northern Rhodesian standards – and evictions also proceeded apace as Livingstone town grew, as farms were pegged out along the river frontage and the line of rail, and as other prime sites for tourist development were reserved as crown land. Mukuni and Sekute's people were forced into the Leya and Toka reserves respectively, along with others, where soils were poor and where they lacked access to the river for gardens, fishing and grazing cattle.[92] The justification for these restrictions was that African settlement and land use spoilt the view and undermined the area's tourist potential: 'the future

5.7 Chief Mukuni and Bedyango, 1945 (Livingstone Museum)

prosperity of the district depended on the Victoria Falls and the river, which forms part of the attraction to tourists and for protection purposes its banks for some distance upstream should remain crown land upon which settlement, if allowed at all, should be strictly limited. Certainly no native settlement shall be permitted.'[93] There was also a parallel process of re-ordering Livingstone town along segregationist lines justified in terms of malaria control. When the Livingstone Native Welfare Association was initiated and began to raise grievances against Europeans in 1930, its meetings in Livingstone's Maramba township were attended not only by urban dwellers, but also by three local chiefs, including Mukuni and Musokotwane, who complained bitterly of how they had been chased out 'like dogs' from the lands of their ancestors.[94]

The Land Commission of 1946 considered that 'a great injustice' had been done to the Leya people who had lived around the Falls: 'Quite apart from [water for their cattle], the natives in question are a riverine people and though not many of their villages used to be actually on the banks of the river, they used to have cattle and fishing posts there, and also a certain number of gardens.'[95] Before they had lost access to fishing grounds, the Leya of the Falls area had a profitable trade in dried fish with the labour force of the railways and Hwange mines, which was banned in 1909. Trade in cattle across the river (which had enabled raided people on the south bank to rebuild their herds) was outlawed in the same year. Although there were opportunities for work at the resort and profits from the sale of curios, this was not necessarily seen as adequate compensation for lost access to favoured lands and the resources of the river.

Oral historical accounts of this process of eviction do not only dwell on the material effects of lost access to land and resources. The waterfall and river had been a religious site, and as we have seen, the waterfall itself was strongly identified with Leya ancestors, and particularly Leya female mediums. Accounts today tell of the desecration of religious sites and the undermining of female powers; indeed this line of argument, focussing on past sacred value and its despoliation, has flourished alongside international interests in the cultural heritage of the waterfall. Narratives of desecration look back to the early history of the resort, when the building of the bridge over the river defiled one of Mukuni's religious sites, as the supports for the bridge reached down into the boiling pot.[96] They cast the generation of hydroelectric power as a further intrusion: beginning in 1938 on a small scale to produce electricity for the town of Livingstone, this involved engineering works in the form of canals, concrete pipes and part of the generator passing through the cleansing site at the top of the waterfall.[97] On the south bank, narratives likewise focus on the desecration of the boiling pot and other sacred places, such as the Big Tree, a large baobab associated with ancestors of important Leya spirit mediums and with pools in the river (see photo 5.6).[98]

The current re-validation of traditional sacred sites, however, has taken place subsequent to a long history of decline, as religious uses of the waterfall were undercut in the period considered here not only by reduced access and engineering works, but also by the penetration of Christianity, education and modern medicine. Once the Leya had been expelled from Southern Rhodesia, those who lived closest to the waterfall were people who had little interest in it, as the Nambya looked primarily to the ruins of their former nineteenth-century capital for ritual purposes rather than to the river, and the Ndebele had no interest in the waterfall. New religious movements ignored old religious sites, and around Livingstone, the Watch Tower movement attracted workers and the young in the early part of the twentieth century.[99] Ceremonies at the boiling pot ceased probably in the 1950s, and the light within it is said to have gone out, while sounds of past communities in the form of drumming, cattle lowing and children playing could no longer be heard.[100]

But Leya communities' loss of access to the waterfall and profound social changes do not entirely explain why the religious associations of the waterfall were so totally obscured from public view in the resort's early history. After all, some people still visited the sites for religious reasons, particularly on the north bank, where the process of undercutting Lozi authority was coupled with the recognition of independent Leya chiefs. One reason why the new Leya chiefs appointed under the Northern Rhodesian administration did not initially stamp their cultural presence on the public life of the resort was undoubtedly that they lacked the status of Lozi royalty. Furthermore, although there were Leya chiefs in the reserves, the Leya workers in the cosmopolitan town of Livingstone were simply one of a large number of ethnic groups, and in urban politics they often identified as part of a larger Tonga block. Moreover, much Leya ritual at the Falls was not in the public domain: water was collected from a site which was kept a secret even from most Leya themselves, while cleansing rituals involving immersion in the water were designed for the participants rather than observers. Even the more public aspects, such as throwing offerings to the boiling pot, involved little display and were not intended for onlookers.

But there is one further explanation why Leya ritual and historical connections with the landscape remained hidden from view. This was because the male chiefs, whose powers were shored up by their official roles in Northern Rhodesia, lacked close connections to the landscape, as such relationships were upheld by female authorities. As chiefs Mukuni and Sekute elaborated traditions for Native Commissioners, they publicized versions of their own genealogies, which led away from the landscape of the waterfall to a mythical first ancestor from somewhere else. They developed stories of how they came to the Falls through a long series of migrations from elsewhere.

They elaborated battles between chiefs and controls over war medicines, and of magic drums and ritual stones that mythical ancestors had brought with them on their travels. Thus Mukuni claimed to be originally of Lenje origin from Kabwe District and Sekute claimed Subiya roots.[101] Mukuni's ritual stone was said to have been brought from Kabwe, whilst Sekute's magic drums came with him from Nqui. In these histories, produced in an era of indirect rule and an elevated administrative use of culture and tradition, Mukuni gained the reputation of having arrived first in the Falls area and became the more prominent of the chiefs close to Livingstone. The influence of the female Leya authorities and mediums was undercut in this period, particularly in the case of Sekute.[102] Mukuni in contrast, noted in 1957 that the chief 'ruled his country together with a woman called Bedyango', whose duties were the distribution of land (see photo 5.7).[103] However, unlike the stories told today, in which Bedyango represents the indigenous Leya people, the then Mukuni emphasised her relationship to the chief's lineage. He mentioned in passing her responsibility for cleansing at the Victoria Falls, but otherwise did not dwell on the symbolism, religious values and links to the landscape of the Falls.[104] The book returns to the Victoria Falls later to explore the context in which narratives of the desecration of the waterfall and ancestral connections to the locality have assumed such an elevated importance.

My aim here, however, was to explore the political uses of the landscape promoted through the tourist industry at the Victoria Falls from 1900. These masked the violence of early colonial extractions, reinforced an expansionist imperial ethos, built confidence in white settlement and celebrated colonial science and modernity. They initially made some space for the Lozi, but excluded the Leya and others both physically and symbolically. The acute competition for land and processes of eviction that accompanied developments around the Falls meant that early colonial rule was experienced rather differently than in the increasingly isolated areas downstream. Being closer to the developing colonial economic, transport and administrative infrastructure also meant that the border – or perhaps more accurately the power of the two states – came to matter more in everyday life than it did downstream. There was still contact across the border after the Leya had been totally expelled from the South bank, but Leya and Nambya increasingly came to be seen, and saw themselves, as Northern and Southern Rhodesian subjects respectively. Notwithstanding their roles in funnelling migrant labourers from a broad region, Livingstone and Victoria Falls developed as distinctly Northern and Southern Rhodesian colonial towns, and the divergent character of the two states was reflected in public representations of local history and culture, and the experience they offered to tourists.

The final chapter of the book will discuss the more recent politics of landscape, heritage and tourism at the Victoria Falls, including new claims to the waterfall, new interpretations and uses of history, and the deployment of old idioms about the river to new political and commercial ends. But before moving onto the post-colonial context, the book turns to the politics of the next monumental state intervention along the river in the colonial period, and its consequences for those who lived downstream, in what had been isolated and marginalized parts of the river valley.

Notes

1 For a discussion of tourism in the wake of the development of an overland route to India, see Barrell 2000.

2 Tourism was important in promoting the Cape to Cairo idea, though Merrington (2002:140-157) ignores it in explaining its persistence.
3 Cannadine 2002.
4 Pratt 1992:201-227, Carruthers and Arnold 1995.
5 Husbands 1994.
6 Phillipson 1990a: chapter 8.
7 See African Concessions Syndicate Ltd. *The Victoria Falls on the Zambezi River* (London, 1902). On power at the Falls, which was not generated until 1938, and then only on a small scale, see S 246/278 and S 482 134/48-9.
8 *South Africa Handbook*. No. 6. [n.d. ca.1903]. Held in Rhodes House, Oxford.
9 *The Directory of Bulawayo and Handbook for Matabeleland 1895-1896* (W.A. Richards and Sons, Cape Town, 1896), citing government notice of 22 September 1894.
10 Letter from Hugh Marhall Hole to BSAC Administrator, S. Rhodesia, Siloh, 18 July 1903, A 11/2/17/2.
11 Lord Curzon, *The Times*, 14 April 1909, S 142/13/21.
12 He had been a trooper in the British regiment which helped BSAC forces put down the Ndebele uprising in Matabeleland in 1896, and was author of *With Plumer in Matabeleland* (London, Constable, 1898).
13 F. Sykes, 'Suggestions for the Conservation of the Falls', 8 July 1903. See H. Marshall Hole to Administrator, Southern Rhodesia, 18 July 1903, A 11/2/17/2.
14 Letter from Hugh Marshall Hole to BSAC Administrator, S. Rhodesia, Siloh, 18 July 1903, A 11/2/17/2.
15 See F. Sykes 'Suggestions' and H. Marshall Hole to BSAC Administrator, S., Rhodesia, 18 July 1903, A 11/2/7/2.
16 See Dubow 2000.
17 Mohr 1876:145. Mohr, who visited the waterfall in 1870, was repeating the ideas of earlier travellers. Livingstone had attributed the waterfall to a sudden crack across the river, contrasting it explicitly with the formation of the Niagara Falls, which was already understood to have been formed by the gradual wearing back of the rock. Baines, Chapman, Mohr and Selous did not challenge Livingstone's view. See Molyneux 1905:44-63, Lamplugh 1908:133-52, 1907:62-79.
18 Armstrong and Jones (1936) outline the history of archaeological discovery and interpretation at the Falls. The idea that the relics predated the Falls geological retreat was first suggested by H.W. Fielden in 1905 in a letter to *Nature*, and was elaborated by Lamplugh (1906:159-60)) and Balfour (1906).
19 Presidential Address, *Proceedings of the Rhodesian Scientific Association*, 4 (1903).
20 C.E.F. Allen, 'Notes on the Falls Botany', in Sykes 1905: section xiii.
21 W.L. Sclater, Director of the South African National Museum, 'Ornithological Notes', in Sykes 1905: section xiv.
22 Clark 1936:240, 245.
23 *Livingstone Mail,* 23 June 1906.
24 Crewel 1994.
25 *Traveller's Gazette,* 9 April 1904, 'By Rail to Victoria Falls'. Available in the Thomas Cook archive, Peterborough.
26 Sykes 1905:2-3 (emphasis in original list).
27 Phillipson 1990b: 97. The construction of the bridge is described in greater detail in Watt n.d.:32-33.
28 Croxton 1973:91.
29 *The Financier*, 22 July 1905.
30 *South Africa*, 15 July 1905.
31 *The Globe*, 13 September 1905.
32 Sykes 1905:7-8.
33 Cited in Dubow 2000:72.
34 *Travellers Gazette,* October and November 1905.
35 Croxton 1973:97.
36 For descriptions of the opening, see Croxton ibid.:93-99; 'Opening of the Zambezi Bridge', *Travellers' Gazette*, November 1905; Clark 1936:202-204; Watt n.d.:35-6.
37 *Livingstone Mail*, 7 November 1908.
38 Clark 1936:212.
39 The *Evening News*, 15 July 1905.
40 Phillipson 1990b:103.
41 See also Ranger 1981.
42 Clay 1968:127.
43 Turner 1951:90-91.
44 E.g. Luck 1902:.62, note 79.
45 Descriptions of the Kuomboka can be found in Jalla 1928:90-91,Stirke 1922:121, Gluckman 1951:11.
46 *Livingstone Mail* 20 August 1910.

47 *World Wide Magazine* October 1910.
48 *Livingstone Mail* 12 November 1910.
49 *Livingstone Mail* 12 September 1910.
50 Clay 1968:129.
51 Ibid.
52 Turner (1951) notes that Nalikwanda came to be of European manufacture. On DCs adoption of aspects of Lozi barge design, see photo of DC on his barge at the Kuomboka taken by Gluckman and caption in Schumaker 2001:50.
53 Phillipson 1990b:98.
54 Stokes (1966:261-301) describes the undercutting of Lozi political power outside the reserve over the period 1898-1911. See also Caplan 1970.
55 Caplan, ibid.:94.
56 Sekwaswa notes that the Lozi who flocked to Livingstone for work 'taught local natives basket-making, trays and mats – stool bowls and woodwork'. Government messenger Sekwaswa, 'History of Livingstone District' (unpublished ms, n.d., ca. 1936:16), Livingstone Museum, G102.
57 Per Rekdal, 'Traditional Carving and the Curio Trade', in Phillipson,1990:108-115.
58 Percy Clark, 'A Native in the Wilds of Rhodesia', in *Guide to the Victoria Falls* (4th edition, ca. 1920:25).
59 Livingstone settlers' complaints over the dominance of the south bank appear regularly in the *Livingstone Mail*, see for example, 21 April and 6 June 1906. See also Watt n.d.: chapter 7.
60 From the 1930s, research conducted from the Livingstone Museum under the guidance of J. Desmond Clark was reflected in north bank guides, which came to incorporate considerable archaeological and ethnographic content. See Clark 1952.
61 Alexander 2006:chapter one.
62 Schumaker 2001:54.
63 On Young's ideas, and the origins of the Livingstone Museum and Rhodes Livingstone Institute, see Schumaker, *ibid.*:54-5.
64 *Ibid.*
65 David Livingstone Unveiling Ceremony, *The Address* and *Programme,* August 1934. Both held in Rhodes House, Oxford.
66 Elizabeth Goodall, the government's sole ethnographer in the 1950s and 1960s complained repeatedly of the lack of interest and funding for the collection or display of African cultural artefacts. Munjeri 1991:444-57.
67 This unsuccessful attempt is described in Mansfield 1911.
68 Because of Cook and Son's long-standing presence in north Africa, the trips were organized from Cairo to the Cape rather than the other way round.
69 *Travellers Gazette*, November 1930.
70 See collection of Victoria Falls guides held in Bulawayo Reference Library, including Dept. of Publicity, Salisbury, *The Victoria Falls of Southern Rhodesia* (S. Rhodesia Gvt printer, 1936), Jonah Woods and Stuart Manning, *A Guide Book to the Victoria Falls* (Bulawayo, 1960).
71 Agatha Christie, *The Man in the Brown Suit* (Bodley Head, 1924). Thanks to Elizabeth Colson for drawing this book to my attention.
72 Annual report, Wankie, 1903, NB 6/1/3-4.
73 Lunn 1977:107-8.
74 ANC de Lassoe, Monthly report, Sebungwe, December 1902, NB 6/4/4.
75 Monthly Report, ANC Wankie, May 1903, NB 6/4/5-6.
76 ANC de Lassoe, Monthly report, Sebungwe, December 1902, NB 6/4/5-6.
77 For cases of murder, assault etc. against employers see *R.* v. *ganger Lawrence O'Neill*, 24 January 1911; *R.* v. *ganger Henry Richard Marshall* 25 July 1919; *R.* v. *loco foreman John Carruthers Robley* 15 May 1919; *R.* v. *(truckbuster) Prya* 21 December 1911. Criminal cases, Wankie, D 3/37/1 and D 3/37/5. The files also contain cases of workers assaulting gangers, of railway compound fights and arson.
78 Detailed in Lunn 1977:120.
79 See discussion in *R.* v. *John Carruthers Robley* of 15 May 1919 – a case involving sickness, injury and death through neglect. The magistrate reviewing the case noted that 'it maybe the case that the railways are greatly to blame in not providing for medical attendance on, or supervision of their natives in so far as Wankie is concerned, as the magistrate finds, but unless the accused is proved to be "an employer" he is not liable to punishment'. Criminal cases, Wankie, D 3/37/5.
80 W.E Farrer, NC Sebungwe to CNC Bulawayo 23 March 1906 NB 3/1/6. See discussion in Phimister 1994:22-3, and van Onselen 1974:275-89.
81 Lunn 1997:120.
82 Testimony of native constable, in ANC Randolph, 'Report on the Application of Mr A. Giese JP to have a Degree of Control Over the Amasili of the Western Border', 26 October 1912 [hereafter, 'Report on Giese'], Wankie Correspondence, S707. This is a sizeable file containing the results of the ANC's investigations,

including many transcribed witness statements by farmers, constables, tenants and others. Discussed also in R. Holst 1993.
83 ANC Randolph, 'Report on Giese'.
84 Interview John Nicholas Mlamba Gumbo, Monde 27 March 2000; Dorcas Ndlovu, Monde, 28 March 2000.
85 This system violated the pass laws, and was revealed through official investigation in 1929. See Lunn 1997: 117.
86 van Onselen 1976:107.
87 See Annual Reports Wankie, National Archives of Zimbabwe [NAZ] S235 /513. People were also moved from outside the Victoria Falls Game Reserve (which is not clear in the archival record) from land to the west then being discussed as an 'Extension' to the Reserve, from the Kazuma Pan Reserve and what subsequently became the Zambezi National Park. They were also evicted from land to the east and south of the Falls (now part of the Hwange Communal Land), Fuller Forest and commercial farms south of the Falls.
88 Interview, Mafikizolo Silus Ngwenya, Monde, 30 March 2000, and other interviews with elderly Leya conducted in February and March 2000 in Hwange and in September 2001 in Mukuni Village, Zambia.
89 Annual Report, Wankie, 1940, S 1563.
90 Extract from the Medical Director's memo reproduced in Sec. Dept of Agriculture and Dept. of Lands to Director of Public Works, 10 October 1938, S 1194/1614/1/3.
91 Cited in Ncube 1994:154–5.
92 Report of the Land Commssion, 1946, entry for Livingstone District. Available in Rhodes House, Oxford.
93 Report of the Land Commission, 1946, entry for Livingstone District.
94 Cited in Rotberg 1966:126.
95 Report of the Land Commission, 1946, entry for Livingstone District.
96 Interview, chief Mukuni, Mukuni Village, 14 September 2001.
97 Interview, chief Mukuni, Mukuni Village, 14 September 2001 (and other interviews Mukuni village).
98 Interview, Esinath Mambaita Kasoso, Milonga, 11 April 2000.
99 On Watch Tower in Livingstone, see Muntemba 1970:39.
100 Interview, Maxon Munsaka Ndlovu, Chidobe, 27 March 2000.
101 See Mukuni 1957:5.
102 Sekwaswa, 'History of Livingstone' does not mention a female authority.
103 Mukuni 1957:2.
104 Ibid.

6

Damming the Zambezi at Kariba
Late Colonial Developmentalism

The Kariba dam was the flagship project of the complex political entity of the Federation of Rhodesia and Nyasaland (the Central African Federation), which was set up by the British government in 1953 with the support of the majority of white settlers in the two Rhodesias in disregard of African opposition.[1] The dam was a monumental intervention that transformed the valley beyond recognition, providing the energy necessary for post-war industrialization, creating the largest man-made lake in the world and displacing those who had lived along what had been a marginalized part of the valley. It was also the focus of a high-profile, internationally funded animal rescue operation. As such, Kariba was important both in terms of the politics of landscape and the making of the border. Moreover, its legacies have been far-reaching in shaping post-colonial politics in the valley.

Like the Victoria Falls bridge, Kariba was always more than simply an infrastructural development. The dam was a symbolic initiative that captured the imagination of global publics, fuelled an expansionist confidence in Southern Rhodesia and was used to justify late-colonial rule. Federal politicians used it to rally support for the Federation and its political ideology of interracial 'partnership', to foster a sense of pride in its achievements, and cultivate a sense of historical continuity with a lineage of white ancestors in central Africa. In the words of Southern Rhodesian journalist Frank Clements, Kariba stood 'as a monument to the white man's genius' – it represented tangible evidence of the benefits of white settlement, and the culmination of a long history of European endeavour to conquer the river.[2] In settler stereotypes, the Zambezi valley in the mid-twentieth century was still a frontier, a primitive border wilderness where time had stood still and the Tonga people, if they were known at all, were cast as 'still leading much the same life as they had when the Livingstones pushed up the river in 1860'.[3] The valley's reputation for backwardness made its transformation through the cutting edge of global technology the more striking and threw the Tonga into the public eye as an icon of the primitive.

This chapter is about these political uses of the dam and transformed landscape. It focuses on the Southern Rhodesian side of the border, where settler interests predominated, and in the wake of the dam's construction, the valley was turned over to a racialized access regime that gave privileged place to white commerce, conservation and tourism. The dam stimulated a large volume of journalistic coverage and other popular writing in both territories, which I draw on here, as it provides insight into official and settler discourses, and the ways these differed in the two Rhodesias.[4] I also draw on the experiences and counter-narratives of those displaced, as expressed in oral histories and other sources. The popular writing worked to promote confidence in the role of Europeans in central Africa at a time of acute challenge, furthered sentimental concerns for the welfare of African nature and fed popular racism, through stories that juxtaposed the achievements of dedicated European engineers, Native Commissioners and conservationists with infantilizing caricatures of the Tonga. Its authors invoked a lineage of white writing about the river, but they reflected and intervened in specifically mid-century debates over modernization, conservation and race relations, which are notable both for their distance from the attitudes accompanying colonial interventions described in previous chapters, and for their continuities with strands of post-colonial discourse, particularly in some romanticized invocations of the 'primitive', 'natural' and pre-modern.[5]

The dominant discourses on Kariba that were promoted by Federal politicians and white popular writers were, however, far from undisputed. David Hughes has argued that the popular literature performed a 'white cultural agenda' at Kariba by facilitating its appropriation to spaces of leisure and the wild, but his focus on Europeans' 'hydrological heritage', and the way the lake 'answered to settlers' longing for water' tends to essentialize white views and downplays contemporary controversy among settlers and officials.[6] Nor does it give space to African perspectives, as he argues that the displaced Tonga were rendered 'invisible' by the new uses of the landscape. Yet conflict over the dam and the subsequent use of the valley is important for the narrative in this book as it provides insight into diverging cultures of colonial authority and settler nationalisms in Northern and Southern Rhodesia, contrasting histories of African nationalism and their effects on processes of identity formation among the displaced border people. The counter-narratives to the dominant settler discourse on the dam also require further elaboration, as they have mattered in post-colonial contexts.[7] Colson and Scudder's studies of the short- and long-term consequences of resettlement and broader processes of social change were developed through close interaction with the resettled Tonga and have been influential both in shaping international policy towards development induced displacement and in the subsequent history of development assistance in Zambia.[8] On the Zimbabwean side of the border, the appropriation of the landscape to settler interests might have effaced Tonga views more fully in the short term, but it is important not to lose sight of Tonga perspectives; as we shall see in following chapters, they resurfaced, as the story of the dam continued to be told among the villages of the resettled, in very different ways from those expressed in popular white writing, and to very different political ends from those intended by Federal politicians.

In order to investigate the emphases, omissions and political effects of settler discourses in Southern Rhodesia, as reflected in popular writing on Kariba, it is necessary first to revisit the political context in which the dam was built, and then to reconstruct the process of the resettlement on the southern shores, both as it was represented in official discourse and in the recollections of those affected.

Federal Politics & Kariba

The idea of intervening in the course of the Zambezi river has a long history, as we have seen, but the drive for industrial development after the Second World War provided a new impetus focussed on generating hydroelectric power. The potential of two dams stood out – Kariba and Kafue – backed respectively by enthusiastic lobbies of Southern and Northern Rhodesian settlers. It is important to briefly review the dispute over which dam to build, as it highlights the contested political dynamics of the Federation itself. The issue at first appeared settled in favour of Kafue, as the two governments had signed an inter-territorial agreement to proceed before the Federation had been inaugurated, on the grounds that it was cheaper, smaller and would meet the pressing energy needs of the Copperbelt mines more quickly; that decision had been endorsed by the Federal parliament in early 1954.[9]

But construction on Kafue did not begin. This was largely because Federal Prime Minister Sir Godfrey Huggins (later Lord Malvern) was in favour of Kariba rather than Kafue, and used 'the authority of a PM as a dictator' to reverse the decision.[10] Instead of starting work on Kafue, he redefined Kafue as a 'territorial' rather than 'Federal' project, (which blocked international finance), and re-opened the debate by setting up a new Federal working committee and commissioning a team of independent international consultants to review the decision, led by the famous French engineer André Coyne – a move that was perceived as a delaying tactic even in London.[11] Lord Malvern was Southern Rhodesian as well as Federal premier, and in his home territory settler opinion was wholly behind Kariba; indeed, it had become a 'prime argument in favour of Federation' as the finance needed to build it was beyond the scope of Southern Rhodesia alone. Moreover, in the context of white immigration and a booming economy, there was an expansionist enthusiasm, and as Kariba would meet the needs of long-term growth, it was felt such an ambitious project 'would show the world what a young colony could achieve', and by spanning the river that divided the Rhodesias it would also cement ties between them.[12]

The French consultants and new Federal working committee reported in favour of Kariba and Lord Malvern used their view to announce his decision that the Federation would proceed with Kariba, based on 'the best independent and up to date technical advice'.[13] But the response to the announcement from Northern Rhodesia was 'an outburst of feeling' of an intensity that threatened to split the Federation.[14] The fall-out began immediately, with a public meeting in Lusaka, at which settlers formed a 'Kafue Protest Committee' led by Lusaka's mayor, agreed they had been 'Sold Down the River' and drew up a petition to the Queen contesting the constitutional basis of the decision.[15] J.H. Lascelles, Northern Rhodesian member of the Federal Hydro Electric Power Board, publicly tendered his resignation, on the grounds that the HEP Board had not been allowed to comment on the French technical reports, had been ignored in favour of advice from Malvern's own civil servants, and that the decision was 'indefensible' and 'detrimental to the interests of the Federation'.[16] An ensuing Northern Rhodesian Assembly debate expressed unanimous 'disquiet and disappointment', over the abrogation of a legally binding agreement and the lack of consultation.[17] The rift was about much more than simply a dam, as Mr Franklin, Member for African

Interests, appositely summed up during the debate:

> We could have developed Kafue on our own... We are not allowed to raise a loan; we have already spent £1,000,000 on Kafue and now, without any consultation as provided for in the constitution, and in a dictatorial fashion to which we in the north are not accustomed, we are coldly told 'you cannot have Kafue' .... The Kariba demarche has become symbolic amongst the people of this country. It is much more than an argument over HEP, the country feels that if a solemn agreement can be curtly dismissed, that if the powerful Southern Rhodesian elements in the Federal Parliament can secure such an action, there is great cause to worry about the future of Northern Rhodesia.[18]

At issue were not only the 'facts and figures' – the so-called Meshi Teshi calculations and the Kafue hydrological records that the French team deemed inadequate (which were not made public, and were contradicted by other consultants' reports) – but also the issue of cost.[19] Critics felt initial financial estimates for Kariba were surprisingly low, and when these were revised significantly upwards in November 1955, Northern Rhodesian Chief Secretary Sir Arthur Benson 'accused the Federal government of calculated deception'.[20] The depth of Northern Rhodesian suspicion appears wholly explicable in the light of other (unrelated) Federal decisions that clearly showed the predominance of Southern Rhodesian interests, such as the choice of Salisbury as Federal capital.[21] But the enthusiasm of the time for monumental technological projects was also significant, as the idea of damming such a famous river and creating the largest artificial body of water in the world appealed to international engineers and financiers alike, most of whom were entirely uninterested in Kafue. Funds for Kariba were readily forthcoming from the mining companies, the BSAC, Barclays and the Standard Banks and the Commonwealth Development Corporation, while the International Bank for Reconstruction and Development (the World Bank) later invested US$80 million following their own appraisal and report of June1956.[22] Sir John Caldicott, then Southern Rhodesian Minister for Agriculture and Lands thought (typically for Southern Rhodesian politicians) that, 'there was no political motive in this whatsoever ... facts and figures' decided the matter: 'Kariba was bigger than Kafue ... and better from the point of view of attracting international finance, as the Zambezi was a famous river unlike the Kafue, which no one had ever heard of.'[23]

African opinion in the two northern territories also opposed Kariba; indeed the decision was seen as a fulfilment of long-standing criticism of the Federation as a cynical device to extend white power, heralding 'segregation and oppression'.[24] Mr S.H. Chileshe, one of three African members of the Northern Rhodesian Legislative Council argued, 'the feeling of most thinking Africans towards the Federal government's decision to proceed with the Kariba scheme was that their fears towards Federation were now coming true, they are saying "look how right we have been all along"'.[25] Although the founding of the Federation initially deflated nationalist momentum, it was quickly regained in the northern territories, as the Federal government lived up to expectations – by blocking further African representation on the Legislative Council, giving extremely limited powers to the African Affairs Board, perpetuating the colour bar, taxing Africans disproportionately and pushing for dominion status.[26] ANC leaders condemned the Federal ideology of multiracial 'partnership' (which Huggins described, memorably, as the 'partnership of a horse and its rider'), as 'the partnership of the slave and the free',[27] and increasingly Northern Rhodesian and Nyasaland were convulsed by strikes, boycotts and protests.[28]

Indeed, Kariba facilitated this recovery of nationalist momentum. In Northern Rhodesia, the ANC had its heartland of support in Tonga/Ila country, and mobilised around the issue of the displacement. Kariba was typical of big dam projects of its era, in that the decision to build was based on technical calculations related to the generation of energy, and the displacement of people was considered only a marginal cost rather than a social issue.[29] Moreover, initial plans grossly underestimated the numbers affected, and figures of 30,000 had to be revised upwards to 57,000. As the details of the move were being worked out, NR Congress leader Harry Nkumbula petitioned the Queen, in the name of 'an assembly of several hundreds' of valley Tonga, opposing the Native Authority's acceptance of Kariba and requesting a commission of inquiry:

1. To determine whether it is just that the people should be dispossessed of their land...
2. To determine whether the power that is to be generated by the Kariba HEP scheme could not be better generated by nuclear energy and thereby make unnecessary the removal...
3. To determine whether the compensation payable... is adequate...
4. To determine whether the lands to which the people are being moved are equal in value to those from which they are being moved....[30]

When the Colonial Secretary refused to meet Nkumbula in London, on the grounds that the petition was a 'mischievous document presented for political reasons, seeking to enhance the power and influence of the ANC', the ANC staged one of a number of protests.[31] By September 1955, before construction started, 3,000 Tonga in Gwembe District were said to have joined the NRANC.[32] As Congress activists toured the valley, sold party cards, and recruited local party representatives, they campaigned on the grounds that membership of Congress would halt the construction of the dam and prevent their displacement. Their influence expanded within the valley on both sides of the border, as Southern Rhodesian nationalists did not take the issue up, partly because the various nationalist organizations were in disarray and fragmented in the early years of the Federation, partly because the valley was so marginal.[33] By the time Southern Rhodesian nationalist activism had regained momentum in late 1957 with the founding of the Southern Rhodesian ANC, it was too late to influence developments in the Valley – the Southern Rhodesian resettlement had already taken place.

The dam site itself was also politicized. The Federal government had initially met labour needs by appealing to the Nyasaland government for 8,500 labourers in January 1956, for fear that a local labour force would not be disinterested, and because complaints of poor conditions were circulating among workers.[34] But later the same year, labour supplies were disrupted after two African members of the Federal Parliament from Nyasaland sparked controversy by claiming labourers conditions on the site were 'horrible' and 'disgraceful', even 'akin to slavery'.[35] European and African journalists, and a delegation of Nyasaland chiefs, were encouraged to investigate and their reports did little to quell suspicion: a *Rhodesian Herald* journalist felt that comparisons to slavery were 'preposterous', but found there were some legitimate grievances over accommodation – new workers' housing schemes were excellent, but others were still in 'tents' and 'shanties of hessian sprayed with cement' and 'the place is ridden with rats and sanitation was poor'.[36] The Nyasaland chiefs reported even more ambiguously that conditions were 'generally good', but that the treatment of recruited men was indeed, 'more or less slavery.'[37] There were also rumours of mass deaths from sleeping sickness, malaria and other unknown causes.[38] Despite salaries three times the norm in Nyasaland, the persistent rumours and complaints stemmed the flow of

Nyasalanders, forcing the Federal authorities to turn to Mozambique, Tanzania, Angola, Congo, Botswana and elsewhere.[39]

Yet unrest among workers continued, which the authorities blamed on Nyasaland agitators, especially as events in that country got increasingly out of hand. The most serious disruption of work occurred in February 1959, and appeared to be triggered by an accident in which three whites and 14 Africans were killed underground. On the following day, 500 underground workers began a strike demanding a pay rise, were joined by surface workers and within two days the entire African workforce was out and the First Battalion of the Royal Rhodesian Regiment were flown to Kariba 'as a precautionary measure'.[40] Impresit, the Italian firm that had (controversially) won the contract for constructing Kariba, responded to the troops' arrival with an ultimatum of '50% pay rise or dismissal, said to be directed to the 1,300 Nyasaland workers held to be the ringleaders of the strike', and subsequently 700 or more were repatriated.[41] The Southern Rhodesian authorities denied that the Kariba walkout was anything more than an 'industrial dispute', but connections with Nyasaland nationalism could not be ruled out, and the Southern Rhodesian Minister for Labour justified the troops on the grounds that 'we will not countenance illegal action or intimidation', from 'agitators' disrupting work on the dam. By the time the state of emergency had been imposed in Nyasaland, and later in the other two territories, however, work at Kariba had resumed.

The Politics of the Resettlement: Debating the Morality of State Force

Although Congress movements and unions increasingly tried to mobilize across borders to disrupt the dam, and territorial and federal governments were coordinating actions against them, state planning in the Valley proceeded as a wholly territorial affair. The resettlement was handled in very different ways in the two countries, revealing a stark disjuncture in cultures of state power, particularly regarding the issue of state coercion, which also shaped patterns of resistance and compliance among those to be moved on either side of the border.

'Absolute control' over the resettlement was devolved to the two governments, who could apply for only restricted expenses from the Federal Ministry of Power.[42] The British government had made 'adequate arrangements for resettlement' a condition of approval for financial assistance, but could only exert influence over procedures in Northern Rhodesia.[43] In Southern Rhodesia, the 23,000 to be displaced (mostly Tonga, but including some Korekore and Goba around Kariba itself) were to be removed from unassigned crown land, which ruled out any possibility of compensation for lost land.[44] Though they could have claimed Federal funds to support other aspects of the resettlement, Southern Rhodesian officials only applied for resources for food and two years' exemption of adult males from the £2 poll tax.[45] High-ranking native affairs officials dismissed local NCs' concerns about the fate of the Tonga on the grounds that national interests and the benefit to the Tonga of contact with the modern world overrode any potential hardship. Typically, Provincial Native Commissioner Yardley put down his juniors' complaints with the argument, 'I do not share the view that resettlement of the BaTonga will have any serious repercussions. They will be brought into close contact with civilization for their benefit and for that of the colony'.[46] Moreover, it was argued that the Tonga were backward and had never exploited the river to its full potential: Lord Malvern felt they were 'among the most primitive in the Federation and are living in one of the unhealthiest parts; resettlement can only be to their advantage'.[47]

This was, of course, the official discourse on the move, but Northern Rhodesian officials were more sympathetic. When the decision had first been announced, Stubbs, Secretary for Native Affairs in Northern Rhodesia, emphasised the 'enormous importance and enormous distress' to be caused by the move; though their 'sacrifice' would be offset as 'they would be among the first to benefit through electricity in their areas'.[48] Northern Rhodesian Ministers made formal complaints to the Power Board about the limitations on the compensation that could be claimed – they wanted funds for long-term development, to offset the inadequate quality and extent of resettlement lands, and to mitigate the effects of population growth and economic change. The 37,000 people to be moved on the northern bank lived on native land, and compensation claims were negotiated by the Gwembe Tribal Authority, headed by a university-educated man from the valley, Hezekial Habanyama, and District Commissioners, who strove to secure the maximum they could in return for the Tribal Authority's acceptance of the move.[49] In the end, the Federal Power Board spent £3.98 million on the resettlement (out of a total project cost of £77.61 million), amounting to £134 per person on the north bank and £59 on the south bank – but even the more generous north bank payments have been judged retrospectively as 'grossly inadequate'.[50]

On the south bank, those responsible for the move were Patrick Fletcher (Minister of Native Affairs), Hostes Nicolle (then number three in the NAD) and Native Commissioner Igor Cockcroft (known in the Valley as Sikanyana), who did the overwhelming body of the work on the ground, operating first from Gokwe and then from the new district of Binga, created in 1957. Unlike the Northern Rhodesian government, which decided to fund its own agricultural development plan in 1956 to 'remedy the disastrous situation inherited from resettlement at substantial cost to its own finances',[51] Southern Rhodesian planners considered it 'too expensive' to survey resettlement lands (costed at £60,000). The state opted for what Rupert Meredith Davies (then Assistant Director of Agriculture) regarded as 'rather a slipshod eventual solution', and left decisions about where to put those resettled entirely in the hands of Cockcroft. In Davies' estimation, the resettlement lands were 'inhospitable, waterless, arid, on which only baboons, antelope and rhinoceros could live'.[52]

The fact that many of those resettled on the south bank were relocated very far from the lakeshore reflected the aridity and rockiness of land close to the lake, but also Southern Rhodesian planners' desire to keep them away from potential new developments around the lake.[53] Early impressive (if ungrounded) projections of the fishery's potential encouraged much commercial and scientific interest, inspiring the two governments to invest in expensive bush clearance (designed to prevent petrified trees from inhibiting fishing), and to devote some of the new shoreline and islands to biological research. Yet other attempts at joint planning – through the Lake Kariba Fisheries Committee set up in 1955 – were undermined by unreconcilable interests. Southern Rhodesian officials upheld the segregationist principles of the Land Apportionment Act, and planners divided the shoreline into white and black areas. They planned European domination of the lake's commerce, even in African areas, which were opened to a European concessionaire who was given a monopoly over the purchase and marketing of fish.[54] Their conception of African fishing as a disruption to commercial and scientific plans for the lake was in stark contrast to the privileged access granted to Tonga fishermen in Northern Rhodesia.[55]

The Southern Rhodesian authorities drew on the usual legal mechanisms to evict those living along the river, instructing the Tonga straightforwardly that the move was a government order. A high-profile team (comprising the Minister of Native Affairs, Minister of Agriculture and Lands and NC Cockcroft) presented this order to the people, in a tour of the south bank villages. The press in Southern Rhodesia and Britain followed the patrol to the valley and repeated official discourse about the lack of resistance: 'The Batonka are gradually accepting that by magic means the white man will create an inland sea because he wants more power for mines and other industries'.[56] With the exception of Cockroft, officials' reports are cold, even hostile: Hunt, for example, who accompanied Fletcher in the valley for a short period, criticized the Tonga migrants who had returned to the valley 'ostensibly to help out with the move, but actually to live a life of idleness and ease', and thought (unlike Cockroft) that 'measures should be taken' against their receiving maize handouts.[57] Cockcroft also recorded people initially accepting the instruction, 'with fatalistic acquiescence which is typical of their law abiding attitude, though the impending move will mean an uprooting from their beloved river, the forsaking of their holy places, the abandonment of the graves of their ancestors, in fact the departure from all they hold dear, they philosophically accepted the "Government's Order" without question'.[58]

Though public reports on the move and press coverage ignored all forms of resistance, presenting the move as a swift and efficient operation, and people as 'in general happy to go', the district records make it clear there were problems.[59] In January 1956, ANC Hanson asked 'that proclamation be gazetted concerning the coming movement so that legal compulsion may be exerted should it become necessary'.[60] Parliamentary Secretary H.J. Quinton recalled sitting in Parliament when 'a note was put in front of me which said that the Batonka had refused to move and that there was a lack of discipline and could disciplinary measures be taken and if necessary one or two be shot, and I said "No". I appealed to the Speaker to adjourn the debate and said to the PM "I'm going to Binga tonight, I'll report to you when I come back".' He described going to Manjolo and addressing a crowd of one thousand resisters led by the chief and headman, who refused to move because of the *malende* shrines and trees associated with ancestral spirits, which would be submerged under the lake. This was not an isolated incident. In Wankie District, the NC also reported that 'the Batonka through chief Siansali' were giving 'trouble over … removal from the dam'.[61] The current chief Siansali recalled his predecessor's refusal to move, how people threatened to cross to the north bank (and some subsequently did so), and how the final evacuation was made in haste only when floodwaters were lapping around the huts.[62] His neighbour, chief Siachilaba, did likewise, moving only at the last minute and refusing relatively well-watered lands far from the river in favour of more arid lands close to the new lake.[63]

Bush clearing operations in the Valley were also disrupted. The vehicles of D.G. Vorster (who had the main contract for bush clearance) were repeatedly brought to a standstill through obstruction from Tonga villagers trying to prevent the knocking down of baobab trees linked to *malende* shrines. Vorster recollected having to call Native Commissioners out on several occasions to pacify the resisters – and on each instance wait idle for days while they travelled the 200 miles from Gokwe. [64] In many cases, such as Manjolo, Siachilaba and Siansali, this resistance was nothing to do with nationalist mobilization. Direct nationalist influence was not necessary to provoke opposition to the move; people did not want to leave their homes and fertile lands by the river to move to the arid tsetse- and game-infested hinterland, and there was widespread incredulity and scepticism about the possibility of damming the Zambezi, and much suspicion that the prime motive was to turn Tonga land into European farms. They saw no benefit in the scheme for them.[65]

But the heartland of resistance in both territories was undoubtedly in areas where the NRANC had influence. Congress had mobilized quickly in the rainy season of 1955-6, when much of the valley was inaccessible to Southern Rhodesian officials based in Gokwe. Resistance was strongest in those parts of the valley that were better connected to Northern Rhodesian political centres, regardless of the river and border. Thus the strongest resisters and most ardent local nationalists were in relatively well-connected Chipepo District (on the north bank) and in the same part of the valley across the river, in the areas of chief Mola, Sampakaluma and Sinagatenke. Cockroft complained in 1956:

> the NRANC has seen fit to interfere with and attempt to make capital out of the Zambezi movement. Their emissaries crossed the river during the rains, when access was impossible from Gokwe, and persuaded a number of people, mainly from Chief Mola's to join their ranks and contribute to their funds. They attempted to influence the people into refusing to move on various pretexts and had a certain amount of success.[66]

By the following year 'Agents were busy working up River and in but a short time no less than 6 chiefs were induced to join the movement. At first the people refused to cooperate in the movement and even decided to refuse to obey the government [eviction] order.'[67] In these areas of Congress influence, Southern Rhodesian unions also contributed to the disruption. Fletcher recalls how, when he was in chief Sampakaluma's area in late 1956 where the NRANC was well organized 'all transport riders went out on strike'. In order to continue, he had to rush to find alternative transport at the last minute and hurriedly purchased '50 5-ton lorries in ten days complete with drivers' from Duly's in Gwelo.[68] Cockcroft and other government officials tried to counter NRANC mobilization by arguing that Congress was simply exploiting gullible villagers to extract funds. He reported that 'Every opportunity was taken to bring home to the people the true facts that Congress was but using them as pawns in a financial and political struggle with no regard to their personal fate, and that the movement could not be halted by anyone'.[69] Assistant NC Hanson echoed this contempt for activists who 'play[ed] with plausible stories on the minds of primitive and conservative people'.[70] Their arguments to discredit the ANC were bolstered by displays of force in the form of contingents of armed police dispatched to the valley.

The emphasis in Cockcroft's reports on 'willing cooperation' is very different from the stress on forced labour, hasty forced removal, reluctance and fear that is central to the recollections of those displaced. The Tonga term for the eviction is *kulonzegwa* (from the verb *kulonga,* to move, also used as the Tonga translation for the book of Exodus), and conveys the idea of being moved by force.[71] People portray the preparatory work of building roads as 'chibaro' or 'slavery', partly in complaint about inadequate remuneration (some were paid only in food, though others got wages), and partly because the roads were designed for an unwanted purpose.[72] Philemon Munkuli, for example, thought: 'The time of removal was a time of slavery for us because we were forced to clear the roads which would take us away from our homes'.[73] The contingents of armed police that accompanied Cockcroft made a big impact, affecting how the risks of resistance were calculated; indeed, assessments of the likelihood of the Southern Rhodesian state's recourse to violence undermined organised opposition. Elena Mumpande, who lived on the river under chief Sinakoma, recalled, 'When the chiefs heard about the message about the forced removal, they did not resist. They were afraid of being harassed by the police. All the police had guns. Other people did not come to help us because they feared that Sikanyana's police might kill them. We were very afraid of them.'[74] The evacuation itself was often a chaotic and rushed affair. 'We had no time to bid farewell to relatives across the river, the trucks arrived before we were organized, things were hurled into the lorries, and we were herded on board with threats, "if you stay you will drown".'[75] Magoyela Mudimba recalled travelling away from the river 'burdened by our thoughts', willing the driver to stop or overturn and singing:

> Let the driver overturn the lorry
> We are perishing
> We who used to be very happy
> Hit the brakes, driver, we are perishing.[76]

Cockcroft himself is not remembered as a bad man for organizing the move; 'he was not rough', rather he was 'hard-working', assiduously carrying out the orders of a powerful and coercive state.[77] Another song sung during the move lamented:

> The DC Sikanyana and
> His followers are overzealous.
> The DC has forcefully removed us
> The people of the river
> We who were used to fetching water
> We who were used to our seasonable crops
> We who were used to our pools of water.[78]

On the south bank the evacuation was achieved in a very short period, such that by the end of 1957, Cockcroft reported that only '26 hardcore fanatics' remained in their homes in Binga District.[79] The organizer of the most prominent 'hard core' group was Siabiyaka Makaza, a kraalhead under chief Sinakatenge, an NRANC politician and also an influential traditional healer who had worked in the towns and mines of the two Rhodesias and Nyasaland. John Powell (NC for Gokwe), recalled the paramilitary operation he needed to remove him:

> We had one group of people, a group under a leader called Mkasa. And they flatly refused to move and we had to carry out an exercise with a lot of police. We swooped in on them in the early hours of the morning ... They were reluctant to stay [in Sanyati prison] and they weren't cooperative. The police pulled out and I stayed a little while to look after them but the minute I left, they returned on foot to their old area.[80]

Southern Rhodesian officials were condemned by their north-bank counterparts for 'inhumanity' in executing the move in an extraordinarily short period of 18 months and without compensation.[81] But on the north bank, the emphasis on persuasion and dialogue culminated in violent conflict. In Chisamu village in Chipepo, the heartland

6.1 Fishing before the dam (National Archives of Zimbabwe)

6.2 Cultivating the Zambezi margins before the move (photo by Emil Schulthess)

of NRANC resistance on the north bank, people were still refusing to move by mid-1958. Unrest deepened with an attempt to arrest resisters, and when the Governor finally intervened and issued an order for people to board the trucks in June 1958, men attacked a police unit with spears and pangas. The police opened fire, shot eight people and wounded 32 others. Elizabeth Colson described how people in the valley were 'shocked and frightened' thereafter, very aware of the dangers of refusing to comply with government orders. The 'Chisamu war' became a point of reference in moral debates among the resettled, not only about the suffering of the move, but also about understanding the Northern Rhodesian state itself, which had long been considered more benign than its southern counterpart where recourse to force and violence was seen as the norm. The incident shaped discussion of appropriate responses to state coercion, and judgements about the value or futility of making a stand in the face of state displays of force.[82] Though later appropriated to a nationalist narrative of heroic resistance to colonialism, in its immediate aftermath the episode was not discussed in these terms, and those who died did not become 'folk heroes'; indeed these debates revealed an ethics that did not translate into European terms of heroism and martyrdom. Colson elaborates:

> They received no honours, and they were not spoken of as having given their lives for their people or as having set an example of resistance. Instead the common complaint was that the government had been deceitful and men had died as a result. None of them, so it was argued, would have been so foolish as to take part in the charge if they had thought the government forces would open fire. They had called what they regarded as a monstrous bluff, trusting in the restraint of Europeans. The Chisamu dead, five years after their death were regarded as the unlucky victims of miscalculation rather than as men who had preferred death to compliance...
>
> Those who had not been present talked about fleeing at the sound of distant guns and how they had hidden in reeds or bushes until all danger seemed over. They admitted to two emotions, fear and anger. The latter was aimed perhaps equally at government and at their own leaders who had encouraged them into danger. Violent resistance to resettlement collapsed immediately after the Chisamu shooting. In succeeding years, until independence, a resentful and morose people found considerable pleasure in needling European officials and their own chiefs by the display of their disaffection, but they did so with a keen eye for the limits to which they could press their attacks. People avoided protest that would again endanger their lives and property. Each government demand or order was examined for the likelihood that it would be backed by force.[83]

The episode was also a point of reference in moral and political debate among colonial officials and settler publics north and south of the border, in discussion of the merits of authoritarianism, comparisons of Southern and Northern Rhodesian approaches to the move (and Native Affairs more generally), the pros and cons of interventions from London, and evaluations of African nationalism. It also attracted comment in the popular books on the dam. For Southern Rhodesian journalist Frank Clements, for example, the incident revealed not only nationalists' incitement of ignorant villagers, but also the advantages of a firm hand and the constraints of Colonial Office meddling, because 'every action of a Colonial Office servant can become a political issue in London', which undermined 'decisive action' on the part of local administrators.[84] The senior officials involved in the Southern Rhodesian evacuation likewise thought the Northern Rhodesian government had been 'dilatory' over the move.[85] In Northern Rhodesia, Governor Benson upheld the consultative approach, but was quick to place blame on the NRANC and 'irresponsible political agitators'. The official commission of inquiry into the episode did not question the self-defence of the police unit or the best intentions of the District Commissioners and Officers, but dwelt inconclusively on whether NRANC officials had *intended* to make the people of Chisamu oppose the move and why.[86] David Howarth describes the soul-searching of the local Northern Rhodesian colonial officers involved. His detailed reconstruction of the move, based on interviews with these officers and members of the Gwembe Tribal Authority, as well as a reading of Colson and Scudder's scripts, is framed by a broader moral argument. Howarth portrays the violence at Chisamu as inevitable, a moral commentary on the destructive tendencies of modernity, of which urbanized nationalists were only one manifestation. 'The truth was that the intrusion of industry into the valley had overwhelmed tradition. Kariba, the Tonga's innocence, the Congress intervention and the ideal of government by consent: with these four initial causes, the battle at Chisamu had been inevitable. If Kariba had never been built, if the Tonga had had more worldly wisdom, if Congress had refrained from adding its own confusion, or if the government had taken earlier the dictatorial powers it despised; then eight harmless men might not have been lying dead.'[87] Howarth's portrayal of the story of the displacement as a loss of 'primeval innocence'[88] is a common theme in contemporary popular European writing and newspaper coverage and echoes an important strand of official discourse.

Operation Noah

Although this violence, and the fate of the Tonga more generally, attracted some international as well as local attention, the fate of the wildlife of the valley received disproportionate international coverage, and was notably prominent in the volume of popular writing on the dam. This aspect of the story of the dam began – belatedly – in 1959, as the floodwaters were beginning to rise when local journalist Malcolm Dunbar publicized the 'tragedy' of drowning and starving animals trapped on islands, and the story was widely taken up by local and international media. 'Operation Noah' was the response to the consequent international appeal for funds for animal rescue, launched by the London Fauna Preservation Society and Rhodesian interest groups.[89] The operation became a focus for southern Africa's most prominent conservationists and leading international experts in animal tranquilization and relocation, who flocked to the valley to help with the first rescue operation of its kind.[90] Film crews and journalists from Europe and America accompanied the teams of scientists and young Rhodesian volunteers and popularized the 'greatest animal rescue since the ark'.[91] Charles Lagus – cameraman in the popular British TV series 'Zooquest' – covered the episode for the BBC, and also produced a film and book. There were so many journalists that they threatened to outnumber the actors, and the Federal government intervened to limit access and issued a 'warning to writers' that, due to 'limited space', permission was not guaranteed.[92] Latecomers failed to find animal rescue teams they could accompany: Elizabeth Balneaves, for example, was refused permission in Salisbury on the grounds that there were too many journalists, who were filling the boats and leaving no room for the animals.[93] The 'iconic image' for the episode, as Hughes elaborates, was a picture of the operation's leader and folk hero ('Noah') – the Game Department's Rupert Fothergill – 'cradling an impala fawn against his bare chest' (see photo 6.3a).[94]

6.3a & b Popular writing on Kariba – 'Operation Noah' 1959, 'The Struggle with the River God' 1959

The operation was driven by developments in international conservationist ideas of wildlife as an aesthetic and economic resource, a 'world possession not to be despoiled by local whim', and by the surge of interest in the family life of charismatic animals encouraged through wildlife documentary.[95] Though there were continuities with earlier conservationist ideas and with the world of hunting (and related genres of writing), the operation also reflected new sensitivities.[96] The rescue was informed by concerns about biological destruction wrought by dams and evoked sympathy for an innocent animal world. It popularized images of animals in distress and emotional responses from European men trying to help families torn apart or desperate mothers trying to protect their babies. Lagus wrote of his revulsion at the stench of dead animals:

> ...in every direction these pathetic victims floated and gyrated singly, in pairs, in groups in a grisly dance of death. My whole spirit revolted in a flash and I knew from that instant that my approach to Operation Noah could never again be academic, coldly calculating, reasonable: I could only regard this work of rescue emotionally.[97]

The support of the London interest groups and international media was crucial to the whole operation, and without it, local preservationists felt the entire operation risked being labelled 'mere sentimental nonsense'.[98] Albert Rusbridge Stumbles (head of Irrigation and Lands, under which the Game Department fell) felt the operation had been initiated by a 'most scurrilous article' and recalled his surprise at becoming 'almost a world figure – I got correspondence from all over the world about this'.[99] Local preservationist societies represented only one perspective in an ongoing debate; other sections of settler society had a much less romanticized view of the animal world than urban publics overseas – they were familiar also with the idea of wildlife as a 'pest', particularly species such as baboons and others. Moreover, it was common knowledge among the Rhodesian public that the Game Department had invested heavily in slaughtering wildlife in the valley in efforts to eradicate the tsetse fly: controversy over the policy of 'game destruction' had culminated in a public inquiry in the mid-1950s, but was resumed by 1957, as the fly was spreading again. The 7,000 animals saved through Operation Noah paled to insignificance when compared to the figure of 750,000 killed through tsetse control (the figure cited when shooting in Southern Rhodesia stopped in 1961).[100] The idea that the valley should be a game sanctuary had been mooted frequently in the past – and repeatedly resisted by local NCs. Cockcroft himself, for example, was against the establishment of the new Chizarira game reserve in Binga, which was designated a non-hunting area in 1958, and felt that the concern over 'animal casualties' stimulated by Operation Noah was misplaced. Rather, he emphasized human casualties to man-eating lions and depredation of crops by marauding elephant, and argued: 'the fact remains that dangerous game and humans cannot exist together'.[101]

All the international attention focussed on the valley's wildlife and scenery in the course of the animal rescue gave a boost to local preservationist opinion, and affected settler attitudes more broadly. Lt Col. R.A. Critchley (President of the Game Preservation and Hunting Association of Northern Rhodesia) thought it made the white media-reading Rhodesian public 'more wildlife-conscious' than ever before.[102] By highlighting the region's tourist potential, it strengthened the hand of the Game Department enormously, and contributed to the momentum for trebling the size of the Southern Rhodesian wildlife estate between 1957 and 1961, with much of the newly conserved land being in the vicinity of the new lake, and facilitated the Game Department's metamorphosis into the Department of National Parks and Wildlife Management with the hero of the rescue operation, Rupert Fothergill, at its helm. These vast new state assets in the Zambezi Valley included the Lake itself and a five-kilometre strip of shoreline (comprising Lake Kariba Recreational Park), Chizarira and Matusadona game reserves (later National Parks), Sijarira and Kavira Forest area, and Chete, Charara and Sibilobilo Safari Areas.[103] As Hughes summarizes, 'in these institutional ways, Kariba's flood *made* conservation'.[104]

Popular conservationist writers covering Operation Noah usually devoted some space to justifying their interest in wildlife rather than people. They did so by echoing Federal politicians' arguments about the inevitability of progress, the Tonga's backwardness, and faith that the government had the Tonga's best interests at heart. Lagus, for example, wrote:

> The Batonga people (whom I have heard described by white friends as a delightful happy carefree race) have had to be uprooted from their ancestral homes. One can make too much of this I suppose. In all great commercial undertakings of this kind somebody must suffer... Of course a benevolent Government has tackled the resettlement problem with the utmost care and sympathy...'[105]

Robins and Legge end their tale with the prospect that the lake and dam will 'open up new lands of opportunity and wealth in the heart of British Central Africa', with teams of scientists – fisheries biologists, soil chemists, irrigation engineers – working on schemes for economic development, the Tonga acquiring 'a marked taste for "civilization" in the form of filmy underwear, gaily-patterned blouses, and chic hats', and expressing their contentment: '"We who lost our lands to Kariba are happy," one

Batonka elder told us… "The lake has been good to us. We were afraid when they strangled the river the fish would be frightened and go. Yet there are more fish than ever before".'[106]

The emphases and omissions of this body of popular writing on the dam bear further investigation, as they repeat the idiom of contemporary newspapers, and capture the debates, sentiments and caricatures both of late colonial Northern Rhodesia, and of a strand of Southern Rhodesian opinion that was determined not to move with the times.

Popular Writing & Popular Racism

Both the wildlife-focussed texts and other popular stories about the building of the dam evoked a common body of self-justifying stereotypes of the white man in Africa, and of the superiority of European science and technology. At the centre of most stories were the roles of the various European men involved – the engineers and building contractors, zoologists, Native Affairs officials, Game Department rangers and volunteers – who are cast as dedicated, courageous, humble and self-sacrificing. Even the more serious, well-researched popular books on the dam have such figures as their central focus: Howarth's step-by-step account of the evacuation on the north bank, for example, is a valuable, empirically detailed account, grounded in field and archival research and sympathetic to the fate of those resettled. Yet he felt it necessary to heroize the Europeans involved, who were his closest interlocutors, in ways they probably would not have represented themselves, and argues that their humble attitude showed that 'the reserve for which the British are famous probably reaches its fullest development in the Overseas Civil Service'.[107] Other writers did likewise in much cruder ways. The monumental scale of the project underlined the importance of their work and lent itself to the use of biblical metaphors invoking Old Testament floods, Noah, the Exodus or other evocative parallels for a 1950s audience, such as Dunkirk (though many writers also deflated their own metaphors, by invoking a less serious genre of boys' adventure stories and movies, particularly *Sanders of the River*). Other notable points of reference are the ubiquitous comparisons with ancient Egypt, perhaps the ultimate symbol of imperial decadence, and repeated invocations of other famous rivers associated with colonial conquest and/or slavery, such as the Mississippi and the St. Lawrence.

Telling the story of the dam provided an opportunity to remember Livingstone and a lineage of other well-known European explorers and less well-known scientists and engineers who had moved along the river and reflected on its developmental potential. Livingstone was once more cast as the apical ancestor for whites in Central Africa – just before this body of work on the dam, he had been heroized anew with the publication of a series of serious academic studies of his research, timed to coincide with the inauguration of the Federation, which cast him – explicitly in some cases – as laying the basis 'for the beneficent civilisation which the Federation aimed to be'.[108]

The Livingstone Centenary events in Livingstone town, in June 1955, made further connections between the man and the dam, and journalists celebrated the Zambezi as 'David Livingstone's river – an artery for industry'.[109] When Lord Malvern symbolically poured the first two tons of concrete into the dam wall, Sir Malcolm Barrow, 'recalled the debt owed to the long line of pioneers whose imagination was stirred by the possibilities of harnessing the waters of the Zambezi'.[110] Some writers looked back, celebrating Livingstone's foresight, while others emphasized advances made from old misguided understandings. Thus Lagus uncritically repeated Livingstone's geographical explanations for Africa's backwardness (the insufficiently indented coastline, the problems posed for 'river progress' by waterfalls and rapids), while Clements condemned those who imagined the dam as restoring a vast ancient inland lake, or creating a St. Lawrence Highway in Central Africa, as 'eccentrics who live in the fairyland of maps'.[111]

Popular writing on Kariba replaced nineteenth-century representations of the river as a focus for violence with a more romanticized image of the primitive, especially in Northern Rhodesia. Except in Southern Rhodesia, writers of the late 1950s wrote not to encourage colonial occupation, but in justification of what it had achieved. The backward state of the Tonga was now anachronistic rather than typical, quaint rather than threatening; they stood for a disappearing world whose inevitable fate was underlined by the closeness of their contact with icons of 1950s modernity (the dam itself, bulldozers, aeroplanes, animal tranquilization, DDT spraying, malaria prophylactics and the like). The adjectives that went along with primitive were now 'simple', 'innocent', 'gullible', 'conservative', 'picturesque', 'ignorant', 'trusting' 'childlike', and even 'primeval'.[112]

Journalistic coverage of the episode was remarkably consistent in its popularization of the idea of local belief in a river god – once again repeating Livingstone and the nineteenth-century travellers and missionaries who had first popularized the idea, but using it to somewhat different ends. In the Southern Rhodesian press, Nyaminyami was part of a derogatory and cruel caricature of the Tonga developed during the move that made them a 'laughing stock'.[113] This also attracted criticism from white commentators north of the border; Howarth, for example, reflected Northern Rhodesian sentiments in condemning this particularly Southern Rhodesian brand of racism that cast the Tonga as unable to cope with town life, only good for night-soil collection, who did not know that sex produced babies, who worshipped 'a huge serpent that lived in the Zambezi, and whose witchdoctors comically predicted the serpent would be angry with the white man's wall and knock it down'.[114] But Southern Rhodesian journalists Robins and Legge fleshed out the caricature, in describing how the 'primitive and picturesque BaTonga' received the news,

> ...that a 'giant's fortress' was to be erected at the Kariba Gorge... The lake, it was said, would be nearly 200 miles long... [They] listened politely to the white bwana who told this unbelievable story but were sceptical... 'Nyaminyami will protect us', they confidently assured each other... For no man could pit his strength and skill against the River Monster Nyaminyami, the water serpent whose coils caused the yearly floods of the Zambezi and whose whiskers were the leaping, silvery spray of the Victoria Falls... Nyaminyami – through the medium of the muttering, whirling, witch-doctors' scattering of voodoo dice of baboon bones – had spoken.[115]

Lagus was also writing from south of the border, and repeated jokes about the river god that obviously circulated in the European bars of Salisbury and Kariba town (such as exchanges between a *Daily Telegraph* correspondent and Italian engineers over its sex). He cast the exceptional floods of 1957 and 1958 that nearly destroyed the semi-completed dam as the final assault of 'Ol' Man River': the 'roaring, battering monster of a river' was defeated by the final few feet of concrete, 'gradually he grumbled back to his old level and his ravaging could be inspected ... the old primeval river would be a prisoner behind a concrete bar, made to work at the behest of man'.[116] Clements tells the story of Kariba as that of the white man's triumph over a series of obstacles (the quest for international finance and expertise, European cooperation, misguided Northern Rhodesian settlers, unscrupulous African nationalist agitators, disruptive strikes, biblical floods etc.), encapsulated in an overall rhetorical device of a military conquest against a river god representing the primitive world and nature – 'the great elemental forces of Africa' and a continent 'a great part of which is still a survival from the world of prehistory'.[117] His chapter headings make the most of the military metaphor – 'The Great Argument; Establishing Base; Forces Deploy; The First Assault; Behind the Lines; The Angry God; Nyaminyami Bound; A Split in the Ranks', and so on, before finally, 'The Valley at Peace'.

Though the story of the building of the dam was used by some writers to look back on the achievements of late colonial rule, more or less in acceptance of the process of decolonization by then underway, others used it to support the perpetuation of white settler rule. The idea of the dam had cut across the divides of white politics in Southern Rhodesia, and was a project around which Southern Rhodesian settler opinion could find common ground and international acclaim, and liberals such as Prime Minister Garfield Todd supported the dam enthusiastically alongside right-wing, racist lobbies. Yet by the time of its completion, Todd had been expelled by his own party and both Southern Rhodesian politics under Sir Edgar Whitehead and Federal politics under Welensky had turned decisively against any move toward majority rule. Minister of Power Malcolm Barrow described the dam at its opening by the Queen Mother in May 1960 as 'a visible demonstration to all those interested in the Federation that our objectives are sound and that we are capable of carrying them out'.[118] The Southern Rhodesian government composed a song to 'The Wonderful Wall' that joins Northern and Southern Rhodesia, casting it as 'a monument to international and interracial cooperation, a lasting expression of great Federal beginnings and great future promise'.[119] Frank Clements upheld Kariba as tangible evidence that 'partnership' could work (with Europeans in charge and Africans as labourers), and thought it a prime example to counter adverse publicity on the 'insolence and brutalities of life in the Rhodesias'.[120] Clements felt that African workers were as proud of the dam as the engineers, and were 'for the most part happy to be at Kariba' and were 'proud of the wall, by far the greatest piece of masonry constructed in Africa since the days of the Pharaohs and the pyramids.'[121] Conservationists' monitoring and protection of the new lake's natural assets could help rectify any unease over ecological destruction wrought by the dam, while the application of the latest ideas in fisheries biology promised a productive natural resource and new commercial possibilities, through the introduction of new fish species. Moreover, the costs borne by those resettled could be forgotten, as the Tonga were out of the way up in the hills.

The new tourist industry that developed around the dam wall, lake and game parks could reinforce some of the same ideas to white visitors. This began before the dam's completion, with tours of the works by Impresit engineers,[122] but it expanded thereafter. Initially, in the 1960s, the resort of Kariba itself was the main attraction; other resorts such as Binga and Mlibizi developed more slowly and catered more specifically for anglers. At Kariba, white visitors were invited to contemplate the monumental achievement of European engineers, and could view the wildlife encouraged by the creation of this 'vast natural asset' from the bows of *The Ark* ('originally the mother vessel for the animal rescue'). They could indulge in all the activities of a modern waterside pleasure resort, from water-skiing, yachting and angling, to more specifically African entertainments such as visiting one of the new crocodile farms along the lakeshore, or pitting their energy 'against a sporting, fighting opponent such as the long-toothed tiger fish'.[123] Much of the tourist literature simply ignored the Tonga and represented the place as an uninhabited wilderness before the dam. Where they were included, they were a picturesque part of the scenery and relic of an ancient prehistory, originators of a myth about a river god. Kariba was 'Rhodesia's Riviera', one brochure boasted, a 'civil engineering triumph' constructed in a wilderness inhabited by the Batonka 'the most primitive tribe in Central Africa', who had 'turned to Nyaminyami, their God which had the body of a snake and head of a fish, and would never allow the Zambezi to be tamed'.[124]

Popular accounts of the story of the dam and transformed landscape thus promoted a set of meanings to white publics in the Rhodesias, Britain and America which worked to validate the developmental role of Europeans and international finance in central Africa, to further interests in African scenery and wildlife, and to reinforce racist stereotypes of Africans through caricatures of innocent primitive tribes and wily, untrustworthy educated nationalists. This was not, of course, the only way in which the story of Kariba was told. Better known to academic audiences today are the studies written by Elizabeth Colson and Thayer Scudder, which were commissioned by the Rhodes Livingstone Institute (RLI) based in Northern Rhodesia. RLI Director Henry Fosbrooke had tried to present the case for 'resettlement plans to be based on an accurate understanding of the ways of life of those to be moved, since their social order was intimately linked to the ecological niches they had so long occupied' and thought 'a unique way of life ought to be recorded before it was destroyed'.[125] At the time, however, the tight time-schedule created by engineers and financiers did not allow for their studies to inform the Kariba resettlement, even on the north bank, where the authorities might have been receptive, let alone in Southern Rhodesia where other interests prevailed. Colson and Scudder's studies became influential later as foundational texts in the field of forced migration, highlighting the social costs of displacement and helping to alter the way resettlement was considered in terms of large dams and other development projects around the world, specifically in World Bank policy. As we shall see in later chapters, their networks and research were important in moves towards reparations and in securing retrospective compensation for 'dam induced grievances' for those resettled in Zambia.

Yet in Southern Rhodesia too, there were other ways of recounting the story of the move, which were very different from the journalists' and officials' accounts described above. Away from the ears of officials and settler publics among the villages of the displaced, people told and retold the story of the dam and resettlement, trying to make sense of their experiences. This telling did not influence state resettlement plans, but it did come to matter in relation to broader political events in the country. The next chapter turns to the experiences of the displaced up in the tsetse and animal infested hills of Southern Rhodesia, and explores the ends to which the story of the removal came to be used by local political leaders. The new political ends for the telling of the story of the dam in the villages of the resettled are part of a broader history of the development of African nationalism and war in the Southern Rhodesian Zambezi borderlands.

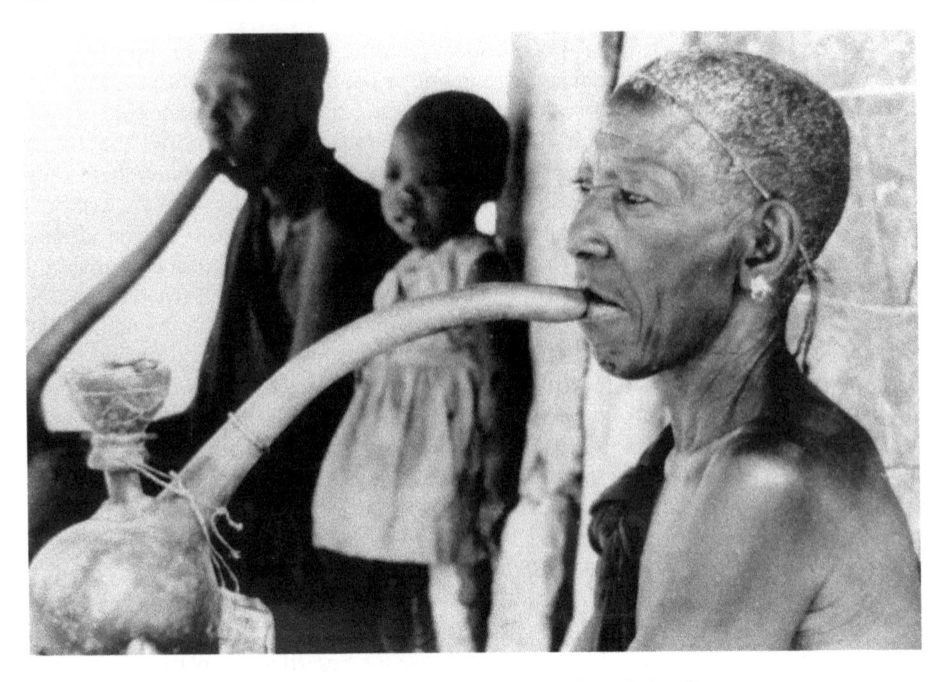

6.4 Tonga women smoking (National Archives of Zimbabwe)

6.5 'The Social Consequences of Resettlement'

Notes

1 The Federal Government was not independent of Britain, but was a superstructure for three very different territories, whose constitutional development it did not control: the white-ruled, self-governing dominion of Southern Rhodesia, and two colonial protectorates – Northern Rhodesia, which had a significant number of white settlers and Nyasaland, which did not. On the history of the Federation, see Wood 1983, Murphy 2005, Rotberg 1986.

2 Clements 1959:13.

3 Hemans 1935:7, 13, *The Victoria Falls Illustrated: A Handbook to the Victoria Falls, the Batoka Gorge and Part of the Upper Zambezi* (Northern Rhodesia, Commission for the Preservation of Natural and Historic Relics, 1952).

4 The press in the three Federation territories as well as in Britain followed the dam's progress closely, observing every aspect of its construction, the process of securing finance, tendering procedures, the evacuation of people and animals, etc. In addition, there was a significant output of popular books and novels of which I draw on the most widely read and circulated: Southern Rhodesian journalist, author and politician, Frank Clements' *Kariba*; American writer David Howarth's *The Shadow of the Dam* (London, Collins 1961) and two books on the animal rescue by British and Southern Rhodesian authors respectively, David Lagus's *Operation Noah* (London, 1959) and Robins and Legge's *Animal Dunkirk* (London, 1959). The South African News Agency also compiled a book (Gillies, 1999) on the dam. For a discussion of J. Davis's novels, which I do not use here, see Hughes 2006.

5 On these connections, see Thomas 1994, Kuper 2005.

6 Hughes 2006.

7 On colonial social scientists role in developing such 'counter-narratives', see Beinart et al. 2005, Tilley 2003:109-30, Mackenzie 2000, Schumaker 2001.

8 For a summary of 40 years' research, see Scudder and Colson 2002:197-238.

9 Although an inter-territorial HEP commission reported in favour of Kariba in 1951, the two governments agreed to go ahead with Kafue on 4 September 1953. Murphy 2005:lxi; on amalgamation, see Wood 1983:chapters 1-3.

10 ORAL FL 1. Sir Patrick Fletcher, then Minister of Native Affairs in Southern Rhodesia.

11 Murphy 2005:lxi.

12 Howarth 1961:36-7.

13 Federal Newsletter 22 March 1955, *Central African Post*, 16 March 1955.

14 B. Curzon, letter to M Rumbold and other correspondence with London, in DO 3/4602, Murphy 2005:lxi.

15 Federal Newsletter, 22 March 1955.

16 Federal Newsletter, 22 March 1955. On Lascelles' resignation, see also CO 1015/946, no.214, Letter, Sir A. Benson to W. Gorell Barnes 1 April 1955.

17 Federal Newsletter, 22 March 1955; *Central African Post*, 16 March 1955.

18 Speech to Legislative Assembly, cited in *Central African Post*, 16 March 1955.

19 Sir Roy Welensky was convinced by the new technical arguments and joined the Kariba camp. The Federal working committee also backed the French consultant. The sole Federal body in favour of Kafue was the Treasury committee, which feared combination of heavy debt and dependence on fickle international copper and tobacco prices. Wood 1983:403, Murphy 2005:lix-lxii.

20 Murphy ibid.:lxii, Wood ibid.:407-9.

21 Northern Rhodesia paid the most into the Federal budget, whilst gaining the least from it (partly because a significant proportion of Copperbelt taxes were retained at Federal level, whilst outgoings prioritised non-African education, privileging the territory with the most white settlers). A. Hazelwood 1967:208-9, Murphy 2005:lix-lxii.

22 World Commission on Dams (WCD) 2000:9.

23 ORAL /CA 5 Sir John Caldicott: 27, 66. On the fascination for grandiose projects, see also Mavhunga 2003.

24 Northern Rhodesian ANC leaders had consistently argued that Colonial Office rule was preferable to them. Confidential Memorandum, 28 December 1948, Sec/Nat/353, cited in R. Rotberg 1966:220; see also responses to the 1939 Bledisloe Commission in Gray 1960:176-7, 191.

25 Cited in *Central African Post*, 16 March 1955.

26 Mulford 1967:36-9.

27 *Central African Post*, 19 October 1956, cited in Rotberg 1966:281.

28 Rotberg ibid.:chapters 9 and 10.

29 WCD 2000:24, Colson 1971a.
30 See copy of the petition, held in DO 35/4605. See also *Central African Post*, 10 October 1955.
31 *Daily Telegraph*, 20 November 1955. See also correspondence between Governor, Lusaka and Colonial Secretary, November 1955, DO 35/4605.
32 *Central African Post*, 10 October 1955.
33 Charles Mzingeli's All African Federation (founded to resist the Federation) had fallen apart in 1954 and nationalist leaders were initially drawn into the politics of partnership; key figures joined the Inter Racial Association, the Capricorn African Society and related Federal political parties. Ranger 1960:29, West 2002:190-202. There were, of course, important localized instances of militancy, such as the Youth League bus boycott in Salisbury in 1956, see West 2002:205-6.
34 See Josiah Maluleke's testimony about his work at Kariba in late 1955, when the casual and disrespectful treatment of deaths at the dam provided his motivation for joining union politics. Josiah Maluleke, cited in Raftopoulos 1997:73-4.
35 *Rhodesian Herald*, 26 September 1956; Fortnightly summary of news, 4-17 July 1956, DO 35/4604.
36 *Rhodesian Herald*, 30 September 1956.
37 *Nyasaland Times*, 2 October 1956, CO 1015/958.
38 The sleeping sickness rumours followed a survey to establish its incidence, while the authorities engaged in spraying against malaria, and denied reports of exceptional deaths. Federal Newsletter (n.d. maybe late 1956); clipping in DO 35/4605.
39 In Southern Rhodesia in May 1957 the Mashonaland Region of the African Trade Union Congress asked the British Trade Union Congress and the International Confederation of Free Trade Unions to investigate. According to Clements (1959:98) the teams of African journalists and Nyasa chiefs invited to visit the site did not confirm accusations of bad conditions.
40 Federal Newsletter, 6 March 1959.
41 Federal Newsletter, 6 March 1959; telegraph, Sir A. Benson to Sec for Colonies, 28 February 1959; *The Times* 28 February 1959; Governor Lusaka, to Sec of State for Colonies, 6 March 1959. CO 1015/1529. Impresit's tender had been controversial, given the conditions that the British government attached to CDC financial support for Kariba, including the 'tacit agreement' that 'a fair proportion of Kariba contracts would come to British contractors'. CO1015/948, Minute by C.G. Wilson, 1 May 1956.
42 WCD 2000:30.
43 See correspondence in DO 35/4605; minute by W.G. Wilson, 1 May 1956, CO 1015/948.
44 Native reserves in Sebungwe were still not demarcated at the time of the move. Annual Report, Binga District, 1959, S2827/2/2/7 vol. 1.
45 WCD 2000:30.
46 PNC Yardley to CNC 3 December 1956, S 2806/1970.
47 Lord Malvern, March 1955, cited in Fabian Colonial Bureau, April 1956 (extract in DO 35/4605). See also, Malcolm Barrow, Federal Minister of Power, in *Kariba: Opening by her Majesty Queen Elizabeth the Queen Mother* (Salisbury, Federal Power Board, 1960).
48 M.W.F. Stubbs, N.Rhodesian LegCo, cited in *Central African Post*, 16 March 1955.
49 Howarth 1961:43-5. They managed to secure resettlement costs, tribal compensation for hardship, inconvenience, loss of tribal lands and customary rights, individual compensation for of loss crops, loss of earnings while clearing new lands (at £5 per acre) and building new huts (at £10 per hut). WCD 2000:30.
50 WCD 2000:31-3.
51 Claim on the Federal Power Board for the Arbitrator, 1956, Northern Rhodesian Government, cited in WCD 2000:55.
52 ORAL/241 Interview with Rupert Meredith Davies, 17 November 1983.
53 This was particularly evident in debates over changes to the tenure of the lands around the lake. See 'Notes on the effects of the Kariba dam on occupation of land' in DO 35/4605.
54 M.F.C. Bourdillon et al. 1985:16-8.
55 Bourdillon et al. ibid.:18.
56 'Tribes Told of White Man's Lake', *The Times*, 2 September 1955. See also 'White Man's Magic', *The Times*, 27 August 1955.
57 Cockroft had 'winked at this'. Mr Hunt's diary, October 1957, DO 35/4606.
58 Annual Report, Sebungwe, 1955, S2827/2/2/3 Vol 2.
59 Mr Hunt's diary, October 1957, DO 35/4606.
60 ANC Hanson, Annual Report, Sebungwe, 1956, S2827/2/2/4 Vol 1.
61 Annual Report, Wankie, 1956, S 2827/2/2/4 Vol 3.
62 Interview, chief Siansali and Sibonde Muleya, 3 April 2001.
63 Interview, chief Siachilaba, 19 March 2001.
64 ORAL 227, John Powell.
65 These views are documented in all studies of the move, on both banks. Colson (1971a:37) argues that local

district officers sympathized with them, and some privately condemned the whole project as benefiting Europeans at the Tonga's expense.

66 Annual Report, Sebungwe 1956, S 2827 2/2/4 Vol 1.

67 Annual Report, Sebungwe, 1957, S2827/2/2/5.

68 ORAL FL 1.

69 Annual Report Sebungwe, 1957, S 2827/2/2/5.

70 Annual Report, Sebungwe, 1956, S 2827/2/24 Vol 1.

71 Tremmel 1994:35.

72 The term 'chibaro' was used by Kachala Mudenda, Sinakoma, 21 March 2001 and in many other interviews. See also quotes in Tremmel 1994:36-7.

73 Cited in Tremmel ibid.:34.

74 Cited in Tremmel ibid.: 33.

75 Intreview, Sibonale Muleya, Kariangwe, 3 April 01; on lack of time, see also Munzaba Mwinde and Philmeon Mukuli, in Tremmel ibid.:32, 36.

76 Tremmel ibid.:37.

77 Interview, chief chief Sigalenke and Siampiza Musaka, April 2001.

78 Jingamulonga Mugande, cited in Tremmel 1994:37.

79 Annual Report, Sebungwe 1957 S2827/2/2/5.

80 ORAL/227. The resort to force in the light of ANC influenced resistance was also detailed in Mr Hunt's diary of the move, October 1957, DO 35/4605. Saina Muntanga remembered the same episode: 'The whites came and told Makaza to go but he refused ... they came several times to try to persuade him to shift, but all in vain.... Then they detained him, dumped him in Sanyati ... they arrested his wives ... but his sons moved up and down at night and returned to the river, going back with the wives as he was detained, then he joined them, having escaped his guards at night. Sikanyana had to send his police and soldiers to find him. By then the river was already covering his villages so he had no choice but to abandon his home to the waters...' Interview, Saina Muntanga, Chuunga, 13 April 2001.

81 ORAL Fl 1.

82 The episode is described in detail in Howarth 1961:chapters 12 to 17 and Colson 1971a:40-41. See also Colson 1971.

83 Colson 1971:24-5.

84 Clements 1959:148-9. Clements was a member of the right wing Dominion Party, whose key members went on to form the Rhodesian Front. On DP politics, see Wood 1983:822-3.

85 Note on the visit to the area of the future lake, Mr Hunt, 3 October 1957, DO 35/4606.

86 Howarth 1961:190.

87 Ibid.:182.

88 Ibid.:192.

89 The appeal involved the Game Preservation and Hunting Association of Northern Rhodesia, the Fauna Preservation Society of London, the Universities Federation for Animal Welfare, the Royal Society for the Prevention of Cruelty to Animals and the Zoological Society of London. Lagus 1959, Haarthoorn 1970.

90 Harthoorn 1970.

91 Robins and Legge 1959.

92 Federal Information Department, 'Warning to Writers' (n.d.), DO 35/4607.

93 Instead, she wrote a profile of one of the tsetse clearance officers stationed amongst the resettled Tonga on the north bank. Balneaves 1971, Lagus 1959:35.

94 Hughes 2006.

95 Wildlife zoologist Dr Fraser Darling, cited in Lagus 1959:163.

96 See Beinart 1999.

97 Lagus 1959:141.

98 Critchley, cited in Lagus ibid.:163.

99 ORAL ST 6.

100 On Southern Rhodesian debates, see Clements 1959:chapter 15. On tsetse control see Ford 1971:320.

101 Cockcroft, Annual Report, Binga, 1958 S 2827/2/2/6.

102 Lagus 1959:163.

103 WCD 2000:94-5.

104 Hughes 2006.

105 Lagus 1959:47.

106 Robins and Legge 1959:176-7.

107 Howarth 1961:8.

108 Holmes 1993:349. The texts include Wallis 1956, Simmons 1955, Debenham 1955, Gelfand 1957.

109 See *The Scotsman*, 18 June 1955.

110 Federal Newsletter, 17 November 1956.

111 Lagus 1959:52, Clements 1959:216.

112 One of the only expressions of concern about the Tonga to come from the Southern Rhodesian side of the border was that of Dr Geoffrey Bond, Keeper of Geology at the Natural Museum in Bulawayo, who felt the move 'was bound to destroy the Mtonga as a tribe and might give rise to discontent', as 'the tribe who wear few clothes, cover their bodies with ochre and put sticks through their noses might not be able to adapt'. Cited in the *Rhodesian Herald*, 14 June 1955, though the editorial the next day was quick to condemn his views, as they could be 'distorted by agitators'. *Rhodesian Herald*, 15 June 1955.

113 Howarth 1961:73-4.

114 Ibid.:73.

115 Robins and Legge 1959:18-19. These two journalists were based in Southern Rhodesia. Legge ran his own international news agency in Salisbury. Before migrating in 1947, they worked respectively for the *Daily Mirror* and *Daily Telegraph* in Britain.

116 Lagus 1959:57-8.

117 Clements 1959:12-13.

118 *Kariba: Opening by her Majesty Queen Elizabeth the Queen Mother Tuesday 17 May 1960*. Federal Power Board Held in Bulawayo Reference Library, Kariba Box.

119 *Royal Occasion 1960: The Kariba Project*, S. Rhodesian Government Printer, n.d. Bulawayo Reference Library, Kariba Box.

120 Clements 1959:209.

121 Ibid.:209-10.

122 'Kariba open to tourists', Federal Newsletter, 16 September 1957.

123 *Rhodesia's Lake Kariba*, Rhodesia National Tourist Board, ca. 1968; Rhodesia Calls, Mardon Printers Ltd, ca.1966; *Kariba: The First Ten Years*, Central African Power Corporation, 1971.

124 Kariba: Rhodesia's Riviera, n.d.

125 Colson 1971a:7.

7

Reclaiming the Borderlands
Ethnicity, Nationalism & War

In the 1950s, new political actors began to claim the Zambezi borderlands, which involved telling the history of state intervention in a very different way from Federal politicians and white popular writers. As a new educated African leadership emerged, they began to construct their own public histories, promoting their own connections to the landscape and protesting their exclusion. These new histories were cast in both ethnic and African nationalist terms; they gave momentum to the crystallization of two ethnic minority identities – Nambya in Hwange and Tonga in Binga – but they were also used in African nationalist mobilization, to legitimize the institutionalization of the Zimbabwean African People's Union (Zapu) in preparation for incursions by Zapu's armed wing, the Zimbabwean People's Revolutionary Army (Zipra).

The intellectuals who led these movements publicizing their history and culture had a powerful sense of their own marginalization, given the predominant discourse of Southern Rhodesia as a country of two African tribes.[1] They also rallied against the stigma of being labelled primitive and accused the state of developmental neglect. Though the caricature of backwardness had been elevated in the popular writing on the dam, it had a longer history and was used by Africans as well as Europeans, as older pre-colonial political hierarchies and ethnic names had been overlaid with understandings of difference that hinged on a developmental continuum.[2] The cultural assertion involved in these movements was encouraged by a shift in administrative policy in the early 1960s towards a traditionalist ethos hinging on the promotion of custom and tradition, and an elevated role for chiefs.[3] Yet it was no accident that those living on the periphery of the state remained marginal in terms of the state's ethnic categories and understandings; rather, as Worby argues, it 'reflects the degree to which the extension of state power and the refinement of ethnographic knowledge are processes that reciprocally reinforce one another'.[4] The exclusion of those who complicated the state's predominant binary formulation of ethnicity was particularly stark in the way that language policy developed, especially when taken out of mission hands.[5]

These ethnic mobilizations in the Zambezi borderlands were important for the emerging politics of landscape. As modernist cultural nationalist movements, they involved essentialized notions of culture that were territorialized and politicized. Their focus on specific evocative 'sites of memory' within lost lands in which culture and the past were instilled made implicit or explicit claim to ownership and access, and demanded compensation and development. The Nambya ethnohistories produced in this period centred on connections with the ruins of the pre-colonial past (by now within the Hwange National Park) and the dishonouring of an agreement that Nambya descendents would benefit from the coal beneath their lands. For the Tonga, the prime memorial site was the displacement and broken promise of development made in the course of the Kariba resettlement, which became the prism for looking back at life with the river, and could likewise function to assert ownership of the new lake and make claims to state resources. Arguably, in Nora's terms, language also operated as a 'site' in these movements, in the sense that it too was reified, rendered static, tied to a singular notion of culture, cast as threatened and in need of rescue. Though these movements to promote ethnic identity reflected a modern historical consciousness, this did not so much efface older ritual means of relating to the past, as Nora suggests, but rather revalidated and entangled them with new understandings. The new ethnohistories were produced under the influence of broader African cultural nationalism. Unlike earlier interests in history, they were not about instrumental bids to chieftaincy, rather they were partly about restoring dignity – nostalgic, patriotic, quasi religious, akin to today's notion of 'heritage'.[6] The African intellectuals who developed these new public narratives of their identity opposed many aspects of Southern Rhodesian state-building, but the notions of culture, tribe and history, as well as the demands for development they incorporated, reflected interactions and mutual influence with European missionaries and Native Commissioners.

These ethnic assertions also reflected a new sense of belonging to Southern Rhodesian political space, and as such were important in the making of the border and border identities. In trying to hold the state to account, local leaders presumed a national community of tribalized Africans with whom they demanded equal status. Although ethnicity was used by the administration during this period for divisive reasons, and the newly reified cultural boundaries created tensions with African others, many of the key players in the ethnic mobilizations of Northwest Zimbabwe were at the time, or went on to be, nationalist activists. For them, the new ethnohistories and the grievances they contained were entirely compatible with broader nationalist messages of freedom, political rights and restored lands. The 'idiom of oracy' deployed in the notion of broken promises was politically weak and focussed on the paternal authority of marginal colonial figures.[7] Nonetheless, it popularized the idea of the colonial state as illegitimate and dishonourable.

This chapter explores the character of these two Zambezian ethnic mobilizations and their relationship to the spread of popular African nationalism in the borderlands.[8] It examines the content of the new politicized narratives of the past that local intellectuals developed, exploring the attitudes towards ethnicity, the state, the border and development they reflected. The Tonga and Nambya movements differed in their historicity and the depth of past they invoked, and in their utility to institutionalizing nationalism (the Nambya movement proving more controversial than its Tonga counterpart). They were also shaped by contrasting perspectives on the border (which Tonga leaders only reluctantly accepted). This process of nationalist mobilization in the borderlands is notable for its similarities with other Zapu/Zipra areas, where civilian party structures were set up in advance of guerrilla incursions.[9] It is important for the narrative here because the spread of nationalist support renewed the strategic importance of the border and recharged the politics of crossing, as well as establishing popular expectations of the post-colonial state.

The chapter begins with the movement to promote Nambya culture and history, before turning to Tonga cultural nationalism, which brings the narrative back to the river and the aftermath of the Kariba dam.

Hwange & the Nambya

The growing ethnic and nationalist consciousness in the Zambezi borderlands reflected the emergence of a local mission-educated elite. Such consciousness developed first in Hwange because missions had a longer history of operation in the vicinity of the colliery and extended into the Binga rural hinterland of the Zambezi valley only belatedly in the wake of the dam and resettlement.[10] Hwange was in many ways an unpropitious location for a movement promoting a singular ethnic identity: in rural homes people spoke both Nambya (the Shona-related language introduced by the migrants who founded the Hwange dynasty) and Tonga (spoken by those who were called Dombe, Leya or Tonga).[11] Over time, Nambya, Tonga, Dombe and Leya had become thoroughly mixed through marriage, and the idea of a single mother tongue tied to a single culture was decidedly misleading. The district's towns added further complexity, as they were dominated by the languages of migrant labourers from Zambia and Malawi – Nyanja was the main language in urban Hwange, including in mission schools, whereas Lozi predominated in Victoria Falls. Furthermore, the district fell under Matabeleland, and the Ndebele-oriented Native Department used Ndebele as the main vernacular, and in the mid-1950s a significant group of Ndebele evictees, cleared off white land around Bulawayo, were dumped in rural Hwange, close to the Victoria Falls.[12]

Despite the long state presence in Hwange, associated with the discovery of coal at the turn of the century, official knowledge about the rural areas was scant, as the area had the reputation of a remote, hardship posting, and there had been a high turnover of Native Commissioners. The educated Africans who emerged through the Hwange missions were aware of being stigmatised by Europeans and Africans alike as 'primitive', 'uncivilised', 'low, 'docile' and 'quiet', people who had 'little use for modern ideas'.[13] These characterisations partly reflected the district's marginality, state developmental neglect and the history of mission failure outside the town, and were reinforced in the 1950s through contrast with the Christianized, educated and politicized Ndebele evictees and central African colliery workers.[14] The Ndebele evictees brought nationalist politics with them to rural Hwange; they continued to present the authorities with problems over contours and land-use planning, as well as provoking conflict with locals by labelling them 'amahole' (slaves) and 'backward', refusing to accept the authority of Nambya chiefs, and eventually being allowed their own Ndebele chief, Mvutu.[15] But labels of modernity and backwardness were also mapped onto the urban and rural; by the 1950s, the Colliery had long been transformed from a 'byword for brutality' into a beacon of modern living[16] embodied in the educated Nyanja elite, who dominated skilled work and urban politics. This Nyanja educated class likewise looked down on rural locals, and their reputation as troublemakers was reinforced in 1954, when they led an all-out strike over wages and rations (motivated partly by comparisons with Copperbelt conditions), and the authorities suspected nationalist orchestration by one of the Federation's northern Congress movements.[17]

When local people from the native reserve went to town to deal with the authorities, or to buy or sell goods in Hwange (or Victoria Falls), they tried to hide their identity because of the stigma attached. They used Ndebele versions of their names, which were on their pass documents, and spoke in the high-status Ndebele or Nyanja languages. The first generation of educated local people felt themselves 'invisible' in the Nyanja-dominated town. They found a niche for employment in government service rather than as colliery workers, and had an acute sense of their marginalization. A retired teacher recalled: 'We had very little, say ... when job opportunities came [in the colliery] they were not given to us because those with important positions were from Zambia and Malawi, and our language was dying ... we were being forced to speak their language. So we tried to say we should also be heard, our language should be recognized, we should be on the map...'[18]

The first activists who tried to put the Nambya on the map were a police officer, Sergeant Vika Marumani, and the district's and colliery's most senior interpreter and head of the Colliery Welfare Association, Ndoswi Sansole. Their positions gave them a proximity to, and influence with, the authorities.[19] In 1946, they persuaded the Native Commissioner to use the identity 'Nambya' on individuals' identity and pass documents, which had previously used the derogatory Nanzwa (the Ndebele name for the Nambya).[20] They also achieved representation for 'the Nambya' as one of the tribal groupings recognized in the colliery workforce and administration, but were unsuccessful in having 'Nambyaland' excised from Matabeleland at the time when provincial boundaries were being drawn in the 1950s.

Recovering Nambya history was particularly important in this new Nambya assertion, and it played a central role in trying to promote a new sense of dignity and pride in Nambya identity and language, and encouraging its public expression. Marumani and Sansole collected oral histories from elders, and formalized a version of dynastic history that began in 1737, with the original groups of Rozvi migrants led by Sawanga.[21] The ethnohistory spelt out the migration and sequence of subsequent Hwanges who ruled from a succession of stone ruins, recounted the myth of the fall of Bumbusi and the flight across the river detailed at the outset of this book, and a wealth of other stories. This process of historical reconstruction was facilitated by the way in which memory of the past had been perpetuated partly through ritual invoking ancestors associated with the ruins – of particular importance was a 'national' rainmaking ceremony that brought together leaders from throughout Hwange in a pilgrimage of the ruins.[22] During this pilgrimage, the dynastic past was invoked through memorable epithets, praises, song, dance and other performances – the names of the original cast of migrants were recited, the different Hwanges and other important figures were honoured, the Leya were respected as autochthons with powers over fertility and the defeat of Tonga and others was re-enacted.

The Nambya ethnohistory also extended into the colonial period, not only through the line of chiefs, but by narrating (amongst other things) the discovery of coal, and evictions from the flat fertile lands surrounding the stone ruins. This aspect of Nambya history promoted the idea of a broken verbal agreement (or a lost, or even deliberately hidden, charter) between the incumbent Hwange and Albert Giese, discoverer of coal and one time 'father of Hwange District' (in reality a dubious figure from Zambezian frontier life, as we saw earlier).[23] Chief Hwange was upheld as honourably refusing offers of money for land, but was said to have negotiated assurances from Giese that his Nambya descendents would benefit from the exploitation of coal. When I moved around Hwange district in 2000, I was asked repeatedly to try to locate the document detailing the agreement.

The young, educated, Christian men who supported the movement for Nambya recognition took it forward through the Nambya Cultural Association, which was formed in 1960. They organized demonstrations under the slogan 'Wankie Colliery for the Nambya', demanding more jobs and invoking the dishonoured agreement.[24] But learning and popularizing earlier, pre-colonial history was even more important. Former members recalled how Marumani and Sansole 'showed us our history, how the Nambya came into this country and came to take the land', and the Association also produced its own summary of Nambya history and a briefing on Nambya tribal characteristics, customs and material culture.[25] The dynastic history that made past greatness tangible was brought to life by trips to the largest and latest of the ruins – Bumbusi – where the pre-colonial state had both been at its apogee and met its destruction. One former member recalled: 'We had to venture to some of the very important places in history, like Bumbusi... We looked at how Bumbusi related to Great Zimbabwe, and we saw it ... had been a great kingdom'.[26] In making these trips, the Association taught how, as Christians, the site and ancestors should be respected as culture rather than religion. Another former member described how 'We ... went out to places of heritage, like Bumbusi, we took people to see the ruins of the Nambya... One time, in the dry season, we performed a certain ceremony, it was our culture, then heavy rain fell, it was shocking to see such a miracle.'[27]

The movement was important in providing a new role for the ruins as a symbol of ethnic heritage at a time when traditional religion was in decline. Though many local shrines remained important, the pilgrimages to the ruins, which had perpetuated the memory of dynastic (rather than local) pasts, ceased. The last attempted pilgrimage was in 1946, with transport provided by the Native Commissioner, but the ceremony never took place, as the group got lost in the National Park before reaching the ruins, the goat to be sacrificed escaped, and the group broke up acrimoniously amidst disputes over whether modern transport was an offence to ancestral spirits, and whether chief Dunduli had the authority to lead the ceremony, as he had converted to Methodism, and came from the disputed female line that arguably brought about the nineteenth-century dynasty's collapse. Modern life had in any case undermined the full extent of the pilgrimage which should have taken a week or more, beginning with a Zambezi river-crossing from Hwange Chilisa's north bank grave, and continuing with visits to a succession of chiefly graves and related ruins before terminating at Bumbusi.[28]

Rescuing Nambya cultural heritage did not only mean giving new life and meaning to physical 'sites of memory' such as the ruins, but also involved recovering a language seen to be threatened to the point of extinction. To this end, a group of dedicated Christians worked with local missionaries to define Nambya as a language of its own (rather than a Shona dialect), developed a Nambya orthography and grammar, translated the bible into Nambya, and composed Nambya songs. After a Nambya catechism was created in 1960, churches began to use the language. The support of the Spanish Fathers in St. Mary's Mission in Hwange was crucial to the success of this exercise. Two missionaries were particularly sympathetic to the Nambya as a minority, derived from their personal identification with Spanish minority causes in Catalonia and the Basque region, and had gone against their superiors to support the codification of the Nambya language through funds they sourced personally from Europe.[29] Fr Alexander regarded the project as a creative exercise in salvage: 'It was a lot of work

7.1 Nambya Cultural Association, ca. 1960 (courtesy of David Kwidini)

7.2 Noah Musimanga of the Nambya Cultural Association at Bumbusi Ruins

to produce the first grammar and catechism, to convince others and make it acceptable – thirty years of work. I love my work. I feel I have helped to create something. We ... have rescued some of the old words. People like that, they say it is beautiful... We have helped them to be conscious and proud of themselves...'[30] As the mission network expanded into rural Hwange, for example with a new catechetical training centre at Sacred Heart mission in Jambezi, so the public use of the Nambya language also expanded across the district, through its use in churches. A new committee was later established to write primary school texts.[31]

This movement to promote and preserve Nambya history, culture and language was notable for its flattening of old pre-colonial hierarchies, and was not led by the chiefs or others with high-status traditional credentials: anyone who could claim descent from families that had once lived under the Hwange could be Nambya.[32] But it was also selective. By defining a singular Nambya history, tied not only to territorial markers, but to a single culture and mother tongue, so it suppressed, denied or condemned the persistence of the Tonga language and cultural forms which had flourished within the district and within Nambya homes, and had also encroached on Nambya ritual. The focus on the Hwange dynasty meant that only some of those incorporated were honoured – such as Nelukoba's Leya who were respected as autocthons; but in general, the subordinated Tonga and Dombe were remembered as having run away or been defeated.[33] The drumming that accompanied rainmaking at the Bumbusi ruins, for example, included hurling insults at imaginary defeated Tonga men.[34] Only cultural forms that could be traced to the Rozvi migrants were deemed authentically Nambyan; activists were thus concerned about the disappearance of the mbira instrument and small Nambya drums, svikiro spirit mediums, and the erosion of patrilineal inheritance, while the spread of tall Tonga drums, mpande mediums, and matriliny was cast as a process of cultural distortion and loss.[35]

The movement continued through the increasingly disputed politics of the 1960s, particularly around centres of mission activity. Yet the earlier momentum was lost over the decade, partly because nationalist and worker politics increasingly consumed the time and energies of many former activists, drawing them away from the work of cultural promotion, and indeed, away from Hwange town as they moved to Bulawayo, were detained, or fled to Zambia. Though the colliery authorities had stereotyped central Africans as political leaders (indeed when yet another high profile strike in 1964 brought the colliery to a stand still, *Drum* reporters cast Hwange as a 'UNIP town'),[36] local workers were nonetheless intimately and increasingly involved in Southern Rhodesian politics. The colliery was, at least for a period, an important hub for Zapu, as it prepared for the first guerrilla incursions from Zambia in the mid-1960s via river crossing points in Hwange.[37] Some of the Nambya Cultural Association members themselves developed high-profile Zapu and union careers, while others helped set up a local network of rural district and branch committees.[38] As Zapu strengthened its rural base in Hwange, it did so partly through networks of educated Christianized men many of whom had been part of the Nambya movement; the vicinity of rural mission stations that were promoting the language, such as Jambezi, became important centres of Zapu activity in the countryside.[39] Expanding in this way, activists mobilized on the basis not only of the main nationalist arguments – the vote, freedom, land and contours – but also used the range of more specific issues popularized by the Nambya Cultural Association such as the marginalization of the Nambya language, and the dishonoured promise.

Yet this ethno-nationalist line of argument was often downplayed in nationalist mobilization in Hwange, for reasons that are important, as they foreshadow the internal divisions that undercut the movement in the post-colonial context. Perhaps most important, not all Hwange's rural residents defined themselves as Nambya. The Ndebele evictees in the district were also an important hub of nationalist activity, and Zapu committees needed to bridge the divides between Nambya and Ndebele and work together, for which purposes the broader nationalist messages provided better common ground.[40] But the Nambya mobilization had also provoked tensions and unease among people who identified as Dombe. The mission churches were careful to use Nambya only when they felt the language predominated, as in some parts of the district, such as in Dombe areas close to the river, the Tonga language was more common. Some people felt the Nambya mobilization looked down on the Tonga, or simply felt the movement was misleading in upholding Nambya language and culture over Tonga, given the complex mix in most families, including those of the chiefs. In the words of one disillusioned former member, who was also a nationalist activist, 'when someone says "I'm a Nambya" – you look at that person and you find the father is a Tonga'.[41] He continued: 'the Nambya movement was causing confusion, it was too selective, we needed a history that could integrate everyone, otherwise it is not on a strong footing, no one could have done all those things alone, without the support of others. We are all people. With the language too – it can't stand alone, the history is mixed up, they should have created a language that took from all – Leya, Tonga and Nambya.'[42] In Tonga-speaking Dombe communities, such as those along the river, life for many still focused around the Zambezi, and the most important religious sites were not the ruins, but local shrines and associated pools around the waterfall, in the gorges and along the river. Dombe intellectuals had also begun to write their own histories, which focussed on their identity as 'river people' who pre-dated the Nambya, and told stories of Dombe ferrymen and skilful crossings, and established intimacy with the river through association with Dombe ancestors.[43] For Zapu to secure support in such areas, some of which were particularly important as they were next to the river crossing points, reference to Nambya issues could hinder rather than help. The Dombe mobilization was encouraged by a parallel movement to promote Tonga language, culture and history in the neighbouring district of Binga. Before following through the politics of crossing during the war, therefore, the chapter first examines Tonga cultural nationalism.

Ethnicity & Nationalism in Binga

The movement for Tonga cultural recognition was focussed on Binga. Given the late establishment of a mission network in the district, it was led at first by a Hwange-educated Tonga elite, but spread rapidly through the network of mission churches and schools built after the displacement. Like its Nambya counterpart, this movement also involved a politicized ethnohistory, emphasized language as a source of exclusion, and provided a new validation of Tonga culture. Yet the history mobilized was more truncated, focussed centrally on Kariba and the displacement, which became the prism for understanding the remoter past with the river, and there was little if any popular interest in deeper pre-colonial history. Past religious sites – the *malende* shrines – were less important than the Nambya ruins in this movement, reflecting their submergence and the contested process of relocating them, and also the fact that the shrines had entirely local constituencies and could not function as 'national' symbols. Much of the intellectual work in formalizing notions of Tonga culture, language and history was not created afresh, but built on ideas already circulating in Zambia among the larger and better established Tonga elite. Bibles, hymn books, schools texts and readers could be introduced quickly into schools and churches as they could be simply imported across the border, subject to some minor adjustments. Connections with institutions and educated Tonga across the river were important in this cultural assertion, and created some tensions regarding the acceptance of the border.

The grievances and sense of marginalization that fuelled Tonga ethnic and nationalist mobilization developed in the aftermath of the resettlement, and it is necessary to briefly recount the hardships that followed the move to set the context for the telling of a new, angry and politicized story of the dam. Colson's *Social Consequences of Resettlement* captures many aspects of social change following the move in Northern Rhodesia that are also salient for the south bank, and the brief discussion here focuses primarily on the political context in Southern Rhodesia, which differed in important ways. Yet Colson's account of the trauma of the first few years is apposite, and in Southern Rhodesia also, the immediate aftermath of the hasty removal from the river was characterized by famine in 1958, outbreaks of disease, loss of livestock, a discrediting of political leaders and a deep questioning of religious practices. Unlike in Northern Rhodesia, where famine and disease among the Lusitu resettled was brought to national attention by the ANC, prompting state investigation,[44] in Southern Rhodesia the famine went unnoticed by African nationalist movements, official state discourse continued to maintain the Tonga 'were happy in their new areas'[45] and were benefiting from contact with the modern world, and only the Roman Catholic church and Binga's Native Commissioner responded to the crisis. Fathers at the new Kariangwe mission raised alarm over 'starvation' in the mission's vicinity as tsetse had killed goats, mice and birds ruined the millet crop, and desperate people depended on the bitter fruits of the *musika* tree, which they mixed with water and ash to make them more palatable.[46] Cockroft's internal reports to his superiors had a generally upbeat tone, but also described the 'disturbing' nutritional status and problems of vitamin deficiency and disease outbreaks (measles, TB and dysentery); the loss of small stock to disease, the failed harvest and subsequent measures to provide relief, involving additional maize, milk power and iodised salt, which were not distributed free, but were 'sold at half cost in order not to encourage lack of effort'.[47] Water supply was a particularly acute problem; of the new boreholes drilled to cater for the resettled communities, many supplied water that was hard or dirty, and in the first year, more than 60 per cent failed to provide any water at all or dried up in the dry season.[48] Oral testimonies also emphasize hunger, lack of water, fear of dangerous animals, death of domestic stock and loss of relatives through death and separation. Reflecting on the death of nine of his children in the wake of the move, Solomon Mutale said 'I will never forget how the white people hurt us'.[49]

The hardships and lack of compensation had always been legitimized through the benefits to the Tonga of contact with the modern world, and Cockcroft's reports following the move were optimistic about how the Tonga had 'been thrown into the competitive whirlpool of modern development';[50] he predicted an 'attack on the "Darkness" of the valley' through the 'scramble' for investment, and by 1958 he had approved sites for two missions (Roman Catholic and Church of Christ), seventeen schools, and several European stores, and had received a flurry of applications from settlers seeking permits for commercial fishing operations and homes along the European section of Binga shoreline, attracted by the 'marine atmosphere' to be created by the new harbour.[51] Yet much of this publicized development never materialized, or only benefited Europeans, leaving Binga the most poorly served district in Southern Rhodesia in terms of its infrastructure, and way behind the districts on the north bank. Roads remained so bad that European businessmen who had applied for permits for shops never opened them, fearing the ruination of their trucks, and the only stores were opened by missions and African entrepreneurs (mostly from other parts of the country). Although the expansion of mission primary schools and churches over the 1960s was significant, it still left many places with no schools, clinics or shops and the nearest secondary school for those in Binga was still over 150 kilometres away in Hwange.[52] Nor was there significant investment in African fishing – the government only sponsored one fisheries officer and four 'orderlies' to train Tonga fishermen, fishing camps were only 'temporary', and use of the lake was undermined by restrictions, the lack of infrastructure, the distance of many homes from the lakeshore, and the build-up of European commercial fishing and marketing operations.[53] State agricultural services were equally limited; by the mid-1970s, there were still less than 20 ploughs in the whole of Binga District, no dip tanks and only two agricultural assistants.[54] Those who managed to accumulate stock (as tsetse retreated) or other possessions, did so largely by investing income from migrant labour.

Although the many small Tonga chiefs in Southern Rhodesia had not been implicated in promoting the move, and were much less powerful than their north-bank counterparts,[55] chiefly authority was nonetheless undermined in the southern resettlement lands because its legitimacy in local eyes was linked partly to the *malende* shrines submerged under the lake. Efforts had been made to relocate the shrines, through symbolic handfuls of clay or branches from sacred trees carried from the old sites by the river, but the process was controversial and surrounded by scepticism, and it was widely argued that the shrines and spirits 'lost their power'.[56] Some displaced communities found themselves in lands where the authority of other shrines had long been recognized, and protracted disputes often ensued between their respective keepers.[57] Latham thought in 1966 that resettled chiefs had 'little or no power', and that for the older among them – such as Sinakoma and Sinamakonde – 'the adjustment [was] proving too severe a test and they [had] deteriorated into confused, often discontented people, and [had] resorted to alcohol as an escape from the realities of their lives'. He also described the undermining of older women's authority through loss of land rights.[58] At the same time, the young were introduced to new ideas and new sources of authority in the mission churches and primary schools, and were also drawn into the *massabe* possession cults that spread particularly rapidly in places where new developments were concentrated, such as Siabuwa and the Kariangwe mission.[59] The resettlement also discredited the NRANC for claiming it was possible to halt the move, and ended its influence south of the river, as local Southern Rhodesian members were now separated from political centres and party leadership in Northern Rhodesia.[60] In the immediate wake of the move, the only nationalist activity reported was in Chuunga, where Makaza's group had been resettled, and where people refused to pay taxes or cooperate with the government.[61] Makaza made an unusual and direct link between Northern and Southern Rhodesian Congress movements (and the south-bank successors, the NDP and Zapu). He is remembered for telling people in his village after the move, 'we are separate from Zambians now and all people should now follow the Southern Rhodesian ANC rather than Nkumbula. Nkumbula can't help us, now we need to look South.'[62] Although Zapu did come to have a broader presence in Binga later, in ways we shall see, the key actors had no history of involvement with the NRANC.

In the context of these social and political changes, and given the ongoing hardships after the resettlement, people looked back at their life before the move with nostalgia, and cast the displacement as central to both short- and long-term processes of loss and impoverishment. They came to see riverine life glowingly, as a time when 'there was no hunger', when 'we were close to our relatives and ancestors'. It was a time when tradition was intact, and when 'God was near us ... when we had our *malende* shrines down by the river'.[63] This is a feature of my own oral histories and of those collected by Weinrich in the 1970s and Tremmel in the early 1990s. The rupture of the move relegated life by the river to a static past, shut off from the present more completely than in other nostalgic movements to reclaim the past and landscape as cultural heritage, such as in Hwange, where old religious sites were still accessible even if many were within a National Park. In these histories, the story of the displacement is the central event, and has become foundational to modern collective identity as Tonga and as 'people of the great river'. The past before the move is generally related in terms of nostalgic memories of everyday life with the river: accounts of fishing and riverbank gardens, tales of the enchantment of sacred pools associated with ancestors, stories of crossing. Although it is possible to probe histories of the individual chiefs and lineages associated with the *malende* shrines, or to collect tales of raiding and slaving, there is relatively little popular interest in such histories; the details are often unconvincing in stories that have a predominantly moral orientation, and in which accusations of past slave status are used rhetorically as insults against others. Moreover, as we have seen, the pre-colonial past is often remembered as violent, not as a source of pride. Regarding the *malende* shrines, even the custodians do not appear to enjoy reciting the lineage of their predecessors; rather such histories can be shallow and difficult to extract. Moreover, the shrines and chieftaincies themselves are vulnerable to accusations of distorted tradition, if not because of the resettlement, through reference to a time in the early colonial period 'when slaves took over'.[64]

The content of these characteristically shallow histories, with a focus on the rupture of the displacement that tore people from their river, has been shaped by the perspective of a new educated leadership, who became influential in the years after the resettlement. These intellectuals channelled the grievances into a highly politicized narrative of the move, in which anger was directed firmly at the Southern Rhodesian government, embodied in the figure of Cockcroft – whose local name 'Sikanyana' came to stand for all DCs after the move – for failing to provide compensation and for neglecting the development of those dumped in waterless and unproductive lands. The narrative they fostered dwelt on how promises of development made in the course of the move were broken. In this version of the resettlement, Cockcroft is remembered to have promised the resisting Tonga leaders 'that the water would follow', meaning the state would provide them with new sources of water after the move. This narrative of a broken promise is now told throughout Binga District. Michael Tremmel describes how, during the meetings he called to compile oral histories of the displacement for his book, published in 1997, 'one of the women, Simpongo Munsaka, kept repeating over and over again, "We left with our property and our bodies, but we left our water behind. We would like our water to follow us to where we are today. They promised that the water would follow us."'[65] Likewise, chief Siachilaba, whose predecessor had refused to move, gave an account that was typical of many I heard:

> When Sikanyana came saying you are going to resettle ... the old people were refusing... Those old people refused and refused. So Sikanyana came with promises of compensation, but the old people still refused... Now, when the river was closed, Sikanyana came back. He promised people, no the water will be following behind, the water is coming after you. But after we moved we saw there was no water following. Boreholes yes, but the water is hard. That promise of water – nothing has come of it up to now.[66]

This story not only made claims to water development, partly through an implied ownership over the river/lake based on past intimacy with it, but also extended to state developmental neglect more generally. In it Cockcroft can also be condemned for obstructing development, for keeping the Tonga away from the modern world and for desiring to keep them primitive.[67] Andrew Sikajaya Muntanga, for example, recalled: 'There was no development here under Sikanyana – no, they were relaxed here, because people were primitive, dressed in goatskins. The DC encouraged that – he thought we were simple people, development might bring ideas that caused him trouble'.[68] The fact that missionaries built the district's schools is upheld as further evidence that the state (and Cockcroft) 'blocked' development. An educated man recalled trying to get Cockcroft's assistance with building a school in his home area, or at least his permission to allow the community to build themselves. Cockcroft is remembered as having 'blocked that – he didn't want us to develop ... that's why we had to turn to the missionaries'.[69] The emphasis on development in this narrative is perhaps partly responsible for tempering nostalgia in some versions of life with the river, many of which contain a note of pragmatism.[70] But it also reflects a particular elite perspective that has spread in the decades since the move, as 'development' was initially opposed by many Tonga elders, who, for example, refused to send girls to school (even in the mid-1970s, most primary schools had very few girls and at least two had none at all),[71] or obstructed school building.[72] The one man who was upheld as a model of 'progess' by NCs, and was famous as the first to own a car and store, and who produced grain on such a scale that Cockcroft used him as a source of food relief, was regarded much more ambiguously in Tonga villages, where his nickname 'the improved Tonga [umTonga umcono]' had also made him famous for subjecting his eight wives and multitude of children to a life of agricultural slave labour, such that many of them fled his home.[73]

But as education spread, so too did the view that the state was guilty of developmental neglect, and District Commissioners' paternal assurances were empty and disingenuous. Comparisons with the way the move and resettlement was handled in Zambia and the experiences of relatives across the lake were also part of this story. Although they were now separated by the lake, people knew that relatives across the border had received compensation, and that they had access to the lake and a better infrastructure. People began to speak of how they too would have liked lump payments in cash like their counterparts on the north bank 'just in recognition of the fact that we suffered',[74] and they too would have liked access to the lake and shoreline. Solomon Mutale reflected on the injustice that Cockcroft 'went to resettle near the new lake in Binga. We too, would have liked to have resettled near the new river, like Sikanyana, but they would not permit this. Instead they wanted us to live with wild animals'.[75]

The intellectuals who helped direct this angry public narrative linked it to both an ethnic and nationalist agenda. Given that education expanded in Binga only after the move, the influence in the early shaping of this history can be attributed to the influence of a handful of men, among whom two stand out, Francis Munkombwe and Andrew Sikajaya Muntanga. Both grew up along the river before the move under chief Saba at the western extreme of the lake, far removed from the old SRANC centres of activism downstream. Saba's chieftaincy produced most of Binga's first generation of educated leaders as well as its most prominent nationalist politicians, as it was close to Hwange's educational, religious and other infrastructure. Munkombwe and Muntanga both did their schooling in Hwange under the auspices of the Roman Catholic church, and both became involved in nationalist politics in town at the time of the formation of the SRANC in 1957; Muntanga met Nkomo personally in Gonakudzingwa.[76] In the aftermath of the move, they became committed activists. The Roman Catholic church fostered the careers of both men, sending Munkombwe for training as a catechist in Uganda in 1963, and Muntanga to Zambia (St Kanisius in Monze in 1961). Aside from their brief studies abroad and periods in Bulawayo, both worked in the north-west through the 1960s. Muntanga had a particularly close influence on Binga politics as he was a shopkeeper at the new Kariangwe mission, which emerged as a focus for nationalist activism, and maintained connections between Bulawayo and Binga through Tonga migrant workers and their engagement with home through burial societies. Munkombwe was stationed in Hwange District (Makwa and St Ignatius missions) and was elected to the Catholic African Association (later the CCJP) in the late 1960s.

Munkombwe began his interview with me with a lament for all that was submerged under the waters of the lake, including his ancestors' graves, the homes and villages he had known, and his own umbilical cord buried in the soil by his grandmother after his birth. But his account of life by the river was also one of hardship and distress following his mother's injury by a crocodile at the river's edge: he described his struggle to pull her out of the animal's jaws, the suffering of the long donkey-cart journey to Hwange hospital and the difficulties of life for the family after her arm was amputated.[77] Both men spoke of the suffering they had witnessed among the displaced, and of their own and others' anger. Munkombwe's cousin had been a driver at the time of the removal; 'that experience made him very angry, the suffering, the pain he saw, he never forgot that'.[78] Cockcroft's driver, Lasten Monga, was also among Binga's influential early nationalists; his job had taken him to villages throughout the resettlement lands, an experience that had left him similarly embittered and provided his motivation for entering politics.[79]

Muntanga and Munkombwe were key figures in efforts to promote Tonga language and a pride in Tonga culture. Muntanga felt the Tonga were 'a neglected people', cast as inferior and labelled 'primitive and ignorant'; Munkombwe felt the Tonga were 'in a pigeonhole', isolated from other Africans in the Tonga municipal compound in Bulawayo. Munkombwe hated Bulawayo; it was a town which 'could have been a springboard for education if the Tonga people had not been confined in a coop in Makokoba township, put behind a fence'.[80] Muntanga had encouraged Tonga migrants in Bulawayo to form burial societies to avoid the further humiliation of a pauper's burial by prisoners, and the societies raised funds through public performances of Tonga music based on the Ngoma Buntibe drumming and dance teams that traditionally performed at funerals; the societies were later used for political ends, to raise money and mobilize for Zapu.[81] Although the Ndebele were held in particular contempt for their arrogant attitude toward the Tonga, for 'belittling and suppressing us – they treat us as their people'.[82] Hwange was seen as little better. Both men had experienced and lent some support to the Nambya ethnic mobilization, but felt the Nambya also looked down on the Tonga, failed to recognise ongoing Tonga cultural and linguistic influence in Nambya areas, and refused them independence and past status. Munkombwe felt the Tonga had spent 'too much time in the past working for other people's benefit', both in Hwange and Bulawayo.[83]

The promotion of Tonga language and culture was supported by missionaries and by Cockroft and subsequent DCs. Muntanga and Munkombwe pushed Cockroft to allow Tonga language clan names to appear on identity cards, when in the past Ndebele versions had been used. Munkombwe recalled: 'In 1958 we changed all the clan names – the DC helped with that – I changed from the Ndebele version of my clan name Inyati to the proper Tonga name Munkombwe. We got those put on our situpas. We encouraged people to use their Tonga names in public, many were shy to do that.'[84] Cockcroft actively supported various aspects of Tonga tradition – he bought cloth and provided transport for the most famous rainmakers, such as Mawala, criticized Tonga elders when they deviated from Tonga practice, and encouraged the performance of traditional dances. Chief Sigalenke remembered: 'Sikanyana wanted people to remain with their own culture – he would criticize – "No, a Tonga doesn't stab with a knife, only a spear." He wanted the Tonga to preserve their culture and was very interested in traditional dancing.'[85] Muntanga thought this was 'a cunning approach to rule – he would allow people to smoke dagga saying, no it's in their culture'.[86]

But promoting the Tonga language was central to this new assertion, and its introduction to mission church services and schools was much easier than with Nambya, as texts and teachers were simply imported from across the border. The new Binga mission schools taught the Tonga language and used Tonga as the medium of instruction for the first years of primary school, using imported textbooks. The Roman Catholic schools also imported Tonga teachers from the sister mission in Monze across the lake and Church of Christ schools are remembered fondly for having an exclusively Tonga staff (in contrast to the Methodists, who had only a minority of Tonga teachers).[87] The Catholic church introduced vernacular languages in the liturgy in 1965, and Fr Arnaldos, who was at Kariangwe, produced a Tonga catechism in the same year, which was based on Zambian translations, but was refined to suit local spoken Tonga.[88] Some of the imported books included stories that promoted the idea of the Tonga as a single tribal unit spanning the river, valley and plateaux on either side, giving a common history of migration – and arrival as the first Bantu in the region – ideas which were increasingly popular and actively debated among Tonga intellectuals in Zambia, encouraged by 1950s archaeologists' ideas of 'Tonga diaspora culture'.[89] The Zambian Tonga teachers passed on understandings of their tribal identity to the school children they taught and also influenced local intellectuals from Binga. Zambia was upheld as a country which was more responsive to the Tonga, and tolerant of diversity, reflected in the great number of recognized tribes and languages, and increasingly, essentialized understandings of Tonga culture and history drew on the idea of one people split by the border.[90]

These close influences across the border that had facilitated the use of Tonga in schools were undermined quite abruptly when the border with Zambia was closed in 1964, when the Binga ferry ceased and legal crossing was restricted to the official border posts of Chirundu and Victoria Falls. Thereafter, Zambian teachers could no longer be imported, and it was also more difficult to get Tonga language texts, forcing mission schools to rely on Southern Rhodesian teachers, most of whom were Ndebele, could not speak Tonga, and were accused of treating the Tonga and their language as inferior. In this context, Muntanga advocated re-uniting the Tonga by re-drawing the international border to include all Tonga on the south bank. He claims to have been encouraged in making this argument around the time of Zambian independence by influential Tonga politicians close to Kaunda, who argued the same cause. Yet this was not a view that was widely or publically discussed, given the OAU policy towards African borders, and Muntanga also recalled an angry response from Zapu to this line of argument: 'we Tonga were called sellouts for that' and 'the Zapu leadership squashed the idea...The Ndebele guys [at Zapu's helm] couldn't pass on that message. They thought they'd lose their labourers...'.[91] Though the idea of changing the border lacked a broader legitimacy and was never seriously debated in either nation, the view that the Tonga and the south bank were a single people who rightfully belonged to Zambia was clearly used by Muntanga and other nationalists in Binga, such as in advance of the visit of the Pearce Commissoners to Binga in 1971 (part of a nationwide exercise to gather African opinion on the constitutional proposals agreed by Rhodesian and British governments). When the commissioners reached Binga, they met resistance, thanks to politicization throughout the district achieved partly through Muntanga's network of Tonga burial societies in Bulawayo which raised money though their drumming to print leaflets with Zapu messages and had agents deliver them to chiefs throughout the district.[92] 'The [Tonga] people refused to vote on the issue of independence under the settlement terms put before them. They claimed instead that they belonged to Zambia'.[93]

In the same year, the issue of language became explosive in Binga, as schools throughout the country were taken out of the hands of missions and placed under the control of local authorities, or under management by the Ministry of Internal Affairs if there were no chiefs councils (as in Binga). The Ministry of Education abolished the teaching of Tonga and other minority languages in 1974, as part of a broader standardization of the primary school curriculum and ruled that the objective of seven years primary schooling was 'an ability to communicate competently in English and fluently in one of *the two* African languages' [my emphasis].[94] The ruling made no mention of the existence of other languages in Southern Rhodesia and made no provision for their teaching. It provoked a storm in Binga and produced a new consciousness of marginality and exclusion. Tonga chiefs protested to the DC that teaching in Ndebele was 'cultural assassination'; elders refused to pay taxes and threatened to withdraw their children from school on the grounds the Tonga were being 'swallowed up and deserved a rightful place in the country'.[95] Fr Arnaldos at Kariangwe tried to rectify the situation by instructing teachers to give private lessons in Tonga and began work himself on a Tonga primer.[96] The then DC, Yates, was also sympathetic to the Tonga chiefs' protest, and supported the formation of a Tonga Writers committee to draft text books – a project that was stifled first by non-Tonga secretaries who could not read Tonga and introduced numerous mistakes into the typed versions of texts, and thereafter by the intensification of the war.[97]

Given the role of local intellectuals and missions in the new ethnic and national consciousness, it is unsurprising that local centres of nationalist mobilization emerged around the district's missions, particularly Kariangwe. The fathers encouraged open discussion of contemporary affairs among the youth at the mission school. James Munkuli recalled how 'The mission was neutral but on our side'.[98] In 1965, Internal Affairs blamed the 'mission complex of teachers and "bright boys" for contributing to a politically suspect atmosphere' and encouraging people to default on their taxes.[99] When the DC singled out a Tonga teacher as a ringleader, the Catholic Fathers intervened in his support, urging the DC to release him 'because of growing unrest in the area' and re-employing the man as a catechist when he was struck off the list of government teachers.[100] Away from the mission, other centres of nationalist influence included some of the African entrepreneurs who had opened shops in Binga – such as the Mangoro brothers (Shona speakers initially from Chirumanzu), who owned stores in Siabuwa, and some of the early Tonga store-owners, such as the Mungombe brothers, who opened a shop in Saba-Lubanda.[101]

The mounting disaffection with the government in Binga, which Munkombwe, Muntanga and others helped to channel into support for Zapu, combined Zapu's main political message of majority rule with the new Tonga cultural nationalism. James Mwinde, one of Muntanga's group around Kariangwe recalled: 'Muntanga was very influential, moving up and down the district, the main argument he used to get people to join politics – we were promised the water would follow, so look, they broke the promise'.[102] Another activist in the group of Zapu politicians clustered around Kariangwe recalled:

> When opening branches for ZAPU, we'd argue, Smith is oppressing you and we'd preach the gospel of independence. We'd argue with them, what have we benefited from the dam? Where is the electricity for us? Where is the compensation? No, we need to be free like Zambia, Zambians are free, we'd say.[103]

Andrew Mudimba, who was Zapu organizing secretary for Binga district recalled:

> Some people were still angry about Kariba – people also knew of the activities in town and rallied behind us. But Kariba was the main issue – people had been moved and there was no benefit. Also the Tonga language, we used that. We were lucky we had been taught by Zambians at first.[104]

Of course, the extent and success of mobilization in Binga was not just a reflection of local disaffection, deep as the grievances were. It was also a reflection of the strategic importance of the border for Zapu. Where people could not be persuaded to support the guerrillas through argument, they were forced, particularly as the war and fighting intensified from the late 1970s. Councillor Muleya, who was trying to set up committees in Simatelele in 1978, when the war was at its height, recalled: [105]

> Setting up Zapu committees, we persuaded them to be free from the white people, who have beaten us and taken our land. Nkomo was to be our president. If you don't join the guerrillas they'll attack you. We could also mention the dam – you were moved from your motherland up to the hills and the tsetse fly… also why is Tonga not recognized as a tribe and a language? Those were the weapons we used. So people were willing, but were also scared of the killing.

But by the mid-1970s, when Sr Mary Aquina Weinrich visited Binga, the government was engaged in what she felt was a fruitless campaign to rally support, as Zapu mobilization already had a widespread impact throughout the district. Moroever, key aspects of the government counterinsurgency strategy received little acclaim in Binga – transport of sick individuals by military plane to Kariangwe was regarded ambiguously by those unconvinced by western medicine; threats to close schools in places

7.3 Chief Siansali and his youngest wife, 1977 (photo by A.K.H Weinrich)

where people were refusing to pay taxes were celebrated by some Tonga elders. But Weinrich was of the opinion that the government 'had little support in Tongaland and its opponents are not just the young, but include even tribal elders and chiefs.'[106] A number of Tonga chiefs were regarded as particularly suspect, such as Siansali (whose people had beaten up a police officer they disliked), who explained in forthright terms to the DC that people did not respect the government and their agents, and Sinamagonde, who the government deposed because of his support for Zapu. Thereafter, the people of Sinamagonde refused to accept the new appointee chosen by the DC, and claimed allegiance to a Zambian chief instead.[107]

This process of nationalist mobilization in Binga was thus closely linked to a Tonga cultural nationalist agenda in which the displacement from the dam was central. In Binga, the new politicized narratives of the dam and place of the Tonga in Southern Rhodesia were universally appealing in the district, and ethnic and nationalist mobilization proceeded more closely linked than in the neighbouring district of Hwange. We have seen how the two districts had much in common with other Zapu areas in their institutionalization of party committees in preparation for guerrilla movements (though Binga was also distinctive among Zapu areas, in that local Zapu did not have women's committees).[108] But the two districts were also particularly important as they abutted the border, and the chapter now turns to the politics of crossing during the war.

The Border During the War

This is not the place for a full reconstruction of the war – even if that were possible.[109] But it is worth ending this chapter by highlighting the role of the border during the war, as it brought these marginal rural areas into the eyes of the state for different reasons than during the construction of the dam, even as it reinforced official and settler discourse about the exploitation of primitive peoples by nationalist agents and guerrillas.[110] For Zapu and Zipra, the river posed a formidable barrier to movement and communication, and created particular problems for the process of infiltrating guerrillas into the country from rear bases in Zambia. Moreover, for the guerrillas themselves, crossing the river was a terrifying experience.[111] Nicholas Nkomo, Zipra Commander for the northern Front, thought the river was 'the first enemy' a recruit had to encounter. He recalls a song which he and other trainees sang before they crossed (for Nkomo, the first crossing was 1974):[112]

> Zambezi, one river.
> One river to freedom.
> We shall carry our guns and our hand grenades.
> There is only one river to freedom.

He continued, 'We sang the last line, "only one river to freedom" without understanding what "only one river" meant':

> What greeted the new guerrillas was the ... expanse of the Great Zambezi River, the mighty, dark blue, swift running waters. At almost all the crossing points I ever used [there were] herds of giant hippopotamus, absolutely terrifying to many new fighters, many of whom had never seen a live hippopotamus before. Crocodiles were also abundant. So it was during this first stage of crossing here at the Zambezi river that most of the new fighters would be gripped by real fear. The fear of having the boat tossed in the air and then capsized in the deep waters, which, because of the river's mountainous course, might sweep the dinghies down into the numerous rapids which would result in instant drowning; the fear of being eaten by crocodiles or other predators. Then there was the terrible fear of being fired upon by the enemy from the Rhodesian side of the river while we were yet to cross.

One important role of Zapu committees and activists close to the border, was to facilitate and support such crossing. Moreover, at a higher level, Zapu strategists deployed individuals with particular knowledge of the river and borderlands to be responsible for this logistical task. After his flight to Zambia in 1973, Munkombwe assumed a position in Zapu's Zambian provincial executive. He recalls participating in debates over whether to blow up the Victoria Falls Bridge and Kariba dam, but was among the voices of opposition, 'No, it is too easy to destroy... How would we cross? No, rebuilding would take centuries'. Among Munkombwe's duties were logistics and transit across the river for all guerrillas crossing between Kazangula and Feira/Luangwa. He recalled:

> Crossing needed knowledge – you had to know the right places, because those trying to cross were fearful and tense as they couldn't swim. They were in our hands. Wherever a group crossed, locals were there to help. If the river was narrow enough, they could use a rope – those from

> the Livingstone side could throw a rope across and use it to pull the men in. Others had small one-man dinghies made in Japan – they were big enough for one man only and were very unsafe. Then there were bigger dinghies and canoes, so many methods! We placed our men on some of the islands; others were doing reconnaissance as the situation could change any time; each group crossing had its appointed carriers to take the group away from the river as quickly as possible, we didn't want fighting by the crossing places. You couldn't just cross anyhow, you couldn't cross outside the designated places.[113]

The network of branches in Binga organized by Muntanga and his team of activists were briefed on the crossing points, on the need for rapid escort away from the river, so that guerrillas could be hosted in the hinterland.[114]

To prevent such crossing, the government built a new road along the river and lakeshore, positioned army camps at key points along it, patrolled regularly, used floodlights and mined the riverbank from Victoria Falls to Binga; efforts were made to seal off the Victoria Falls African township in the wake of incidents in the resort's Baobab Hotel, to prevent its use as a 'bolthole', supply centre and recruitment ground for guerrillas.[115] Those who lived relatively close to the shoreline, or who occupied the fishing camps, became prime suspects. Binga's Simatelele fishing camp, for example, was one such crossing point, used regularly by guerrillas (who also recruited some of its fishermen), before it was closed by Rhodesian soldiers, who beat, tortured, interrogated and killed its occupants.[116] Rhodesian soldiers set up camp at Simatelele and other crossing points, and were reinforced by a South African army unit and by locally recruited trackers and auxiliaries, until Simatelele was destroyed by Zipra bombing.[117] Similar efforts were made to control other crossing points, in Siachilaba, in Mlibizi resort (where soldiers were deployed around the hotel), and within Kavira forest. Further downriver, at Chuunga under chief Sinakatenge, where the lake was much wider and more difficult to cross, there were also fishing camps and communities who still lived relatively close to the lake until 'the security forces shut down the fishing camps and white soldiers occupied them, as the youth used to cross to Zambia at night, and many joined the guerrillas'.[118] The focus of state counterinsurgency efforts on the river and lakeshore made it particularly important to move guerrillas quickly inland, to keep the exact location of crossing points a secret, to have food and porters ready to assist, and to avoid the detection of local guides from the communities with knowledge of the river. Thus villages away from the lake, further into the hills, experienced more intense fighting than those close to the lakeshore.

Upstream from the lake, in Hwange District, the river was narrower and had some of the most attractive crossing points. Nambya and Dombe communities still lived along the river banks, and had not been resettled at the time of Kariba. These communities had become a threat to the government following the break-up of the Federation, particularly in the wake of early Zipra/ANC incursions in the mid- to late- 1960s which had used Hwange crossing points. In their wake, security forces moved along the banks destroying canoes and prohibiting fishing. People remember burning their boats, sinking them or letting them be carried downstream by the current. One recalled, 'All canoes were captured by the government: before we crossed a lot, but it became political and everything was destroyed.'[119] Another said, 'It was during the time when tensions were running high, we were forbidden to cross. We didn't want to see our boats destroyed by the government, so we launched them into the river, casting the canoes downstream. It was a sad thing to see them go.'[120] There were new levels of militarization by the late 1970s, and in 1978, the Hwange rural population living close to the river between the Deka and Gwaai confluences were rounded up and put into a protected village on a hilltop at Simangani. Chief Wange was brought to the village to control people, and recalled: 'I tried to calm the situation ... and told them "This is a war, it is not a game, just stay and don't fight one another." I gave them the example of Noah. I said to them "maybe this is another way of saving us. Here we are in the Ark. When the waters go down, we shall be released and return to our homes".'[121]

Conclusion

Although war thus created new imperatives for the state to intervene and militarize the riverine frontier, by the time the authorities did so they had also lost the support of the border regions. This chapter aimed to explore how a local educated elite had fostered and channelled this disaffection over the decades of the 1950s and 1960s, popularizing politicized versions of past experiences and tying local grievances both to the crystallization of two ethnic identities and to the cause of nationalist mobilization. I hope to have shown similarities and differences in the processes of ethnic and nationalist mobilization in Hwange and Binga. Both movements were centrally concerned with language as a source of exclusion, developmental neglect and the stigma attached to Nambya and Tonga identities, and both popularized the idea of broken promises on the part of the state. In Hwange, local intellectuals mobilized a deeper history that was centrally concerned with pride in the pre-colonial past focussed on stone ruins as tangible evidence of past greatness. But the attempt to tie the multiple ethnicities, languages and cultural forms of the pre-colonial Hwange dynasty and its legacies to one particular and static cultural and linguistic tradition proved controversial, and in a multi-ethnic district, the new ethnohistories proved divisive and of only limited use in nationalist mobilization. In Binga, by contrast, the promotion of a singular cultural identity, focussed on the suffering and separation of the displacement proved universally appealing, and was central to the shaping of popular nationalism. Anger over the Kariba resettlement was still fresh, and the repeated, public discourse of how the Tonga would benefit was easily turned against a state that dishonoured its promises. In Binga, the tensions and compromises of ethnic and nationalist mobilization were different. They meant accepting the dam and state border as permanent features of the landscape; abandoning the idea of moving the border meant accepting a future in a Zimbabwean nation, as a small ethnic group with a history of marginalization.

The end of the war, independence and electoral victory for a majoritarian government did not bring an end to the expectations and tensions fostered in these two movements to reclaim the borderlands. Rather, it provided a new context for their expression. In the next chapter, we shall see how the unsettled legacies of past colonial interventions and the new expectations fostered by popular nationalism continued to shape post-colonial politics in the Zambezi borderlands.

Notes

1 On post-war ethnicity and the disruption to binary formulations provided by another marginal group, the Shangwe, see Worby 1994:375.

2 Alexander and McGregor 1997.

3 On the traditionalist turn in administrative policy, see Alexander et al., 2000:chapter 4, Ranger 1999.
4 Worby 1994:376.
5 Hachipola 1998.
6 Parsons 2006:669.
7 On the idiom of oracy and its weaknesses, as used by Ndebele chiefs to hold the state to account, see Ranger 1999.
8 To see how local popular nationalisms were created in other contexts, see Alexander et al. 2000:chapter 4.
9 For a discussion of the differences with Zanu/Zanla areas, see Alexander et al. ibid.
10 The dramatic expansion of the mission network in Hwange, dates from 1949, when Spanish fathers took over the Catholic missions in the district, established a diocese independent from Bulawayo and concentrated on the expansion of schools for the first decade. On the earlier history of Catholic and Methodist mission, see Ncube 1994:179; Becerril 1988.
11 Nambya ethnic mobilization is explored in greater length in McGregor 2005a. For a history of Hwange in this period, see Ncube 1994.
12 Most were from Essexvale and Matobo. Annual Reports, Wankie, 1956-8, S2827/2/2/3-5; A. D. Elliott, The Wankie Tibal Trust Land Communities Wankie District. Delineation report, March 1965.
13 Annual Reports, Wankie, 1937, S 1563; 1955, S2827/2/2/3. This was an old administrative stereotype, also used in the past to contrast the Nambya with the Ndebele, see Annual Report, Wankie, 1912, NBH 1/1
14 The district had never had an agricultural demonstrator, and before the arrival of the evictees, had received scant state developmental attention.
15 Annual Reports, Wankie, 1956-8, S2827/2/2/3-5.
16 On the colliery's transformation, see I. Phimister 1994:71-7.
17 Ibid.:chapter 4.
18 Interview, Maxwell Sikulu, 17 March 2001.
19 More than ten former members of the Nambya Cultural Association were interviewed in March-April and August-September 2001, and form the basis of this section.
20 The name is said to mean, 'the ones who were licked', and refers to people appointing a certain Parazhoji as leader on the south bank, after Hwange had fled across the river; but Parazhoji fled or was captured by the Ndebele; Mzilikazi put him in a kraal with a bull to test if he was a witch, but the bull licked him and he was freed.
21 Marumani's research resulted in a draft handwritten book on Nambya history, the typescript notes of which are available in Sinamatella camp. In so doing, these two provided the information for a flurry of articles on Nambya history and tradition written by Native Commissioners and published in the 1970s in the Native Affairs Department Annual, namely, M.E. Hayes, 'The Nambiya People of Wange', Native Affairs Department Annual (NADA),11 (1977), A.G.K. Henson, 'History and Legend of the VaNambiya', *NADA* (1973).
22 There were, of course, also family and neighbourhood rainmaking ceremonies at local shrines, as described in chapter 2.
23 This story features in various versions of Nambya history compiled by the Nambya Cultural Association (though does not appear in accounts published by Native Commissioners). See for example, BaNambya Scientific Association, untitled MS, n.d. held at DA's office Hwange. It was also repeated in interviews with chiefs (e.g. chief Nekatambe, 3 March 2000), NCA members and others.
24 The administration regarded this as a 'nuisance'. Annual Report, Wankie, 1961, S28272/2/8.
25 Interview Joshua Siantungwana, 31 March 2001; Nambya Cultural Association, The BaNambya Tribe unpublished MS (n.d. ca. 1975).
26 Interview, Maxwell Sikuka, 17 March 2001.
27 Interview, Amos Malikwe, 17 March 2001, Maxwell Sikuka, 17 March 2001.
28 Interviews: chief Nekatambe, 30 March 2000; chief Whange, 4 April 2000; F.S.B. Dube, 18 April 2000, Patrick Gwaba, 16 April 2000; N. Musimanga and N.L. Chinyati, pers comm., April 2000.
29 The church authorities had been against the project, on the grounds that the Nambya were too small in number, and Hwange town was cosmopolitan. Interview, Fr Alexander, St. Mary's Mission, 17 April 2000.
30 Interview, 17 April 2000.
31 This was known as Lusikunulo, set up in 1975 with goals similar to its predecessors, to encourage the use of Nambya names and language, and to write school textbooks.
32 The history of the Nambya Cultural Association is explored in greater depth in McGregor 2005a.
33 See for example, the Nambya Cultural Association's, The BaNambya Tribe.
34 As one elder recalled: 'when we drum [during rain-making at the ruins], we pretend there is a Tonga man, and shout "You get out of our way, you, whose fathers ran from the Hwange, run, we shall whip you with a stick" and other insults Interview, Anderson Tatane Whange, 16 April 2000.
35 Nambya Cultural Association, The BaNambya Tribe; interviews with former Nambya Cultural Association members, who subsequently left the movement, partly because of Tonga influences within their own families.

Interviews, A. Chiyasa, 30 March 2000; M. Sikuka, 17 March 2001.

36 Phimister 1994:154.

37 Some of the most prominent nationalists from Hwange were based in Bulawayo, such as Francis Nehwati. Zapu later stopped organizing from Hwange, as tight surveillance at the colliery made it a more difficult place to operate from than Bulawayo. Interview, J. Manyinga, Jambezi, 15 March 2000. There was already a rural network by the time of the Pearce Commission, when headmen Nekatambe and Dingane voted 'No', see Bulawayo Record Office, Box 1/7/5/F Wankie Annual Report, 1972.

38 Interview, J. Manyinga, Jambezi, 15 March 2000; Philip Ngonze, 16 March 2000.

39 Jambezi emerges clearly as a centre of nationalist activity through Wankie Annual reports, 1969-79.

40 On surveillance and security forces actions against these politicized Ndebele evictees, for fear of their role in cross Zambezi incursions, in the wake of crossings in previous years, see Bulawayo Records Office Box 1/7/5F Wankie District Annual Report, 1969.

41 Interview, Alexius Chiyasa, 30 March 2000.

42 A. Chiyasa, 30 March 2000.

43 Interview Matthias Munzabwa, 3 April 2000; see also letters spelling out Dombe history to the DC: 7 September 1975; 13 May 1980, Hwange Chiefs file, DAs office, Hwange. It is difficult to date exactly when the Dombe began to organise in opposition to Nambya cultural assertion, though it was certainly before independence.

44 Colson 1971a:53-6.

45 Gokwe's Native Commissioner John Powell's recollections echo official discourse of the time: he recalled, 'after a season ... I found they were very happy in their new areas'. ORAL 227.

46 *Rhodesia Herald*, 14 November 1957, reproducing an extract from Kariangwe missionary's report to Commission on Social Service Development of the Roman Catholic Bishop's Conference, cited in Weinrich 1997: 26.

47 Annual Report, Binga, 1958 and 1961, S 2827 2/2/6 vol 3 and 3; S2827 2/2/8 vol 1/3/5.

48 WCD 2000:41.

49 Cited in Tremmel 1994:42.

50 Annual Report, Binga, 1958, S2827/2/2/6 vol 2 and 3.

51 Annual Report, Binga, 1958, S2827/2/2/6 vol 2 and 3.

52 Weinrich 1997:36-7.

53 Some would-be fishermen activated old links with family on the north bank and crossed the lake to take advantage of the better access to the lake and training facilities at Sinzongwe, developed with Federal compensation funds. Scudder 1965:6-11.

54 Weinrich 1977:37.

55 Colson (1971a:42) describes the discrediting of the Gwembe Tribal Authority and chiefs.

56 This was a feature of most of my interviews, see also discussion in Weinrich 1977:39-40.

57 See also Colson 1971a, 1977.

58 Delineation Report, Binga, 1966; cited in Ncube, p. 233. On alcoholism after the move, see Colson 1971a.

59 These cults offered a means of dealing with affliction – often for a fee – and were initially dominated by women. Luig (1999:132) has argued that in contrast to older basangu mediums, massabe healers did not endeavour to challenge or banish powerful, feared forces but rather tried to appropriate or accommodate them, transforming them from a threat to a source of protection through re-enactment, mimicry or parody of their behaviour.

60 On the north bank, this disillusion was brief, and support was quickly rebuilt; the ANC made a high-profile intervention to publicize deaths in Lusito, and the fate of the Gwembe Tonga more generally remained on the list of the broad range of grievances nationalist leaders defined their movements as addressing. On nationalist invocation of the resettlement see, for example, Mainza Chona's speech, cited in Mulford 1967:127.

61 Annual Report, Binga, 1958, S2827/2/2/6.

62 Interview, Saina Muntanga, Chuunga, 13 April 2001.

63 Interview, Chief Siachilaba, 19 March 2001, Kachela Mudenda, Siachilaba, 21 March 2001. Similar comments were made by virtually all elders interviewed. See also Tremmel 1994.

64 For example, such controversy hinges over chief Saba's malende shrine. It is widely held that the current chiefly line and shrine was taken over by slaves when first formalised under the colonial authorities; after the resettlement, the relocated shrine was within the constituency of an existing malende shrine, further undermining its status. Interviews A. Chiumu Mudimba,13 March 2001 chief Saba, 12 March 2001.

65 Tremmel 1994:37.

66 Interview, chief Siachilaba, 19 March 2001.

67 Interview, Andrew Chiumu Mudimba, 11 April 2001.

68 Interview, Andrew Sikajaya Muntanga, 1 April 2001

69 Interview, Andrew Chiumu Mudimba, 11 April 2001.

70 This was a feature of some of my interviews, see also some of the comments cited in Tremmel 1994. For a

discussion of this pragmatic ethic, see Colson 1971. There were, of course, quite severe demographic and economic pressures before the move. At least two riverine chiefs – Binga and Sigalenke – had tried to move away from the river shortly before the dam, back to lands inside the tsetse belt from which they had been evicted in 1913, on the grounds that riverine soils were impoverished and yields consistently poor. At the time of the move, they tried to find lands they knew from the past.

71 Interview, chief Binga, 27 March 2001, Andrew Chiumu Mudimba, 27 March 2001; Weinrich 197735.

72 Chief Binga – controversially elevated to the status of paramount by Cockcroft in 1957 – refused to allow a primary school to be built close to his home up to the time of his death after independence.

73 Interviews Siabvwara Mudenda 10 April 2001; Andrew Chiumu Mudenda, 10 April 2001.

74 Interview, Francis Munkombwe 15 April 2001; chief Siachilaba, 19 March 2001.

75 Cited in Tremmel 1994:44.

76 Munkombwe first joined during a brief sojourn in Bulawayo and continued in Hwange, Muntanga became involved in Hwange, where he had gone to work, and gained his education in night school. The profile of Muntanga's political career depends partly on my interview, but also on Bernard Manyena's detailed biography of Muntanga, compiled through interview in July 2007.

77 Interview, Francis Munkombwe, 15 March 2001.

78 Interview, Francis Munkombwe, 15 March 2001.

79 Interview, Andrew Sikajaya Muntanga, 1 April 2001.

80 Francis Munkombwe 15 March 2001, see also Tolani Ndlovu 13 March 2001. Tonga migrant workers to the city sometimes consciously hid their identity as a strategy to avoid the compound, and the stigma attached, avoiding Tonga names and the Tonga language, using Ndebele or sometimes Nambya instead.

81 Bernard Manyena, interview with Andrew Sikajaya Muntanga, 7 July 2007.

82 Francis Munkombwe, 15 March 2001.

83 Francis Munkombwe, 15 March 2001.

84 Interview, Francis Munkombwe, 15 March 2001.

85 Interview, chief Sigalenke, 2 April 2001.

86 Interview, Andrew Sikajaya Muntanga, 1 April 2001.

87 Francis Munkombwe, 15 March 2001.

88 Becerril 1988.

89 On shifting identities, see Colson 1996.

90 Francis Munkombwe, 15 March 2001; Andrew Sikajaya Muntanga 1 April 2001.

91 Interview, Andrew Sikajaya Muntanga 1 April 2001. In interview with Manyena, Muntanga describes earlier causes of friction within Zapu that had caused the Tonga to be labelled 'sellouts', dating to the time of Smith's 1964 indabas to test selected African opinion on the topic of independence (ignoring nationalists and urban Africans); one such was held in Binga in the presence of a British Ministerial delegation including Arthur Bottomley. The Tonga chiefs selected for this purpose supported Smith and denounced Zapu. Muntanga described his subsequent efforts to politicize the Tonga in terms of the need to 'reverse this embarrassment of the Tonga', July 2007.

92 Following the Pearce Commission, Francis Simwinde and Siabbunga were arrested in Bulawayo for their part in the campaign, and sentenced to 30 days' imprisonment. Bernard Manyena, interview with Andrew Sikajaya Muntanga, 7 July 2007.

93 Weinrich 1977:46.

94 See Annual Report of the Secretary for African Education for the year 1974 (held in Rhodes House, Oxford). Interview, Duncan Sinampande, 25 August 2001 (who qualified as a teacher in 1974, as one of Binga's first Tonga teachers); Francis Munkombwe, 15 March 2001.

95 Interview, Francis Munkombwe, 15 March 2001; Andrew Sikajaya Muntanga, 1 April 2001; Duncan Sinampande 25 March 2001.

96 Weinrich 1977:35.

97 Duncan Sinampande, 25 March 2001.

98 Interview, James Munkuli, Kariangwe, 3 April 2001.

99 Weinrich 1977:35, 47.

100 Internal Affairs report of 1965, cited in Weinrich 1977: 46.

101 Interview, Moses Mungombe, Saba-Lubanda, 13 March 2001.

102 Simon Mweemba, Kariangwe, 3 April 2001. See also interview with Andrew Sikajaya Muntanga 1 April 2001.

103 James Munkuli, Kariangwe, 3 April 2001

104 Interview, Andrew Chiumu Mudimba, 25 March 2001.

105 Interview, Clr S.P. Muleya, 29 March 2001.

106 Weinrich 1977:45-7.

107 Ibid.:45

108 Former activists in Binga explained that such a role for women was unimaginable to them: 'How could

husbands allow their women to move up and down organizing independently?' They treated any such suggestion as a joke rather than a serious proposal, an idea which they themselves did not support and which certainly would have met with resistance in the villages Interviews, Andrew Chiumu Mudimba 25 March 2001.

109 During the war, DSAs burnt all the Binga DC's records, and the DC himself was assassinated for sympathizing with the nationalists.

110 The wartime context produced a more extreme genre of white writing about the Zambezi and primitive innocence, in which 'terrorism' was cast as the 'rape of Tonga womanhood'. Hughes 2006.

111 For further discussion of Zipra narratives, see Alexander and McGregor 2004.

112 Niholas Nkomo n.d.:14-18.

113 Munkombwe, 15 March 2001.

114 Bernard Manyena, interview with Andrew Sikajaya Muntanga, July 2007.

115 Civil Defence, Victoria Falls, Bulawayo Records Office Box 5/8/8/F Running file 1975, 1977.

116 One form of torture involved tying up suspects and towing them behind speedboats, head submerged, to the point of drowning.

117 Interview, Clr. S.P. Muleya, Simatelele, 28 March 01.

118 Interview, Saina Muntanga, Chuunga, 13 April 01.

119 Interview, Sungani George Munkuni and Nelson Nengwa Munzabwa, Simangani, 14 March 2000.

120 Interview, Petrus Dixon Shoko, Makwa, 24 March 2000.

121 Chief Wange, Mwemba, 4 April 2000. On the creation of the protected village at Simangani, see Bulawayo Records Office, Box 5/6/9R, Wankie Local Board, 1974-78; Box 4/6/5R LAN 21 1971-77.

8

Unsettled Claims
The Tonga & the Politics of Recognition

After Zimbabwe's independence in 1980, Binga – the only district in the country where Tonga speakers comprised the overwhelming majority of the population – continued as the focus of Tonga identity politics and demands for cultural recognition. This defensive assertion was no more separatist in intent than the previous movements that had delivered the region to African nationalism; its leaders emphasized their difference with the aim of enhancing their inclusion in the new Zimbabwean nation, and reversing a long history of marginalization, discrimination and colonial developmental neglect. At all times, Zimbabwean Tonga politicians have claimed their rights within the Zimbabwean state – the brief flirtation with 'belonging to Zambia' before independence was short-lived and did not become grounds for mobilization in the post-colonial period, even as local discourse continued to take for granted the view that the border was wrongly placed.

Although, as we saw in the last chapter, Zimbabwean Tonga demands for recognition developed initially in the context of administrative traditionalism and the broader Zimbabwean cultural nationalism of the 1960s, they became louder after independence. In the post-colonial context, 'minority' language groups were legally defined for the first time, provoking a storm of criticism. Local leaders in Binga made strategic links with other linguistic minorities to campaign for changes to the law, and to reject the new label 'minority' itself. But they also elaborated specific concerns about the place of the Tonga in the Zimbabwean nation, and ideas about Tonga heritage focused on past relations with the river and the injustices of the displacement from the Kariba dam. This chapter sheds light on the reasons why the idea of being a 'river people' became more rather than less important over time, even as a new political generation came to the fore who had no first-hand experience of life along the river or of resettlement, and examines how the history of the displacement has remained a foundational historical event defining modern Tonga public identity in Zimbabwe.

This movement for cultural recognition is interesting for the light it sheds on the 'politics of recognition', which has attracted growing commentary as it has undergone a resurgence throughout the African continent in the 1990s. This resurgence is generally attributed to neo-liberal reforms and democratization, the retreat of the state and the growing importance of globalized NGO networks, which have provided international validation for discourses of indigeneity, culture and rights.[1] The literature analysing these movements often has a polarized tone, as advocates see a potential catalyst for broader debates about citizenship and the reform of unaccountable states,[2] while critics see further evidence of the inadequacies and schismatic tendencies of post-colonial politics, as authoritarian cultures of power are re-crafted through the manipulation of group identity in neo-patrimonial networks, risking balkanizing the nation by producing separatist movements and other forms of insurgency in disaffected border regions.[3]

Yet I hope to show here that there is no intrinsic trajectory to these movements, no essential politics of recognition that can be evaluated as good or bad; nor should they been seen only in the light of neoliberalism, as the notion of culture they incorporate has older and non-liberal roots. The multiple grievances that fuel these movements – which typically do not only revolve around culture, but demand development and also campaign for an end to various forms of discrimination – are available for incorporation either in liberal reformist movements with inclusionary goals, or in exclusive and divisive tribal politics. As we shall see below, localized struggles for cultural recognition in Zimbabwe have been used both in the construction of patronage networks by the ruling party, and in liberal reformist movements that have reframed 'local' issues in terms of rights. The appeal of liberal reforms grew dramatically as the state failed to deliver on promises of material transformation and as anger deepened over politicians prepared to enrich themselves through the poverty of others. Indeed, as Berman and Eyoh argue, the patronage politics that has flourished in African contexts over the 1980s and 1990s, both before and in the aftermath of neoliberal reforms, is intrinsically unstable in conditions of economic decline, as the resources to feed neo-patrimonial networks diminish.[4] The movement to secure a place for the Tonga in Zimbabwe highlights one of the exclusionary dimensions of post-colonial definitions of the nation, when cast as a country of two African tribes and a white minority. However, it also shows the inclusionary potential of democratic reform movements mobilizing around equal rights and citizenship and the popular momentum this movement had gained before it was abruptly eclipsed after 2000.

In the Zimbabwean context, 'minority' debates have been largely ignored by academics, and have achieved little prominence in national political discourse. Moreover, the label 'minority' itself requires some explanation in relation to Zimbabwe, as it is commonly applied not to the country's several newly formalized linguistic minorities, but to the white community as a racial minority, the Ndebele as an ethnic minority in relation to the Shona, or even to farm labourers from central Africa, or to the Indian and Coloured communities (recently described as 'invisible subject minorities').[5] In Zimbabwe, race is generally cast as the country's primary social cleavage, whereas debates over ethnicity have focused predominantly on the country's two major ethnic blocks – or 'supertribes', as Werbner has described them – on the historical processes through which they were created, and the politics of relations between them, particularly in the course and aftermath of the conflict of the 1980s. Yet the historical construction of these 'majority' ethnicities and the expansion involved, was part of the same process through which 'minorities' were created and marginalized in Zimbabwe's political space. As both Shona and Ndebele languages are formally defined as 'majority' vernaculars, the focus in the academic literature on the two major ethnic identities has ignored the role of language as a source of exclusion for others.[6] In the brief period when civil society flourished and public debate over civic rights opened up over the 1990s, leaders of the country's linguistic minorities began to make their voices heard, as part of a broader questioning of the ruling party's governance, and demands for constitutional reform and civic rights. They might have attracted more attention in national politics had such expression not been eclipsed from 2000 by mounting state repression, the violent stifling of public debate and a renewed emphasis on the politics of race.

Although Zimbabwe's linguistic minorities have grouped together to claim their rights to more inclusive policies regarding language in schools and national broadcasting, and share concerns over stigmatizing stereotypes, exceptional levels of poverty, and access to resources, there are also differences and tensions between them. Most of the country's minorities are in Matabeleland, had a history of support for Zapu, and (unlike minorities in the east) lacked historical relations with Zanu(PF), which affected their relationship with the ruling party in a post-colonial context. Within Matabeleland, minority groups differ in their historical relations with 'the Ndebele' and 'the Shona', and their relations with one another have not always been smooth, as we have begun to see. The politics of recognition inevitably involves drawing boundaries between self and other and thus demands for cultural rights have on occasion resulted not only in conflict with the 'majority' ethnicities and immigrants but also with other minorities. As such these mobilizations raise deeper questions about the predominance of essentialized and territorialized notions of identity.

The chapter begins by considering the first decade of independence, when new levels of state development were combined with demands for cultural recognition in the Zambezi valley in an era before neoliberal reforms, in the context of authoritarian modernization and the violence of Zanu(PF)'s attempt to destroy Zapu, which lost the first parliamentary elections in 1980.

Hunger, Coercion & the Politics of Development in the 1980s

National and international attention was quickly drawn to Binga and other parts of the Zambezi valley in the early years of independence, largely due to hunger and poverty. In the context of the postwar reconstruction programme, and attention to the needs of returning refugees, assembled guerrillas and the internally displaced, a nationwide nutrition survey, funded by donors in 1980, highlighted exceptional levels of distress in Binga, where half the district's children were malnourished.[7] The district retained its leading role in measures of hunger and poverty during the droughts of 1982 and 1983.[8] National media publicized this hunger, running stories of charities pledging food aid, and celebrating prospective state interventions, such as supplementary feeding programmes.[9] The publicity had the effect of attracting an unusually large influx (by Zimbabwean standards) of international NGOs – by 1982 Christian Care, SCF(UK), Redd Barna, Novib, Unicef, the Danish Volunteer Service, America Aid, Voice, and Africare, all had projects in Binga, and many others followed.[10] Local leaders had mixed feelings about this NGO presence; although they brought resources, NGOs also had their own agendas.[11]

This was not the only ambiguous effect of all the publicity given to hunger and underdevelopment in the Zambezi valley. Journalists' coverage of Binga in the national press helped to reproduce the negative stereotypes of the Tonga that had flourished in the public sphere before independence, particularly at the time the dam was built, such that in the 1980s and beyond, Binga, the Tonga and the Zambezi all remained by-words for the primitive in Zimbabwean popular discourse. In 1981, for example, *The Herald* repeated old caricatures of the Tonga's backwardness when it blamed high levels of malnutrition in Binga not primarily on drought, war and the history of displacement, but on 'primitive farming methods'.[12] The newspaper reproduced a 1960s colonial image of people in the valley with only two toes, and told stories of communities so backward that they had not learnt to till the soil and so isolated that they had still not heard of the country's independence by the end of 1981.[13] It is clear from the attitude of outsiders who moved into the Zambezi valley after independence that these caricatures circulated widely within Zimbabwean society: the non-Tonga African public servants who came to work in Binga after independence thought they were going to an uncivilized place where people were less than fully human. 'I expected to find little people wearing skins who lived in trees – I had heard their homes were dirty and unclean,' one teacher recalled;[14] another had been told 'so many stories about the people in Binga. I was told that the people looked like animals and that they did not wear any clothes', and that 'their culture was an obstacle to development'.[15]

The persistence of these stereotypes meant that, from the outset, local leaders in Binga combined their developmentalism with demands for cultural recognition. Notwithstanding the presence of international NGOs, the most important institution for pursuing both developmental and cultural objectives in the early years of independence was the new democratic body at district level – the council. As in other parts of Matabeleland, the Binga district council was dominated by Zapu activists from the 1960s and 1970s. Andrew Sikajaya Muntanga and Francis Munkombwe, who were the first two council chairs, then MPs, had a particular influence on local politics over the 1980s.[16] As we saw in the last chapter, both men were educated and veteran nationalist politicians: in exile in Zambia, Munkombwe had been a prominent figure, and as Zapu's representative in the Zambian parliament he was used to dealing with government ministers and donors.

One of the Binga council's first acts under Muntanga and Munkombwe's leadership was to spell out Tonga grievances and a vision for the future in the 'Lusumpuko Plan' of 1981.[17] The plan explained:

> It is an established fact that of the 55 districts in Zimbabwe, Binga has been the most neglected by past colonial administrations. In nearly every sphere of development, Binga has been overlooked and always unfortunate enough to be placed at the tail end of the receiving line where priorities are concerned. As a result of this, it is drastically lacking in all the essential public services such as health, education, communication and agricultural development. Due to lack of progress, Binga has been forced to rely almost entirely on aid from various charitable institutions, which is appreciated, but is far from being sufficient to fulfil her development requirements. As of recently, sizeable financial amounts have been solicited on Binga's behalf, which is much publicized in the mass media, but, in fact never reaches Binga.

The Lusumpuko plan was ambitious on all fronts, advocating major developments in education, health, agriculture, transport and industries around the lake including fish canning and freezing plants, boat and net manufacture, fresh water prawns and aquaculture. It envisaged the lake as a 'commercial waterway combining cargo with passengers on a scheduled route…', with developments 'directed towards the uplift of the Tonga people' to reverse the actions of 'various regimes of the past [who] have given lucrative financial concessions to a limited few with no consideration about the majority of the Tonga people who by tradition rightfully have a claim to the riches of the Zambezi Valley.' The document also called for Tonga language teaching and the development of Tonga literature as well as a culture centre to prevent 'random buyers exploiting the creative talents of Tonga artists unmercifully, giving them pennies for their articles'. In line with the putative socialist ethos of the time, these developments would be run by the local state and cooperatives.[18]

At the same time as the council tried to raise funds for these development projects, it also lobbied for recognition of the Tonga presence in Zimbabwe, mobilizing an essentialized view of the Tonga while criticizing the essentialism of others.[19] Munkombwe argued, 'the colonial government didn't want to recognize us, but we expected more at independence, we argued no, Zimbabwe can't be a country of two tribes – Shona and Ndebele are fiction. There are more Tonga in Zimbabwe than Ndebele – if you look closely, most Ndebeles are Kalanga or other origins.'[20] The council set up a Tonga Language Committee to source Zambian Tonga language books as well as to compile Zimbabwean Tonga materials, lobbied the Ministry of Education for Tonga language teaching and supported a new Tonga Writers Group and a Tonga Cultural Association.[21] It also pushed for the redrawing of internal administrative boundaries, which had 'divided the Tonga' and 'put them under Shona or Ndebele dominance'. Munkombwe argued that 'Lupane [in Matabeleland] was called Silupale and it was Batongaland ... Mt Darwin [in Mashonaland] was also part of Batongaland...'.[22] What was needed was a new 'Tongaland' province encompassing Lupane, Hwange, Binga, Nyami Nyami and Gokwe.

This activism in Binga was replicated in other districts of Matabeleland, where parallel campaigns were underway to promote the language and culture of Kalanga and Nambya, Venda and Sotho, and cultural associations linked to each group proliferated. The campaign in Binga was unusual in comparison to these other minority mobilizations, in that Tonga-speakers were the overwhelming majority of the district, and demands for cultural recognition had unanimous popular support. Elsewhere the situation was more complex, as in Hwange, where both Nambya and Tonga were mother tongues in the rural areas and the town was a cosmopolitan mix.[23] In the districts of Matabeleland South, the situation was different again, as Kalanga, Venda and Sotho activists who wanted to promote their own languages mobilized in communities where many people identified as Ndebele (as well as Kalanga, Venda or Sotho). Where the promotion of minority languages and cultures lacked unanimous appeal, minority activists were vulnerable to accusations of dividing the population and undermining government rhetorics of 'unity and reconciliation', and could not take council support for granted.

The various minority organizations around Matabeleland soon began to make common cause. In the north-west, Tonga and Nambya leaders from the neighbouring districts of Binga and Hwange, which had longstanding historical links, developed an inter-district education fund. The initiative fell apart amidst acrimony and mutual accusation.[24] Tonga representatives on the committee complained that the Nambya leaders looked down on them and wanted to obliterate Tonga cultural influences in Hwange district, as the unilateral promotion of the Nambya language ignored the widespread use of Tonga in the district. Munkombwe felt that 'the Nambya don't respect us; in the past the Tonga spent too much time developing other peoples areas, such as Hwange and Bulawayo and now we needed to focus on our own district'.[25]

More successful in terms of bringing minorities together throughout Matabeleland was 'Vetoka' (standing for Venda, Tonga and Kalanga). Vetoka was set up on Kalanga initiative and operated from the same address as the Kalanga Cultural Association in Bulawayo. It mobilized over the issue of minority languages in schools, but also tried to raise the profile of Matabeleland's minorities more generally, by criticizing Ndebele cultural dominance. Vetoka's maverick leader, Million Malaba, had actively promoted Kalanga causes since the 1960s. His idea of Kalanga ethnicity was defined in essentialist terms, which cast the Kalanga as indigenous rulers and the Ndebele as immigrants, thus ignoring more flexible definitions of identity and the fact that many people spoke more than one mother tongue.[26] In Malaba's view Kalangas 'will never be transformed into being the Matabeles', and deserved recognition for being in Matabeleland since 1441, a time when 'we were all Kalanga and the nation was led by Mazmuyenazwa Moyo'. This antagonistic definition of Kalanga identity as mutually incompatible with being Ndebele worked against many family histories in Matabeleland and a long tradition of more fluid thinking and organizing, particularly within Zapu. Vetoka campaigned against the street names in Bulawayo town and against the naming of Matabeleland's provinces (Matabeleland North and South), advocating alternatives, such as 'Mavetokaland, or Vetokamaland or Tokavemaland'.

This activism on the part of Matabeleland's minorities prompted some action by the state in the early 1980s, notwithstanding concerns that minority languages were divisive and would 'sow the seeds of disunity in schools'.[27] As early as July 1981, the Ministry of Education and Culture decided to allow 'all schools to use the local mother tongue from grade 1-3 and to switch over to either Ndebele or Shona in grade 4',[28] and a Select Committee was appointed in 1984 to look into the teaching of minority languages. The Committee toured the country, collecting evidence from districts where minority languages were an issue.[29] The outcome of the Select Committee's work was a resolution on the teaching of minority languages passed in 1985 by the Minority Languages Committee of the Ministry of Education and Culture. It stated that 'A minority language should be taught in grades 1-3 *excluding* [ie instead of] Ndebele/Shona'. In Binga, where there was overwhelming support on the council for this measure, the teaching of Tonga was quickly phased back into policy by the district education authorities.[30] Adopting this policy, however, was not the same thing as implementing it.

The concession to minority languages occurred at a time when ethnicity in Matabeleland was highly politicized. During the conflict of the 1980s, Zanu(PF) fanned anti-Ndebele sentiment to undermine Zapu. The ruling party tried to associate Zapu narrowly with the Ndebele, ignoring its history of multi-ethnic mobilization, and claimed that the Zipra guerrillas who fled persecution in the newly integrated army to take up arms in Matabeleland had support from Zapu and Ndebele civilians. As soldiers of the Fifth Brigade massacred villagers in Matabeleland, they justified doing so in terms of revenge for nineteenth-century Ndebele atrocities, and stopped at the borders of Binga and Hwange, where villages were Tonga and Nambya rather than Ndebele-speaking.[31] In this context, Matabeleland's minorities' complaints of Ndebele dominance were used by Zanu(PF) to build a constituency in a region where it lacked support. Senior Zanu(PF) politicians told Tonga leaders 'we have no quarrel with you, only with the Ndebele'.[32] Binga's Zapu leaders also found it useful to emphasize their Tonga identity to distance themselves from alleged support for dissidents and to try to harness state funds. Munkombwe recalled saying to Mugabe and other Ministers visiting Binga in 1981: 'look at the backwardness here, what did we go to war for? For the suffering and poverty caused by the colonial regime... I emphasized that we were Tonga and made a link with Mugabe, because he had taught in Zambia and had seen what the Tonga were capable of.' Councillor Munkuli recalled: 'When people were saying Matabeleland North is supporting dissidents we said, no not in Binga, we Tonga don't support that, we had a record of capturing and killing them. We told the Ministers that because others were trying to include us in that...'[33]

Binga's politicians achieved some state investment in Binga in the early 1980s despite the conflict. They interpreted this not as the state's rational bureaucratic response to objective need, but as a result of their own strenuous efforts, and particularly the strategy of circumventing provincial structures in Matabeleland. Time and again, Munkombwe organized delegations from Binga council to go in person to Harare and appeal directly to ministers and the president. Munkombwe explained: [34]

> I used to go straight to Mugabe, without an appointment, or to the ministers, knocking on their doors, first as council chair and later [after 1985] as MP, there were very few offices we didn't visit, presenting our causes. If the ministers didn't solve them, we'd go to the prime minister himself and he'd give a directive. I didn't go through the procedures, I didn't use the provincial structures, I took a short cut – I'm a politician not an administrator. The cake had to be shared, and Harare was where the cake was. Despite the problems with Zanu, we did well at that time.

These Tonga delegations were frequently arrested en route to Harare, such as in 1983 when trying to secure funds for Binga hospital, and again in 1984 when complaining that the road promised in 1981 had not been built. Yet both delegations were successful on their release from detention; in the latter case, Munkombwe secured a hearing with Mugabe who 'organized a meeting for us with the construction minister and all was signed in Harare'.[35] Realizing these and other projects around the district, once funds had been secured, depended heavily on raising local Tonga labour through food-for-work schemes, which was readily available due to severe droughts through the 1980s.[36]

Zanu(PF)'s strategy for trying to construct support in Binga depended on appealing to Tonga grievances, and claiming to be delivering development and cultural recognition, whilst also making threats against Zapu. The ruling party's key agent was a former teacher from Hwange, Jacob Mudenda, who had a Zapu history but also had long-standing connections to Zanu(PF), developed whilst studying at Kutama Mission in Mashonaland West, where Mugabe and other nationalist figures had done their schooling.[37] In the early 1980s Mudenda crossed the floor to Zanu(PF) and rose rapidly through both state and party hierarchies. He was promoted to a series of senior teaching positions, before being offered a job first as District Administrator in Binga, then as Provincial Administrator for Matabeleland North, and subsequently as Provincial Governor. In the run-up to the 1985 parliamentary elections, Mudenda (by then Provincial Administrator), also stood as the Zanu(PF) parliamentary candidate for Binga, in a constituency that covered a large swathe of the north-west. He courted his potential Tonga voters, claiming to be delivering on long-standing Tonga demands. As the council minutes of his address record:

> Tonga would be taught in every school... After independence in 1980, the district had 6 secondary schools and 52 primary schools in just five years. The development was done by the government and now the government wants representation ... The Tonga suffered enough

> during the racist regimes and cannot continue to suffer in free Zimbabwe. It was not a mistake to be born a Tonga. The translation of Tonga surnames into other languages should cease because that destroyed Tonga culture… Some of you might think Mudenda was tribalistic, but was the Prime Minister being tribalistic when he said Tonga should be taught in schools? From today headmasters should learn Tonga.[38]

But Mudenda also threatened Binga's Zapu councillors with massacres, warning them to:

> Behave like a chameleon and change with the times… Councillors should not support dissident activities… Do not canvas for Nkomo's party as others were doing. In Lupane and Tsholotsho they did it and people were dying… Those who do not change with the times will experience what those in Noah's time experienced. People were told to get into the ark but they refused. The floods came and they drowned... The meeting was advised that time was not on their side, one-party state was coming.[39]

As punishment for influencing voters to return Zapu in the 1985 elections, all councillors were arrested, detained and interrogated.[40] For some, this was not the first time, and more than one had been shot.[41] The majority decided to cross the floor to Zanu(PF), the exception being the council chair and Zapu MP, Munkombwe, who resigned his position on the council. During council elections in 1987, Binga council was closed again, and councillors were both threatened and 'encouraged' by drinks and a meal at Zanu(PF)'s expense at an exclusive tourist venue. These politicized interventions did much to discredit the council and the ruling party and, as Zanu(PF) tried to establish a shaky network in the district, many local intellectuals drew a distinction between their promotion of Tonga culture on the one hand, and Zanu(PF)'s tribalism and coercion on the other.

In the same period, though not primarily as a product of Zanu(PF) strategy, tensions did mount between Tonga and Ndebele in Binga. This was because the process of immigration into the Zambezi valley from other parts of Matabeleland gathered pace in the early 1980s. As the tsetse fly began to retreat and cattle-keeping was possible, Ndebele immigrants flocked into the relatively well-watered parts of the valley suited to commercial agriculture in a movement that was encouraged by flight from the Fifth Brigade massacres. This influx of Ndebele-speaking immigrants and refugees was greeted with hostility: migrants were accused of encouraging armed dissidents into Binga, subjecting locals to violence and enhancing the risk of state retaliation. Moreover, the newcomers looked down on the Tonga as primitive and refused to respect the local Tonga political leadership. The local council repeatedly and unsuccessfully tried to stop the 'flood' of immigrants through council resolutions to evict them on the grounds that the valley and its resources belonged exclusively to the Tonga by historical right.[42]

Although the ruling party had courted Tonga grievances to try to build political support, it did not deliver enough, and there was widespread resentment over the way Tonga grievances were used to further party political ends and personal careers. The council was discredited not only by its subjection to Zanu(PF), but also by scandals over the diversion of donated food and other allegations of corruption. The Unity Agreement of 1987, through which Zapu was absorbed into the ruling party, brought a temporary end to heavy-handed political intervention, but it did not rejuvenate the council. After being forced into the Zanu(PF) fold, individual councillors were accused of diverting food aid, and eating at the expense of those they were supposed to serve, while the district's Zanu(PF) patron was himself amongst the politicians accused of corruption in the national scandal over the sale of subsidized motor vehicles – known as 'Willowgate' – in 1988.[43] Further, the signing of the Unity Agreement was coupled with a narrative of reconciliation between two warring tribes, which once again excluded the Tonga and other minorities. Mugabe, for example, referred to the agreement as 'a charter which would bind once and for all, the two major tribes of Zimbabwe, namely the Shona and Ndebele, into one… The Unity Accord thus forms the bedrock upon which peace, democracy, social justice and prosperity should be built.'[44] Cast in this way, the politics of unity overlooked the ethnic minorities, excluding them from a place in the nation and denying them democracy, social justice and well-being. Yet discrimination against minorities was not only about definitions of the nation, it was also about state bureaucratic practice. All over the country, the elected councils were structurally weak not only in relation to Zanu(PF), but also to the powerful line ministries. In terms of language, the Ministry of Education was particularly important.

The 1987 Education Act, which formalized policy on language teaching in schools, did so in a way that minority language activists saw as a step backwards from the advances of the early 1980s. The Act allowed for the teaching of minority languages not as an alternative to Ndebele or Shona (as stipulated in the 1985 circular) but alongside it. Primary school children in the first years of school were to be bombarded with three different languages all at once. This provoked a furious response from those now legally defined as minorities. The chair of Binga Tonga Language Committee explained: 'Imagine the confusion in a small child who comes to school for the first time and is introduced to three different languages. They end up confused, and disadvantaged. We joined together with other minority groups to argue that Ndebele should be removed… Being labeled a minority is discriminating'.[45] Quite aside from the educational disadvantage, the policy was seen as 'cultural assassination'. As Munkombwe argued, 'you kill a child culturally as well as educationally if you teach in Ndebele – the Ndebele want to do to us what they did to the Kalanga, to swallow us up. We're not being tribalistic, we can't rest until we are recognized, we need to have a rightful place in this country as Tonga.'[46]

Vetoka also campaigned against the Act, casting tribes as a biblically ordained natural order and framing their opposition in terms of state rhetorics of 'unity':

> Vetoka language groups are treated as 'language slaves' in Zimbabwe… The Vetoka language groups are for *political unity*. We do appreciate and cherish the political unity signed by our leaders… we would however, strongly object and resist any process devised by anybody towards the extinction of our Vetoka languages in the guise of Unity. For we are not man-made or man-formed like political parties. We are divinely made or divinely formed and should never be dominated, discriminated and suppressed disguisedly in the name of political unity in the land of our ancestors… The creation by God of his declaration through Jesus Christ before his Ascension that people should speak different languages was not by error. This is supported by the world body that 'Each man is entitled to his language and culture and its development.'[47]

A further problem was the rate of progress in introducing the minority languages (whether instead of or alongside the 'majority' ones), the obstacles to which activists explained not in terms of party politics, but tribalism within the Ministry of Education, whose teachers, headmasters and officials were seen as intolerant of diversity.

As one Binga councillor argued, 'Some people at the top [of the Ministry of Education] are Ndebele and Shona, they are so tribal they don't want Tonga to be taught. Mugabe is not the one blocking that... They came with a lot of technical arguments – teachers, books, what not...'[48] The Binga council deemed progress since 1985 to be slow to the point of 'failing'[49] and organized a petition against the local District Education Officer and headmasters, and sent a delegation to the Regional Director of Education in Bulawayo. Vetoka resolved to print books themselves, given the slow progress on the part of the Ministry.

In 1990, the Binga District Council conducted a survey which showed 75 per cent of Binga schools were still not teaching Tonga in grades 1-3, due to 'resistance by some heads of schools, the shortage of trained Tonga speaking teachers and the shortage of teaching/learning materials.'[50] The Curriculum Development Unit of the Ministry of Education reached very similar conclusions, was highly critical of the pedagogic effects of introducing three different languages simultaneously in the first three years of school, raised questions about why texts had not reached schools in sufficient quantity and pointed out widespread ignorance about the new policy on the part of teachers and headmasters. The CDU recommended delaying reading and writing in Ndebele and Shona 'until the learners have mastered these skills in their respective minority languages. At this stage the teaching of Ndebele/Shona should be restricted to oral/aural skills.' The report also outlined the need for further in-service training for teachers and clarification of policy grey areas for district education officers, headmasters and PTAs, and recommended greater cooperation and support for local cultural associations and district writers groups by education officers.[51]

By the end of the first decade of independence, the campaign for development and cultural recognition of minorities was still far from reaching its goals. The unity agreement had changed the national political stage, but had not rectified the marginalization of the country's ethnic minorities. Moreover, the discrimination attached to the label 'minority' was now legally enshrined in the 1987 Education Act, and introducing the languages in school had met with a range of seemingly intractable obstacles. The construction of schools, clinics and a hospital, though welcome, was far removed from the ambitious schemes of the Lusumpuko plan and still left Binga along with other districts of Matabeleland North lagging behind the rest of the country. Binga continued to rank poorest in surveys of poverty; many of its inhabitants continued to suffer hunger, and remained dependent on food handouts.[52] The movement to gain recognition for the Tonga and development for Binga district thus continued in the 1990s. It did so, however, in a changing national political and economic context, which demanded new institutional strategies.

NGOs & the Politics of Recognition

The politics of recognition changed in important ways over the course of the 1990s in the context of neo-liberal reforms, which drastically curtailed central government funds to local councils, forcing them to seek new sources of finance locally through an unpopular combination of elevated taxes, user fees and partnerships with the private sector. NGOs achieved a new prominence nationwide and, in Binga, where they had always been a significant presence, their role was similarly elevated. New donor agendas of state retreat, democratization and civil society were influential, shaping priorities and projects on the ground. Other changes also affected the campaign for development and cultural recognition in Binga: in the 1980s, there had been no more than ten Tonga speakers in Zimbabwe with O level qualifications or above, but by the second decade of independence, the secondary schools constructed in the 1980s had begun to produce Tonga graduates. The 'pioneers' of Binga secondary school were a highly motivated group, who had begun to involve themselves in the development of the district even when still at school, through the Binga Youth Progress Association.[53] By the 1990s, a significant number of these pioneers who had gone on to do A levels, diplomas or university level qualifications returned to Binga to help their own people, taking up professional jobs in the local state or with NGOs. Efforts to deliver on long-standing grievances were thus rejuvenated by the energies, concerns and strategies of a new generation of educated Tonga activists.

The new solution to Binga's lack of development was an 'indigenous' development organization – the Binga Development Association (Bida) – which was founded towards the end of 1989. Bida very rapidly achieved a high profile within the district and began to usurp the central place that the council had occupied in the previous decade. The rise and fall of Bida over the decade of the 1990s is a fascinating and sad story, with lessons for those who romanticize local NGOs as an alternative to local state institutions or international NGOs, and see them as necessarily more accountable to grassroots constituencies and less amenable to politicization and corruption.[54] The conflict over Bida was in many ways a microcosm of unfolding national politics in Zimbabwe, as challenges to its increasingly corrupt management mirrored challenges to the ruling party's authoritarian politics of patronage on the national stage through a popular counter-mobilization around issues of civic rights and democracy.

Bida was registered as a non-profit organization and run by an elected committee of voluntary executives. The first chair, Joseph Machina, worked for Ministry of Social Welfare in Binga, the secretary worked for Christian Care, and the remaining officers were employees of other international NGOs, council executive officers or teachers. Bida combined developmental and cultural goals, aiming to help the district 'catch up for 100 years of neglect'.[55] At its opening ceremony, Bida's secretary explained that despite its abundant resources, 'Binga district was one of the worst underdeveloped districts in the country… Nationally, the mention of Binga in public raises stares and even among the Tonga themselves … [who] used to hide their identity as Tongas.'[56] Bida's logo featured 'blue: symbolizing the lake; green symbolizing agriculture and red symbolizing the suffering experienced in the development process of the district, plus the Tonga double handed hoe.'[57] Its projects were diverse but water was the first priority, focused on a 'Zambezi water scheme'.

In terms of its mode of operating, Bida combined an elected, voluntary decision-making structure, with salaried professional managers and a grassroots membership; chiefs were co-opted onto committees and senior figures were brought in as advisors and patrons, such as Matabeleland North Governor, Jacob Mudenda. Bida's structure in many ways copied that of the council, though its functioning was envisaged as diametrically opposed to it.[58] Bida's membership paid an annual fee of $5 for the employed or $1 for the unemployed, and its constitution provided for regular elections to committees at all levels. Bida defined its way of working as bottom-up and apolitical, in contrast to the council, which was criticized for its top-down implementation of party/state directives.[59]

Bida was stunningly successful in securing funds from NGOs, which were fed up with working through the council. Channelling funds to an indigenous development organization fitted broader shifts within the donor community in support of activist NGOs, which it was believed would strengthen civil society and contribute to democratization. It provided an attractive alternative to attempts to work directly through or to reform the state. By 1991, Bida had sent out more than 30 project proposals to donors, most of which were financed, and some were notably successful.[60] Bida was first housed in the council offices but a Novib grant allowed the Association to move into a purpose-built building, and allowed for the future construction of a community hall, a resource centre and library.[61] The salaried staff also grew quickly through donor support, to include not only a full-time coordinator and other managers, but also a network of paid 'community mobilizers'.[62]

However, the rapid development and success in securing external resources quickly led to problems.[63] As early as 1994, there were already conflicts between individual Bida staff and between Bida and the council, which (accurately) felt 'sidelined'.[64] Bida's internal problems began when a group in the management committee tried to take over unconstitutionally and attempted to alter the bank signatories, which they succeeded in doing by rigging the 1996 AGM elections.[65] Though the Department of Social Welfare declared the elections invalid, the new management continued in place, and began to transfer Bida assets to individual pockets, inflated staff salaries threefold, and appointed relatives and friends to new Bida posts (including a new director, who left a well-paid job in town to take up Bida work).[66] This state of affairs did not go unchallenged, as 'disgruntled members' became increasingly organized: they pressed the Ministry of Social Welfare to nullify the 1996 Bida elections 'to save the organization from loss of donor confidence and support, loss of grassroots support, loss of experienced personnel and plunder of the association's resources', and pointed the finger at three individuals – two of them local Zanu(PF) politicians, plus the Provincial Governor.[67] They mobilized Bida members to sign a petition calling for properly supervised elections, access to a critical donor evaluation, details of appointments and an external audit.[68]

The two groups came into head-on conflict at the 1997 AGM. The incumbent management committee were accused once again of manipulating the elections, by providing food and transport only to members from areas where they had support, refusing to allow critical questions, and using eulogies to the Governor – involving singing, clapping and stamping – as a means of stifling debate. The 'disgruntled members' (plus 227 supporters) tried to boycott the vote, but the meeting turned to chaos and violence after the local Zanu(PF) MP locked the gate to prevent their departure, and the police had to intervene to restore peace.[69]

The conflict over Bida had a greater significance than simply the fate of the organization itself, according to the disgruntled members, because the same small yet well connected group that had taken over Bida were also trying to take over the council and the local Zanu(PF) structures.[70] The scale of the asset-stripping was also significant: by 1997, the management had transferred to a 'ghost welfare fund' two tractors and tractor accessories; $50,000 from the Bida education fund; $80,000 from Bida's main account; and the proceeds from businesses based on Bida refrigerators (kept in individuals' homes) and Bida retail outlets.[71] A review concluded that the internal and external conflicts had cost the organization around Z$15 million worth of development funds.[72] The critics of this use of Bida for private gain succeeded in securing donor withdrawal and the organization was paralysed by the two competing groups.[73] Yet the conflict had broader effects – it had stimulated critical popular debate over governance, democracy, constitutions, transparency, members rights and the proper management of elections. These issues were discussed in villages throughout the district in efforts to mobilize the grassroots membership, and extended beyond the issue of Bida itself to a critique of the operation of the local council and ruling party.

With Bida's demise, development and cultural initiatives in Binga continued in a less coordinated manner, with local and international NGOs running a diversity of projects, sometimes in partnership with the council or other local groups, and sometimes independently. International NGOs continued to provoke criticism from the council, for 'pursuing hidden agendas and goals'.[74] Local intellectuals argued that NGOs were uncoordinated, operated from distant headquarters, and ran projects with short life-spans staffed with outsiders.[75]

The council, for its part, became dependent on revenue from wildlife industries and tourism, particularly through the donor-funded programme known as Campfire. These and other commercial ventures were boosted in 1990 by the tarring of the road linking Binga to Matabeleland's major transit artery from Bulawayo to Victoria Falls. Yet the income from tourism, wildlife and other natural resources was curtailed by adverse systems of revenue sharing, such that profits from state land including National Parks, Safari Areas and land designated for tourist development accrued to central rather than local coffers. Partly for this reason, the council was suspicious of the flood of applications that came in for tourist lodges, hotels, safari leases and the like, and tried turn down proposals which did not take the form of joint ventures with the council or did not provide for significant profit sharing.[76] Their caution led visiting Minister of Local Government Msika to berate the council for obstructing new developments.[77] The council chairman complained the council was 'spending a lot of its precious time on state land issues but without getting anything in terms of revenue... all monies for rents and title deeds were channeled directly to Central Government'.[78] Moreover, many of the developments that were given the go-ahead on council land resulted in protracted disputes not only over profits, but also over conflicts with local communities and workers about land, resources and pay. As we shall see in the following chapter, the issue of who benefited from developments around the lake was particularly sensitive.

Despite these drawbacks, tourism brought jobs and finance into the district, and also provided new opportunities for the display and marketing of Tonga heritage to visitors. Aside from the various informal activities and the proliferation of roadside markets, a Tonga museum and craft shop was organized through the council (with assistance and funds from Danish volunteers), and the sale of Tonga crafts outside the district was increasingly well organized. A local NGO, the Kunzwana Trust, developed a 'Tonga tourism project', which popularized the tragedy of a people who, 'once deported from their habitat, were abandoned', yet retained a 'vibrant, exotic, and ancient culture.'[79] Private lodges and safari operators also increasingly marketed the cultural as well as the 'wilderness' attractions of the Zambezi valley. Just as the white popular writers of the 1950s elevated the Nyaminyami myth in their story of the dam (and seemed to need it more than those who were displaced), the icon and related tales were now marketed to tourists along the length of the valley. Travel Africa, for example, promoted Nyaminyami as the Tonga River God, and compared the mythical creature to the Loch Ness Monster,[80] Safari Africa details the story and symbolism of

8.1 Nyaminyami carving (courtesy of BaTonga Museum, Binga)

Nyaminyami, 'protector of the people of the Zambezi Valley', with

> a body like a snake and a head like a fish and no one knows how big he is, for he never showed himself in full display. But he is very big! The people of the Zambezi valley in Zimbabwe were protected by Nyaminyami, their ancestral spirit... The people pledged their allegiance to him by performing ceremonial dances ... One season ... the white man came to build a wall. It took five long years to see it through because Nyaminyami did not want to be disturbed. He caused some floods and loss of life, but at last he was kind enough to let the wall to be all complete. It is also believed that the occasional earth tremor felt in the lake surroundings is caused by this spirit. It was the work of the Tonga elders and their medium spirits to persuade the Nyaminyami to allow the Zambezi to be tamed.

In terms of Tonga heritage and its promotion to local, national and international audiences, 1997 was a particularly important year. It was the fortieth anniversary of the displacement from the river, and was marked by a series of commemorative events. In Binga itself, local leaders from the council and Bida organized a cultural celebration to remember this foundational event in modern Tonga identity, which involved speeches by politicians, and performances by local drumming teams.[81] Due to NGO funding, the year of the displacement was remembered not only in Binga, but also outside the Zambezi Valley, in Bulawayo, Harare, and Europe. The Kunzwana Trust organized 'The Nyaminyami Festival', celebrating 'Forty Years of Cultural Survival', which involved the display of paintings and artifacts, and the performance of Tonga drum teams at the Bulawayo Gallery and the Zimbabwe International Book Fair in Harare.[82] Also on display was an exhibition created on the initiative of the Bulawayo

Museum and the Choma Museum in Zambia entitled 'Batonga Cross the Waters'. Thanks to Austrian funds, displays and a local drumming team also toured Austria, and an internet site was constructed to reach broad international publics.[83] Alongside the exhibitions and performances, there was also a human rights seminar, hosted in the Alliance Française, entitled, defensively: 'The Tonga Have Ten Toes'. The seminar flier announced:

> The rights of cultural minorities are often underplayed in Zimbabwe in the interests of promoting national unity and the political requirement that all Zimbabweans conform to one set of cultural values. The Valley Tonga are the third largest ethnic group in Zimbabwe, but even so, their number amounts to far less than a couple of hundred thousand... The Valley Tonga have no voice and they suffer all the setbacks of being a small number of rural people in a remote corner of a country. They are the victims of myths, ignorance and prejudices which portray them as dangerous, deformed and primitive.[84]

The emphasis placed on past relations with the river and the history of displacement in these cultural projects is striking. This was not just a backward looking exercise remembering what was lost, nor was it only a celebration of cultural survival. Rather it was also a reminder of the broken promises by Native Commissioner Cockcroft and the colonial state, of the failures of the post-independence state, and of Tonga rights to compensation and development. Memories of past relations with the river and of the broken promises have been invoked in more than rhetorical gestures and performances such as the cultural events described above. For example, the Catholic Commission for Justice and Peace conducted an investigation of the problems leading to the closure of a small irrigation scheme in chief Siansali's area of Binga District. They reported that when the Lungwalala dam was built in 1992 by the government, the people regarded it as compensation for their removal in 1957. For the first two years they had not been charged for water, but then payments were demanded, and 'people could not understand why, as the DC had promised them that water would follow them to the new areas where they were made to settle'.[85] This attitude had contributed to the refusal to pay rates, and the failure of the scheme.

The use of the history of the displacement as the basis of a right to development was boosted through recognition by interstate and international bodies. In 1995, the Zambezi River Authority (ZRA) the interstate body responsible for managing the dam, recognized that 'there was insufficient time allocated to the planning and implementation of the resettlement programme, and that insufficient resources were made available for the massive exercise'; it acknowledged the absence of compensation in Zimbabwe and the 'grossly inadequate' sums paid in Zambia.[86] The ZRA undertook to raise compensation funds, through golf tournaments, handbooks and videos on Kariba history and the resettlement, which raised Z$0.5m by 1999. The money was used to buy grinding mills in Kariba and Binga districts, though the amounts involved were seen as an insult by local leaders. Thereafter, the Zambezi Valley Development Fund was launched to raise funds through donations and a 1 per cent levy on water bills for water used in generating power. Perhaps more important than the sums raised, was the recognition these initiatives gave to past injustices as the source of rights to development and reparations for 'dam-induced grievances'. A major evaluation of Kariba by the World Commission on Dams welcomed the ZRA initiative: 'What counts in this initiative is not so much perhaps the size of the fund, but the recognition of the need for reparation and the very act of initiating a reparation fund. One hopes that both the financier, the World Bank, and the governments of Zambia and Zimbabwe who were parties for the Kariba Hydro-Electric Scheme, will follow suit with even more meaningful projects which will have significant positive impacts on the lives of those displaced.'[87] Although the Zambian authorities and the World Bank recognized these historical arguments as the basis for action, and initiated a new US$26m rehabilitation project for the resettled on the north bank, the Zimbabwean state did not, and development initiatives on the south bank – such as Campfire and poverty alleviation programmes – were implemented on the basis of national ministerial directives and council policies.[88]

In the light of the Zimbabwean state's inaction, Tonga grievances over developmental neglect and lack of recognition were as deep as ever. Given the weakness of the council and the paralysis of Bida, the most important local forum for discussing them was a new Binga office of the Catholic Commission for Justice and Peace (CCJP), founded in 1995.[89] The CCJP had a long interest in human rights and development issues and also had broader roots in society in the north-west thanks to the prominence of the Roman Catholic church. Its Binga office was staffed with Tonga intellectuals, including some of the 'disgruntled critics' who had tried to put Bida back onto a developmental path, and who were still determined to find a way to redress the combination of poverty and stigma that the Tonga faced in Zimbabwe. The CCJP began to facilitate discussion at community level around the district, encouraging people to conceptualize their problems as human rights issues and set up a range of developmental and advocacy initiatives, from water and community development projects to new advocacy campaigns.[90]

The CCJP office rejuvenated lobbying on the issue of minority languages and education.[91] Its research in Binga showed huge problems: staffing levels in schools were critical as qualified teachers continued to 'shun the area because it has been over publicized as underdeveloped and "remote"', and enrolment in secondary schools, which had been increasing, was reversed towards the end of the 1990s. Moreover, 73 per cent of the district's schools still did not teach Tonga at all, and revisions to the Education Act in 1996 had not changed the law on minority languages. CCJP researchers cast these grievances as human rights issues, framing them in terms of the constitutional debates then raging in Zimbabwe:

> Section 62 of the Education Act may contravene section 23 of the Constitution of Zimbabwe… The 'minorities' (the Tonga included) are being discriminated against on grounds of tribe. They are being subjected to a restriction (of learning their languages to a limited level, that is, only in primary schools) to which other persons of another description (Ndeble/Shona) are not being subjected. Furthermore, the government has accorded to persons (Ndebele/Shona) a privilege or advantage which is not accorded to the first mentioned description (the 'minorities' – the Tonga inclusive). It is in this view that some legal experts argue that section 62 of the Education Act may contravene Section 23 (2) (a) (b) which prohibits the discrimination of people on grounds of race, tribe, etc. The restriction of the teaching of the 'minority' languages to primary schools, while the 'majority' languages are learnt up to universities, is itself a discriminatory move. It is therefore suggested that the Tonga together with other 'minority' language speakers in Zimbabwe should lobby the government to amend the education Act so as to accommodate them.[92]

Framing the issue of minority rights in this way gave the arguments more weight, by appealing to universal principles and thus a broader constituency than simply the five districts with significant minority populations. The CCJP was a professional human rights organization with salaried staff, was well connected to the increasingly organized network of human rights, civic and advocacy organizations operating in Zimbabwe and was a key player in stimulating national debates over citizenship. It provided the funds to facilitate a series of workshops to bring together the country's minority groups in 2000-2001. As a result of these meetings, resolutions were passed to lobby parliamentarians on a range of education issues, from the need to amend section 62 of the Education Act, to the banning of the term 'minority' and its replacement with the term 'indigenous languages'. A new national organization was also formed to pursue issues regarding the country's indigenous languages, named the Zimbabwe Indigenous Languages Promotion Association (Zilpa).

From the late 1990s, the constitutional debate stimulated by the National Constitutional Assembly (NCA) provided a receptive context in which to air grievances over the law on minority languages. Joint action on the part of the minority groups had always been weak in the past as, even together, they could muster no more than five MPs who had any interest in the issue.[93] But framing their concerns in terms of rights and citizenship provided a link to a national movement to construct a more inclusive definition of the Zimbabwean nation and to build more accountable national institutions. The NCA had tried to incorporate rural areas and marginalized social groups through an outreach programme that deliberately targeted special interest groups, such as women's and gay activists, as well as 'indigenous language' organizations, and printed its information leaflets in the minority languages and not only in Ndebele and Shona.[94] In order to be seen to be responding to a popular cause, the government tried to take the lead on the constitutional debate through its own Constitutional Commission, which initiated a rival process of public consultation throughout the country, including in remote rural areas such as Binga. Yet the final draft constitution did not accurately reflect the public opinion the Commission had drawn upon, revealing the government-led process as disingenuous. Its exclusions were many: restrictions on presidential terms of office did not apply to the incumbent president, the labour movement's demand for the right to strike was ignored and minority activists' call for the abolition of the label 'minority' and the recognition of 'indigenous languages' in schools and the media was excised (despite appearing in earlier summaries). The rejection of the draft constitution in a national referendum in February 2000 was both a vote of no confidence in the government-led process of constitutional reform and a more general protest against an unpopular government. [95]

This disaffection translated into support for the new political opposition, the Movement for Democractic Change, formed in late 1999 out of the labour and constitutional reform movements. Like the NCA, the MDC appeared receptive to minority activists' complaints and took care to publish election materials in minority languages. In Binga, the local leaders who had led the campaign against corruption had initially tried to work through the ruling party's own political structures, particularly the Ministry of Youth. But by 1999, they were thoroughly disillusioned with the possibility of reform from within, and felt that the MDC would be more likely both to respond to the district's specific needs and to deliver political and economic reforms more generally. In the June 2000 parliamentary elections, Binga delivered the highest opposition vote of any rural constituency in the country. Those who campaigned for the MDC in Binga did not have to mobilize from scratch, but could capitalize on their history of campaigning for improved governance in the council, and for cleaning up Bida and reforming Zanu(PF) from within.

In the wake of the parliamentary elections, and in the run-up to the 2002 presidential elections, the ruling party once again embarked on a heavy-handed effort to bring local authorities under Zanu(PF) control.[96] These interventions were more far-reaching than similar measures in the 1980s, and extended well beyond interference with the district councils. From 2001, local Zipra war veterans based in Binga were used by the ruling party as a proxy force to initiate a process of purging the local state of individuals accused of supporting the MDC – this involved threats to the lives and livelihoods of councillors, council executives, and officials in the Ministry of Local Government and line ministries, as well as teachers and nurses, and public employees at all levels down to community workers, cleaners and security guards. Local NGOs were also threatened, and the CCJP office was singled out for closure, due to its effective role in civic education and creating 'an informed and alert electorate'.[97] Although Zipra war veterans were instrumental in staging the 'demonstrations' through which offices were closed and individuals accused, they worked together with local party structures, and police and the intelligence services, with back-up from provincial level party structures and the Matabeleland North Provincial Governor, Obert Mpofu. The war veterans tried to give their actions popular appeal by raising a range of genuine local grievances against the local institutions they targeted. Thus they locked the gates of the council on the grounds that it was becoming 'the headquarters of the MDC', but also raised issues of corruption, demanded the revisiting of a controversial council audit, complained about the employment of non-Tonga executives, raised concerns over tendering procedures, and complained about the distribution of revenue from lucrative wildlife and tourism projects run by the council.[98] But the political goals of these demonstrations, closures and expulsions were transparent as key MDC activists were the primary targets, and the interventions had counterproductive effects in relation to the problems they purported to address. Rather than redressing concerns about corruption and poor local governance, the expulsions entrenched relations of patronage: qualified professionals lost their jobs through illegal procedures and were replaced by members of the ruling party, war veterans and their relatives.[99] The district lost its cadres of educated Tonga public servants, many of whom were forced into the diaspora, and the capacity for professionalism within the local state was drastically undermined.

Zanu(PF)'s emphasis on the politics of land and race also failed to win local support in Binga, as it had little resonance with local concerns. People in Binga had not lost land to white settlers, but to state conservation bodies; they did not want to be resettled on former commercial farms but wanted access to state resources and reparations for unsettled grievances relating to the dam and displacement. In the words of a report authored by Bernard Manyena, one of Binga's intellectuals, Zanu(PF)'s land reform was notable for 'missing the Tonga'.[100] One of the report's informants explained: 'the people of Binga have got different needs from the rest of the country. We are still suffering from the shock of the traumatizing forced resettlement from the present day Lake Kariba. Our farm is Lake Kariba. We have unfinished business in Binga.'[101] Zanu(PF)'s politics of land and race gave the Tonga new reasons for feeling excluded, and made matters worse through its negative effect on the national and local economy. Punishing Binga's residents by interrupting and withholding crucial supplies of food had the effect of deepening poverty and hunger. The ruling party's politics of patriotism had also eclipsed the work of activist NGOs such as the CCJP, which had tried not only to transform material deprivation in the valley, but also to stimulate debate over citizenship and rights.

Conclusion

This chapter has described the motivation and strategies of activists who have tried to create a place for the Tonga in Zimbabwe. The politics of recognition over more than twenty years since independence involved demands for a remarkably consistent set of issues: investment in development infrastructure and other initiatives to redress poverty and hunger, and recognition for Tonga language, culture and history in the Zimbabwean public sphere, particularly in school curricula and the national media. By 2000, the articulation of minority rights appeared to be gaining some headway through the work of human rights NGOs and the incorporation of local issues into broad national debates over citizenship, civic rights and the constitution stimulated by the NCA. Binga had become a globalized hub, through networks of communication developed by NGOs and tourism, providing opportunities to promote Tonga history and grievances, as well as publicize Tonga culture and heritage. Yet thereafter progress on the combined material, historical and cultural concerns of Tonga activists was abruptly brought to a halt. The ruling party's emphasis on 'patriotic history' and the politics of race marginalized other debates, and silenced discussion of rights, whilst developmental initiatives were undermined by the politicization of the local state and increased controls on NGOs. The national economic crisis was experienced locally through the elimination of tourism, the withholding of food, and the reduction of Binga's inhabitants to hunger more far-reaching than at any time since the war of the 1970s and the aftermath of the displacement from the dam.

The silencing of debate over minority rights in the context of patriotic history is not the product of any intrinsic conflict between the promotion of local cultures and an authoritarian, militarized patronage state promoting the values of Africanism and counter-imperialism. Indeed, as we have seen in the course of the chapter, ruling party patrons have at different times tried to build constituencies by courting Tonga grievances. Yet patronage politics is inherently weak and unstable, especially in a context of economic decline or in marginalized and politically insignificant constituencies. In such contexts, it cannot deliver the material transformations that are a crucial component of social inclusion, and patrons lack the political weight and motivation to influence national law and policy.

This chapter has focused on the motivations of the elite district leadership, the politics of the networks linking local, national and international institutions, the shifting character of the local and central state, competing definitions of the Zimbabwean nation and the role of language as a source of discrimination. We have already begun to see how the politics of land and race 'missed the Tonga' and left them with 'unfinished business' regarding access to the lake. The next chapter reveals what this meant in terms of the political economy of fishing and daily life along the border at a time of mounting economic crisis. As we shall see, the governance regime over lake and border can cast further light on the nature of the state and the way it was reconstituted from the late 1990s.

Notes

1 Englund and Nyamanjoh 2004, Cowan et al. 2001, Berman et al. 2004.
2 Solway 2004, Werbner 2004.
3 E.g. Berman (1998) on Kenya.
4 Berman et al. 2004.
5 Muzondidya 2005.
6 The only Zimbabwean text on this topic is Hachipola 1998. On language and exclusion in other contexts, see Bamgbose et al.,1994, Bamgbose 2000.
7 *The Herald*, 12 and 16 October 1980.
8 See district drought assessment for 1982, discussed in Binga council meeting of 5/3/82. More generally, see drought committee file, Binga, 1982 onwards. On chronic hunger through the 1908s, see Reynolds 1991:xxiv, 19.
9 See for example *The Herald*, 24 January 1981.
10 On early NGO projects, see RDC council records, including Social Development File, Meeting of organizations working in Binga, 21 April 1983; district team minutes, 1981-1982. One outcome of NGO funding in support of culture and history, initiated in 1986, was the children's book, Reynolds and Cousins 1993.
11 Over and over again, the council minutes record anger at NGOs distributing food and setting up projects without consultation. In 1982 the Binga council resolved, typically (and ineffectually) in the wake of one such debate, that: 'all donations to the district must be made open to the council. The donor's priority might not be the council's priority, in which case the council reserves the right to channel the funds to the urgent projects'. District team minutes, 26 January 1982.
12 *The Herald,* 24 January 1981.
13 *The Herald*, 19 October 1981.
14 Ndebele teacher who had worked in Binga in the early 1980s, interviewed in Lupane 25 February 1995.
15 Teacher, Sianzyundu, cited in Dahl 1997:223.
16 Andrew Sikajaya Muntanga handed over to Francis Munkombwe having been voted out in 1981 on allegations of financial mismanagement.
17 Binga District Lusumpuko Plan. Compiled by the Lusumpuko planning committee, and proposed by full council 11 December 1981.
18 Lusumpuko Plan, 11 December 1981.
19 The campaign for greater access to natural resources in the district, particularly around the lake, is covered in Chapter 9.
20 Francis Munkombwe 15 March 2001.
21 NGOs also helped: the Zimbabwe Trust, for example, tried to engage the University of Zimbabwe's Department of Linguistics in the production of Tonga textbooks. BDC, Social Development file, minutes of meeting held on 21 April 1983.
22 Binga Distrist Council minutes, 12 September 1983. On Tonga settlement in Matabeleland, see Alexander et al. 2000:chapter two.
23 On the Nambya Cultural Association, see McGregor 2005a.
24 Interview, Francis Munkombwe, 15 March 2001.
25 Interview, Francis Munkombwe, 15 March 2001.
26 On debates over ethnicity and nationalism in Matabeleland, see Ranger 1994..
27 Letter, Minster of Education and Culture to Kalanga Language Promotion Society, Bulawayo, 6 July 1981, Vetoka MS 936. Concessions were made on the issue of language, though not on the teaching of different local history in different schools, as 'all significant local history should be incorporated into a general History of Zimbabwe which will be taught in all schools'.
28 Letter, Minster of Education and Culture to Kalanga Language Promotion Society, Bulawayo, 6 July 1981, MS 936.
29 In Binga, the council presented the Select Committee with a list of demands: for a full-time Tonga announcer on Zimbabwean TV and radio, and for the teaching of Tonga at all levels not only in Binga, but in schools nationwide. Aside from the issue of the Tonga language, councillors complained that there was discrimination against Tonga people, as there was still no school in the district teaching form 4, the Ministry of Local Government imposed non-Tonga staff, prevented the deployment of Tonga civil servants, and left schools partially built and understaffed. Council minutes, special meeting discussing points for the Parliamentary Select Committee, 28 May 1984.
30 Curriculum Development Unit, Ministry of Education and Culture, *A Report on the Survey of the Teaching/ Learning of Minority Languages in Zimbabwe,* April 1990. Copy held by CCJP, Binga.

31 Alexander et al., 2000:chapter nine.
32 Interview, former council chair, Binga, 12 March 2001.
33 Interview, former councillor James Munkuli, 3 April 2001.
34 Interviews, former councillors and council chairmen. See particularly, interview with Francis Munkombwe, 15 March 2001.
35 Interview, councillor Saina Muntanga, 12 April 2001.
36 Interview, Francis Munkombwe, 15 March 2001.
37 Mudenda had further links with important Zanu(PF) figures from his time in teaching at the Marist Brothers' college in Hwange in the 1970s, when one of his colleagues had been Canaan Banana, Zimbabwe's first President.
38 See Binga District Council records, minutes of council meeting 22 February 1985, 12 September 1985.
39 Binga District Council, minutes of 22 February 1985.
40 Interview, former councillors, 15 and 26 March 2001, 12 April 2001.
41 Interviews, former councillor, 26 March 2001.
42 See for example, council special report on illegal settlers, December 1984 (held in file of council minutes). Conflicts between Tonga and Ndebele settlers are detailed in Dzingirai 2003, Dzingirai and Bourdillon 1998.
43 The allegations against Mudenda were not proved in the courts, and his political career with the ruling party survived, indeed flourished, in the 1990s and beyond.
44 This was repeated in Zimbabwean history textbooks. Cited in Teresa Barnes 2004:142.
45 Interview, Duncan Sinampande, 7 September 2001. A councillor recalled his own experience: 'I suffered that confusion, as a child, the experience of a child who cannot understand the teacher and the teacher who cannot communicate with the child, I remember how English and Ndebele were all one language to me at first.' Interview, councillor, 26 March 2001.
46 Francis Munkombwe, 15 March 2001.
47 Vetoka Society to Fay Chung, Minister of Education, 15 May 1989.
48 Interview, councillor James Munkuli, Kariangwe, 3 April 2001.
49 Council minutes, 15 May 1987.
50 CDU, Ministry of Education and Culture, 'A Report on the Survey of the Teaching/Learning of Minority Languages in Zimbabwe', April 1990. Copy held in CCJP office, Binga.
51 Ibid.
52 See for example Warndoff and Gonga 1994.
53 Interview, Fanual Cumanzala, 15 April 2001.
54 An excellent account of the problems with this view of local NGOs is in Rich-Dorman 2005.
55 Interview, former Bida management committee member, 17 March 2001.
56 Bida minutes 2 September 1989. Held in Bida file, DA's office, Binga District [hereafter Bida file].
57 Bida minutes 11 November 1989, Bida file.
58 Bida's central district management committee delegated to specialized sectoral committees (for culture, education, water, etc.), and below district level, operated through a hierarchy of regional committees and below them forty area committees.
59 Minutes 25 May 1990, Bida information sheet.
60 Early successful projects, including the construction of Chipangawodi at Binga hospital (accommodation for relatives of those admitted), teachers' houses and scholarships for a handful of students who today are university graduates.
61 Bida minutes 21 April 1990, Bida file.
62 Aside from Novib, other funders included SCF(UK), the Binga RC church, SNV, CADEC, Christian Care and SIDA. Details in the Bida minutes, 1989-1994, Bida file.
63 This narrative of Bida's internal problems that follows is based on the accounts provided in several letters and petitions: a letter from ousted Bida treasurer, G. Nyathi to the Provincial Social Welfare Officer, 10 March 1996, a petition to the Minister of Public Service and Social Welfare, 10 February 1997, and an undated letter from Joel Gabbuza, Bida chair until September 1996 (ca. August 1997). All held in Bida file.
64 Council minutes 28 March 1994; Letter from CEO to Mr Muntanga, Bida 14 August 1996, Bida file.
65 Petition, 10 February 1997, Bida file.
66 Letter, former Bida chair, 14 August 1997, Bida file.
67 Letter of 10 March 1996; petition of 10 February 1997, Bida file.
68 Petition to Minister of Public Services and Social Welfare, 10 February 1997.
69 This description is based on a letter from the boycotters to the chair of the Binga council, 5 April 1997, Bida file.
70 Interview, former 'disgruntled member', Binga, 15 March 2001. Letter from the boycotters, to BRDC, following the AGM of 4-5 April 1997, Bida file. Letter detailing problems stifling Bida (n.d. April 1997).
71 Letter from boycotters following the AGM of 4-5 April 1997; letter detailing problems stifling Bida (n.d,

April 1997), Bida file. Two of those involved in running Bida's finances after the takeover had already served prison terms for stealing finance from other Binga institutions, according to critics See also letter to the *Sunday News* 13 April 1997.

72 Report on the Bida review and planning workshop, 7-9 October 1997.

73 Letter, disgruntled members to Bida Director, re meeting on conflict resolution facilitated by the Zimbabwe Council of Churches, 7 July 1999. They argued the management had a culture of 'imposing ideas on how the problem has to be resolved to your advantage as demonstrated at the 1997 AGM where you chased and even fought some people to secure your positions and interests. Your fascist calculations are not the best way of helping the marginalized people of Binga. No nyanga will convince us to change our position that the Bida assets and funds were misappropriated .... We believe that you are control in the situation since you achieved to make Bida your private limited enterprise. What would you ... need from the disgruntled members? We will assure you that the people of Binga will not lose anything by your continuing running your private limited company ... we know this is not genuine since your agenda is to use Bida for your political campaign.' Bida could not be re-activated even through a series of donor funded 'conflict resolution workshops', which were boycotted by the disgruntled members.

74 See council minutes, 13 December 1996.

75 Interviews, former council executive officer, 12 March 2001, F. Cumanzala, 15 April 2001. Interview, SCF employee, Binga, 14 April 2001. Binga's SCF(UK) office was criticized for bringing in unskilled labourers from outside the district, and decided (apparently without local consultation) to phase out a community water supply project in favour of a child rights programme, even though water problems around the district were far from solved and the sustainability of their own water project was debatable. These issues were raised against the organization in the politicized assault on the local state undertaken by war veterans in the course of 2001, described below.

76 See, for example, debates over proposals in council minutes 8 June 1989, 29 June 1990, 14 December 1990.

77 Council minutes, 13 March 1992.

78 Council minutes, 13 March 1992.

79 http://www.werfus.at/argezim/tongatour/ (accessed 6 November 2002).

80 Travel Africa Magazine, May 1999; for interpretations of Nyaminyami, see also http://www.safariafrica.co.za/nyaminyami.htm. (accessed 7 April 2005).

81 Bida minutes, 14 August 1997.

82 See http://www.servus.at/argezim/nyami/meeting.htm, http://www.servus.at/argezim/tougatour and related sites (accessed 6 November 2002).

83 Later, the project expanded into 'Tonga online', an IT development programme involving a newsletter and discussion forum run from Binga schools at the website http://www.mulonga.net ('on the river'.net).

84 Minority rights seminar, 5 August 1997. Details on http://www.servus.at/argezim/nyami/minority.htm, (accessed 6 November 2002).

85 CCJP, Report on the Problems That Led to the Closure of Lungwalala Irrigation Scheme', October 2000. Copy held in File AGR 16, Binga Rural District Council.

86 This paragraph is based on WCD 2000:44.

87 Ibid.:45.

88 Ibid.:45.The Zambian initiative was a project funded jointly by the World Bank and the Zambian Electricity Supply Corporation (ZESCO).

89 For a full discussion of CCJP Binga, see Conyers and Cumanzala 2004.

90 F. Cumanzala, Binga, 15 April 2001.

91 See Mumpande 2000. This work was facilitated by an officer funded by Silveira House's advocacy programme for minority languages, initiated in 2000.

92 Ibid.

93 Former Binga MPs, involved in a Parliamentary Select Committee for minority languages emphasized this problem. Interview, M. P. Siachimbo 5 September 2001.

94 See Kagoro 2004:247.

95 Ibid.

96 This process is detailed in McGregor 2002:9-37.

97 Conyers and Cumanzala 2004.

98 McGregor 2002:26-27.

99 In the Ministry of Youth, for example, expelled officers and community mobilizers were replaced by war veterans and the relatives of local party officials.

100 Manyena 2003.

101 Ibid.:15.

9

Surviving in the Borderlands The 'Unfinished Business' of Lake Kariba

Lake Kariba and the Zambezi state border both achieved a new prominence as the Zimbabwean economy declined and then plummeted after 2000, and people turned to fishing and cross-border trading to offset the effects of mounting inflation and collapsing formal employment. Rapidly developing social networks criss-crossed the landscape of the lake and border, disrupting the image of uninhabited 'wilderness' long cultivated through state controls privileging conservation, large-scale commerce and tourism. Though a punitive regime of conservationist regulations on the use of the lake was still in place and intermittently enforced in 2000-01, state capacity was crumbling rapidly.

This chapter is about fishing, trading and the reconfiguration of authority over the lake, as revealed in the activities and perspectives of Binga's gillnet fishermen, who work the inshore waters of the lake on a daily basis. Although the landscape has provided a continuous theme for the book, my focus here is on perspectives arising from day-to-day material interactions with the water, rather than memories. The chapter uses the fishermen's own accounts of their daily life to explore the shifting material relationships governing fishing and trade, which link the micro-ecologies of the south-western shores of the lake through a web of social relations to distant Zimbabwean urban markets as well as across the nearby border with Zambia. These relationships were under strain, as inflationary pressures had contradictory effects in Binga, and one of the ways they were experienced was through an increased isolation from Zimbabwe's main urban centres, as soaring fuel costs cut off transport links and undermined the viability of small-scale fish trading, in turn reinforcing the need for a stake in cross-border trades. As livelihoods became increasingly risky, so disputes multiplied, pitting fishermen against each other, the authorities, and other users of the lake, such as the large population of wild animals – particularly crocodiles – that had flourished through half a century of state protection.[1] Indeed, the mythologized figure of the crocodile could stand as a metaphor for the state itself, embodying state conservationist priorities and Tonga fishermen's marginality, as well as invoking older ideas of power. By examining changing livelihoods and authority on the lake at this time of transition in the borderlands, I hope to throw further light on why the lake was considered 'unfinished business' among Zimbabwe's Tonga leaders and fishing communities, but I also hope to illuminate a broader process of the remaking of state authority at this time, rather than invoking misleading metaphors of 'collapse' or 'failure'.[2]

The burgeoning irregular activities on the lake were normalized to the point of being considered 'illegal yet licit', to use Janet Roitman's terms.[3] They were openly discussed with outsiders such as myself, and were justified in terms of the economics of survival and invocation of Tonga traditions of crossing the river and entitlement to the lake as river people who had borne the costs of its creation. The expansion of informal activities did not, however, mean that these new economic spaces were outside or beyond state power. Roitman's analysis of informality marks an important shift away from previous romanticized accounts of informal economies as autonomous, equated with resistance and state collapse.[4] Rather, she sees the clustering of irregular economic activity around seemingly isolated borders as constitutive of, rather than undermining, state power, indicative of its re-crafting rather than its collapse. She finds it useful to distinguish between 'state regulatory authority' (as expressed in bureaucratic controls) and 'state power' (expressed partly in 'brute force' but also in 'command of central parts of the bureaucratic apparatus' and the 'capacity to constitute the field of "the state" through forms of power that exceed the state bureaucracy or its central institutions').[5] The distinction can help account for 'the contradiction between the expansion of unregulated activities, which seems to indicate a loss of state control, and the continuity of state power in spite of it all'.[6] Although, in Das and Poole's words, such reconstituted forms of the state can be notably 'illegible', they are nonetheless enmeshed with bureaucratic forms of statecraft.[7]

In the Zambezian context discussed here, the legal-bureaucratic framework for governing the lake remained important in a multitude of ways. Aside from being the basis of intermittent punitive action, it underpinned the social hierarchy and inequalities of power among lake-users and had marked biological legacies in the large population of state-protected wildlife. It also shaped the language of fishing itself. Unlike other fisheries, such as Mweru-Luapula, where traditionalist 'oral charters' such as that of *Nachituti's Gift* continue to circulate and invoke ritual and chiefly authority, on Lake Kariba the old sacred Zambezian landscape, marked by *malende* shrines and pools linked to chiefly powers, was submerged and deemed impotent. The narrative of broken promises invoked in Tonga cultural nationalism claimed entitlement based on the suffering of the displacement and justified breaking the rules of the lake, but also implied a breach with the past and alienation from the contemporary waterscape. Like so many other fishing communities, Kariba's gillnetters are stereotyped as unruly and irresponsible, a reputation reinforced by the unregulated flows of cash, goods and women through lakeshore camps located far from the wealth of towns in the midst of an otherwise impoverished rural hinterland. But whereas youthful fishers on Lake Malawi break traditionalist sanctions by invoking urban identities and metaphors, such that their fishing grounds and villages are 'emblazoned by urban lights',[8] Binga's gillnetters subvert state regulation of Kariba's waters by invoking their identity as 'river people' with a heritage of fishing and crossing, and by mimicking state authority, such that irregular fishing expeditions take on the status of 'patrols'.

At the time of the research, the official regime over the lake was being modified to undermine such undisciplined behaviour through new 'participatory' co-management that sought to involve lake users in policing themselves.[9] But the reformed structures were grounded in the abstractions of resource economics and theories of collective action and involved creating incentives for conforming to the rules that could not function because they did not address broader relations of power. They also ignored

9.1-9.2 Fishermen diarists: Paul Muleya, Laxon Mutale

9.3-9.4 Diarist Elias Munkuli; Fisherman's home, Malala fishing camp

fishermen's practices and their need to invest in networks of social relations beyond the fishing economy, and the close connection between broader politico-economic changes and apparently remote local fishing communities. By assuming there was a 'tragedy' of over-fishing to mitigate, the reforms also failed to incorporate new ecological understandings of the lake itself as naturally fluctuating and resilient, which located the source of biomass and productivity in the lake's hydrological regime and the annual pulse of fertilizing nutrients washed in by the rains.[10] The power of water to restore fish stocks despite intensive exploitation cast doubt on the whole architecture of control and punishment.[11] On the Zambian side, little of the shoreline had ever been closed to fishing, fishermen's numbers had not been controlled, restrictions on mesh sizes were removed in the 1960s and managers had long considered use of the lake to be essentially unregulated.[12] As this has not affected the lake's productivity or biodiversity but simply made catches more dependent on rainfall, ecologists have now come to see the Zimbabwean fishery not as over-fished but as 'underutilised'.[13]

My main sources for discussing fishermen's perspectives and strategies are the texts of two sets of 25 one-week diaries, commissioned from men living in five of Binga's forty fishing camps, in March/April and again in August 2001, combined with interviews and my own observations.[14] I asked the diarists (selected by the fishing camp committees) to write about their daily experiences as fishermen, and to report on gossip in the fishing camps.[15] Although the authors were literate, often relatively privileged among the broader gillnetting community, they did not write romanticized accounts of their daily life. Rather they told stories of being harassed by wildlife, of exhaustion and capsizing, combined with descriptions of the tedium and frustration of fishing. The presence of significant numbers of literate fishermen in the camps reflected the contracting opportunities for school leavers, the retrenchments and decline in formal employment, politicised expulsions of suspected MDC activists from state jobs,[16] as well as positive evaluations of the freedom of working without a boss and the attractions of money. The diarists were, of course, selective in what they chose to describe, and the diaries themselves were of varied quality. Nonetheless, both the texts and the process of soliciting them were enlightening and, in a context where research was difficult, allowed for freer discussion around topics fishermen were comfortable to talk about, raising issues I had not planned to investigate – such as crocodiles (which dominated the March batch of diaries) and thieves (which dominated those written in August). Before elaborating these themes and their capacity to shed light on the remaking of state power and fishing livelihoods, it is important to explore the history of governance of the lake and its effects.

Governing Fishing on Lake Kariba's Southern Shores

Fishing on Kariba's southern shores is shaped by a complex interaction between the ecology of the lake and the political economy of access and trade, as it developed in the decades after the dam. Although the law was de-racialized after Zimbabwean independence, there were notable continuities in the lake's regime. Conservationist interests grew stronger as tourism boomed, and white-owned companies continued to dominate both the highly capitalized multi-million dollar *kapenta* business of the offshore fishery and the commercial trade in fish from the inshore waters – despite an explicit and unsuccessful 'indigenization' programme in the 1990s.[17] Access to the lakeshore continued to be restricted, and all aspects of African gillnet fishing remained regulated by the state through the Department of National Parks and Wildlife Management (DNPWLM), whose officers demarcated fishing grounds and closed areas, located temporary fishing camps, controlled the number of permits, methods of fishing and net sizes, monitored catches and prohibited permanent homes and lakeshore agriculture (the latter justified as a strategy to 'professionalize' fishing).[18] Efforts to reduce the high death toll further regulated movement on the lake by outlawing dugout canoes and demanding the purchase of expensive life-jackets. Gillnetters caught breaking the rules remained subject to draconian punishment, involving the confiscation of boats and nets (while transgressing *kapenta* fishermen and tourist operators were treated leniently and subject only to fines).[19] The DNPWLM also operated a 'shoot-to-kill' policy against wildlife poachers, which caused significant loss of human life in the National Parks in the valley, particularly but not exclusively of Zambians.[20]

The state regime over the lake has created a series of contradictions for the gillnet fishermen. Perhaps most important, the places where they are allowed to fish lack good fish stocks, while many of the best fishing grounds are closed to fishing, either because they are designated breeding grounds or because they abut islands and shoreline reserved for conservation and 'wilderness' tourism. As fishermen have increasingly used these prime fishing grounds despite the restrictions, the rules have had the effect of obfuscating and criminalizing fishing practices, exposing fishermen to enhanced risks and reinforcing managers' stereotypes of fishermen as irresponsible. Further contradictions were created by the lack of transport and infrastructure in remote parts of the lake, such as Binga, where fishermen also face the predicament that even when fish are plentiful and catches good, traders can be scarce and transport unreliable, making consumers in urban markets inaccessible. All of this has made gillnetting a risky business, with profits tending to be invested outside the fishing economy. Scudder showed how Zambian fishermen capitalizing from the one-off 'boom' caused by the nutrient rush following the initial impoundment put their new-found wealth primarily in cattle and the agricultural economy;[21] Murphree found that Zimbabwean fishermen in the 1980s did likewise.[22] Thus, gillnetters typically also tilled fields away from the lake, tried to accumulate livestock, made artefacts for tourists, rented out boats (if they had them) and ferried traders. Those 'gillnetters' in the fishing camps making the most money clearly often did so not from gillnetting itself but from a stake in the other more lucrative trades, such as the booming informal *kapenta* business and cross-border trading.

The access regime that has helped produce these contradictions has been the object of persistent protest on the part of Tonga leaders, whose elaborate new blueprint for the development of Binga developed in the early 1980s – the *Lusumpuko Plan* – focused centrally on the resources of the lake. The plan aimed to re-orient business to achieve 'the uplift of the Tonga' through major council-led infrastructural development, fish canning and freezing plants, boat and net manufacturing industries, and fresh water prawns and aquaculture, with the lake serving as a commercial highway along the valley.[23] None of this materialized, and the council's one *kapenta* rig, named evocatively to invoke Tonga connections to the Zambezi – '*Kasambavesi*': 'only those who know can cross' – was eventually sold off. Councillors fruitlessly passed resolution after resolution to increase the number of permits for Tonga gillnetters, campaigned against license fees, protested the eviction of fishing camps in the interests of tourism and conservation, and opposed police actions in which whole camps were periodically rounded up and punished indiscriminately.[24]

Yet their inability to increase access significantly while also administering the permit system for DNPWLM, and their increasing implication in other aspects of the lake's business convinced gillnet fishers that the council was 'against fishermen' and interested only in 'selling' the lake to the highest bidder.[25] Fishermen moved on by tourist developers complained: 'we are not recognized as legitimate users of the lake ... [we are] forced to move from camp to camp as the camps are sold by the council officers. We black men will be forced off...as the lake is purchased by the white man.'[26] The successful establishment of some African *kapenta* cooperatives did not disrupt the old hierarchy, as they too considered themselves superior to the gillnetters.[27]

Social scientists also criticized the Kariba regime.[28] In the 1980s Michael Bourdillon, Angela Cheater and Marshall Murphree undertook the first social science research on the Zimbabwean side of the lake and questioned the assumptions of over-fishing and declining catches, the rationale for closing 20 per cent of the shoreline to gillnetting, and the emphasis on 'professionalization' and 'co-operativization' as a means of developing attitudes of responsibility.[29] Jeremy Jackson examined the voluminous data on daily catches, collected obsessively by managers over previous decades, but 'minimally analyzed' and never discussed with fishermen themselves.[30] He argued: 'It is fair to say that after nearly 30 years of monitoring and research we know more about fish than about fishermen'.[31]

The new Combination Lakeshore Masterplan and fisheries joint management scheme, finally drafted in 1998,[32] fell short of radical proposals for reform, as the 'exclusive' gillnet fishing zones to be controlled by committees of fishermen were not delimited by fishermen themselves and did not involve powers over other users of the lake, while the punitive aspects of the old regime persisted.[33] Malasha argued that the new committees representing gillnet fishers developed their own interests in the workshops, travel allowances and other perks brought by participation in the management regime, which took them away from their colleagues and control over everyday issues in the camps.[34] Gillnet fishermen speculated at length on the 'real' motives behind the rules, interpreting irregular spates of enforcement as motivated by 'grudges' against particular individuals and communities, or as career advancement strategies on the part of ambitious young police and National Parks officials trying to impress superiors. They complained of a new Officer-in-Charge demonstrating he was 'clearing up' after predecessors accused by the Parks authorities of illegal fish and cross-border trading. But as party politics became polarized and the security arms of the state were politicized after the MDC's success in the referendum of 2000, so gillnetters believed they were being punished as disloyal opposition supporters.[35] The new decentralized management and user committees thus had little impact on the legal structure of resource rights, or fishermen's daily strategies, developed over previous decades in circumvention and subversion of state rules.

Patrolling Kariba's Waters

The strategies that fishermen have developed for exploiting the lake during decades of punitive bureaucracy are best explored in their own terms. Fishermen regarded 'proper fishing' with a boat and gillnets to be a man's job and contrasted it to inferior and feminized forms of fishing, such as 'hooking' from the shore and illegally 'driving' fish (*kutumpula)*, in which an entire cove is netted off, and the shallows are beaten to scare the fish into nets. Increasingly, 'proper fishing' necessitated going beyond the bounds of unproductive legal fishing grounds – dubbed 'bathtubs' by fishermen who spoke to Malasha in the late 1990s.[36] Such illegal expeditions were called 'patrols', and often took the form of trips of several days. 'Patrolling' alters the economics, social relations and risks of fishing in important ways that have not received the recognition they deserve in the existing literature. The vocabulary for describing working relations between fishermen has also been shaped by the lake's regime, particularly the practice of allocating permits preferentially to cooperatives and allowing two workers for each individual licensed fisherman. As a result, the term 'cooperative' was bureaucratic fiction in most Binga camps, as cooperatives filled the role of legitimising fishing and access to other assets (such as donor-funded boats) rather than describing cooperative action among members.[37] 'Worker', however, could describe someone legitimately resident as such, but as fishermen did not employ permanent workers and controls on residing in the camps were only irregularly enforced, the term had a more general application to anyone without a boat or nets of their own, doing casual work for others, 'hooking', or borrowing equipment to fish for themselves.

The discussion below begins with an investigation of the network of relationships fishermen construct to navigate the range of ecological, social and political factors that give fishing its characteristic 'boom-and-bust' nature. Sociological studies tend to treat fishermen's negotiation of such relationships narrowly in terms of calculations of 'effort', with 'increased effort' being measured in terms of an increase in the number or size of nets being worked, which is cast as the main means by which economic or ecological pressures are overcome.[38] Here I hope to add complexity to such discussion by illuminating the effects on the material relations of fishing caused by a changing economic environment, and to show that 'effort' cannot be divorced meaningfully from risk, itself reflecting both new forms of state power and the residue of older practices of conservation and bureaucratic control. Aside from seasonal variations in the availability of fish and the effects of intermittent policing, fishing relationships are also shaped by the capacity of full moons to suppress catches, or the action of wind, waves or wild animals in making fish stocks temporarily inaccessible. Fishermen compete directly for fish with the large population of crocodiles as well as with other fishermen. A further set of problems relate to nets being torn in the waters by wind or drifting mats of *Salvinia* weed, getting caught on the many submerged petrified trees, snagged by other boats or ripped by crocodiles. Nets are also frequently stolen from the waters when they have been laid out.

Against this diverse assemblage of human and non-human adversaries, and the periodic crises they produce, fishermen tried to construct networks of support, and to ensure protection from ancestors and God to guard against the hidden powers they considered to lie behind success or failure, which were often revealed only in the context of disaster. As the old *malende* shrines and 'places of power' in the river were redundant, fishermen depended on an eclectic mix of trying to ensure benign intervention from immediate family ancestors, purchasing charms and consulting healers and diviners, praying and going to church and investing in social relations in and beyond the fishing camps.[39] All the fishermen were enmeshed in complex webs of obligation and credit that developed over time from their first engagement in fishing as workers or dependents of others (generally brothers, fathers or other senior male kin), and, as they accumulated assets that allowed them to fish independently, incorporated clients of their own.

Although the risks of patrols were considerably higher than fishing the bathtubs, the possibility of bumper catches provided a clear economic rationale for patrolling. Patrols were returning after three or four days with catches of over one hundred kilograms of fish (one returned with 300kg; another was considering firing his workers when they only returned with 15kg when other patrols were bringing back 100kgs).[40] Although returns were distributed among the crew they were still considerable, indeed they compare very favourably with the 1980s catch figures discussed by Jackson.[41] The same cannot be said for fishing legally in the bathtubs: in Simambo, for example, fishermen with nets in legal grounds in the peak season regarded as 'OK' a week in which the two highest daily catches were 6kg, despite returning most days with nothing more than relish. Several of the diaries in March and August recorded entire weeks with no significant catches at all, with fishermen returning daily with half a fish, or a single small fish, or at most 2kg. The August batch of diaries from Mujere camp coincided with a full moon, and none of the diarists caught anything significant the whole week. Indeed, normal returns from legal gillnetting compared unfavourably to 'hooking' – one diarist spent a day in March 'hooking' and secured a very reasonable '5 heaps', i.e. 5-6kg, which was directly comparable to 'OK' returns from gillnetting legally.[42] Only if a fisherman was prepared to take the risk of going into illegal waters, and was organized and also lucky, could gillnetting lay the basis for an income or accumulation rather than simply 'fishing for relish'.[43]

These kinds of comparisons meant that fishermen were increasingly going on patrol despite the risks. By breaking the law, patrolling created the risk of punishment from the authorities, raised the chances of losing assets or craft for other reasons, exacerbated encounters with life-threatening wildlife, demanded significant bodily exertion and meant sleeping rough, exposed to mosquitoes and other dangers.[44] On return from patrol, fishermen described physical exhaustion from rowing long distances across the lake. One Kaluluwe diarist described being 'forced to go on patrol' to an island some 20km away, involving more than four hours paddling in the middle of the night: 'Feel absolutely tired and my fingers are painful due to the long paddling', he wrote. Returning was no better: 'The wind and waves left me stranded all day. I had to wait until sunset and then battle against the wind… There was no plan except to paddle. I forced myself and my body. My body was wet all over and hunger was my friend because of overworking.'[45] The following day he was ill due to this 'over-paddling'. The bodily discomforts of patrolling help to explain why boat-owners sent others rather than going themselves, though there were also advantages in terms of dealing with the authorities (if the patrol was intercepted, workers could argue the assets were not theirs, and owners could claim innocence of sending workers to fish illegal waters).

Risks from wildlife were more acute on patrols because fishermen laid their nets in crocodile breeding grounds, such as in the river estuaries, and worked and slept with inadequate shelter within National Parks and safari areas or on islands. The fishers were unsurprisingly hostile to the lake's wildlife, an attitude that was reinforced by the community based programme, 'Campfire', which they saw as devaluing human life and serving others' interests.[46] In Mujere, for example, fishermen complained of the place 'being controlled by the safaris, together with Campfire. Elephant, buffalo, crocodiles and hippos trouble people every time. Ten or more people can die within a year but when we send the report, they say they haven't yet found the client to buy the animal. And when they address the meeting they teach us that animals are better than people because animals are being sold, whereas a human being brings no money.'[47] Among the animals that took human life on the lake, crocodiles featured particularly prominently.[48] Crocodiles had benefited directly from state protection as an endangered species and indirectly from measures to restrict access to fish breeding grounds in the river estuaries (where crocodiles also breed). But their numbers had also increased through the flourishing business of crocodile farming, as the farms supported wild populations by releasing a proportion of those bred in captivity back into the wild at an age where their survival was more or less guaranteed.[49] The estimated 1980 population of 40,000 on the southern lakeshores had thus continued to grow, particularly but not only in conservation areas. Although fishermen tried to control crocodile numbers illegally by destroying or poisoning nests, and some crocodiles drowned or were speared after being tangled in gillnets, their efforts did not prevent the population increase.[50] The amount of fish crocodiles consume is considerable – ecologists estimate that they eat as much as 10 per cent of the total annual fished off-take from the lake.[51]

The fishermen's diaries provide a wealth of detail on daily harassment from crocodiles, which emerged as by far the most important cause of damage to nets; indeed, 21 of the 25 diarists recorded crocodile damage to nets in the week they kept the diaries, some reported such damage on every day of the week.[52] In more than one instance, crocodiles completely destroyed all a fisherman's nets: one diarist reported returning from patrol 'stranded – no nets, no plan' after such a disaster and had to plead for help from his matrilineal kin;[53] a second was reduced to hooking.[54] The fishermen tried to minimize damage through a range of strategies that they considered tiresome, and which greatly increased the effort and dangers of fishing. Instead of leaving nets in the water unattended overnight, for example, they stayed out on the lake, checking nets on an hourly basis, fending off crocodiles by throwing stones at them. 'I spent a sleepless night on patrol' recorded one, 'losing my sleep just to protect against crocodiles'.[55] Another summed up at the end of the week, 'It was a bad week, the life of a fisherman is hard and risky. I am bored with daily encounters with crocodiles and torn nets. We'd kill them all, given the chance.'[56]

The fishermen also feared the animal. One diarist chose to document little other than crocodile movements and sightings, his anxiety apparently stemming from a narrow escape the previous year.[57] A second capsized during the week he kept the diary and recorded 'a most fearful day in my life, when the canoe I was in capsized in the strong winds. What I had to do was swim for the nearest dead tree in the lake thinking all the time of crocodiles and leaving the canoe to float away. I clung to the tree and shouted for help, to be rescued by two white men fishing.' Later in the week, he nearly capsized again, in the wash from a speedboat – 'I reflected that my canoe will disturb my life', he wrote.[58] After sleeping rough one night on patrol, a third diarist described his fear upon waking to find that during his sleep he had moved in such a way to have exposed himself to crocodiles:

> ... I looked for a place to sleep where the mosquitoes would not bite me, I moved and moved and eventually found a place between the rocks by the lakeshore. Finally I slept, but I woke with a shock because I found my feet were in the water and I was so scared because a crocodile could take me. I looked for my nets, still thinking of crocodiles, but all I could see was floating

> corks, so I paddled quickly to check them. When I got there, I found all the nets had been cut by crocodiles, there were only pieces remaining. One piece had half a kg of fish in it. That's worth Z$17.50, so I bought half a loaf of bread... I boiled tea and I reflected on the problem of low catches and crocodiles. If there were no crocodiles I would be eating margarine, I reflected.[59]

Stories of infamous crocodile attacks circulate in the camps, at least some of which are founded on actual events. In Simambo camp, fishermen claimed that crocodiles had killed five people in the previous four years, mostly poachers (who were said to have deserved their fate). They also reported being harassed by one particular crocodile, which had been accustomed to following a ferry and being fed by tourists, but began troubling fishermen when the ferry and tourists ceased abruptly in 2000. A particularly devastating series of deaths had caused chaos in Kaluluwe camp (and were also known to the authorities), when three young fishermen were killed in the single hot season of 1999-2000; one was said to have been eaten after capsizing, the second man's disappearance was attributed to crocodiles when clothes were found on the riverbank and the third was apparently taken by a crocodile whilst trying to rescue a boat that had gone adrift.[60]

The disruption caused by attacks was heightened by occult understandings of the creature as a witches' familiar, and interpretations of assaults as acts of witchcraft (*bulozi*).[61] Fisherman claimed witches were able to move about in the form of a crocodile, and that people with powerful medicines sent them in attacks against others. Patterns of accusation, which were reinforced by diviners, were typical of those in Tonga communities more generally, as they pointed almost exclusively to elder male kin.[62] Thus, surviving colleagues of the three Kaluluwe deceased accused elderly male relatives of killing their own sons, motivated by jealousy of their economic success, and n'angas reinforced their suspicion.[63] In the wake of the attacks, a sizeable group of young men vacated the camp in preference for a neighbouring camp where there were no old men.

Accusations of witchcraft were not the only source of tension disturbing the fragile trust and threatening the social relations that are so important to fishing. Accusations of theft also proliferated and bore some similarity to witchcraft accusations in the frequent lack of evidence and the manner they were handled – including by consulting diviners and with recourse to hasty retribution against suspects. Unlike witchcraft accusations, however, the blame for theft did not fall on old men, but targeted workers, 'poachers' and other outsiders, such as itinerant prophets. [64] The uncertainty of knowing whether thieves were at work or whether nets could have been simply swept by weeds or taken by animals lead to rapid action to beat up suspects. Moreover, the authorities were rarely consulted or considered to have the capacity to intervene. Thus, when a victim of a nasty beating took his case to the police, the officers were more interested in the wrongful retribution than the initial theft, which was embedded in an impenetrable web of social relationships.[65] If the interactions with crocodiles revealed the legacies of official management, the thefts revealed other aspects of state power and its ability to infuse the expanding informal economic networks that undermined official regulation, as well as illuminating the ramifications of hyperinflation on social relations in the fishing camps.

The population of workers in the fishing camps swelled rapidly as the economy deteriorated: in March 2001, the number of 'workers' in one camp alone was said to have more than trebled.[66] The camp committee was sympathetic and could not evict them: 'What can we do? We have to let them stay, we are human, they are our relatives.'[67] These workers lived a precarious life, combining hooking with whatever piece work they could get – helping for a few days on patrols, or being hired for specific tasks such as gutting fish, or cutting firewood. Although boat-owners were fully aware of the risks of patrolling, disputes with returning workers were commonplace. When one owner lost all his seven nets on patrol, his immediate reaction was to fire all the workers without payment; another, who felt his workers were 'playing' when they returned with catches below those of other patrols, likewise dismissed them all.[68] Fishermen's distrust of their workers and the reason why they too could become part of the assemblage of adversaries on the lake was explained clearly by a Masumu fisherman: 'I keep changing [my workers], if one gets used to you he doesn't work hard ... he abuses... After 2-3 patrols, I'll employ others, always like that... They have their own families ... you'll find each trip they come with less.'[69] Confidence was eroded from the outset because employers (who had started out as workers themselves) were fully aware of the various strategies workers use to become independent (at their patron's expense). As the Masumu fisherman continued: 'It's easy to start up as a fisherman ... First you struggle to start as a worker, but it's easy to progress – you come from patrol and you say, "The nets are all broken". You come, saying, "Oh! the nets are taken by crocodiles", but you'll have taken the nets, hiding them somewhere. That's how you do it to start up. So you see, I can't trust my workers.'[70]

Rampant inflation over the six-month period between the two batches of diaries had seen the cost of nets treble, along with soaring costs of mealie-meal and other basics. Yet fish prices had remained virtually static.[71] The consequences were an acute crisis of employment for workers in the camps, a spate of retrenchments, disputes, accusations and thefts, and a related dramatic reduction in the population of workers. Those remaining in the camps in August described 'getting poorer... our group broke up with accusations. No one is employing workers these days. Some are cutting firewood, but most have gone home.'[72] Although accusations of theft targeted 'poachers' and scapegoated Zambians in particular, fishers often took out their anger on the workers: the former were less readily dealt with, as their activities revealed the violence of informal economic networks and the lack of state protection and entanglement of the authorities, particularly the police, in ways that will become clear below.

Incidents of theft heightened the value of the social networks fishermen could command to borrow money or loan equipment to continue and regain their independence. As economic problems deepened, finding assistance to offset such normal mishaps became more and more difficult. A relatively well-off fisherman from Kaluluwe, for example, who had employed several workers for patrols when we first visited, spent the week prior to the August diaries moving between villages trying to piece together the funds to replace stolen nets, and was tense and worried when we saw him again. He described his 'business collapsing', and related in his diary his efforts to resurrect it:

> I explained to my friend I was out of business in fishing, all my [seven] nets were stolen when left under water so I am now desperate, and want to borrow money and will pay it back after two months, just to buy nets. I said I only wanted $6,000. He smiled and told me he was very sorry, but he had no money at the moment, unless next month maybe, but he had some old nets. I really needed them. Oh, I was very, very happy because he gave me for free. From there I had to go back with pleasure and joy.... But fortune is not with me this week. Those nets that were saving me were destroyed and taken by crocodiles...So now I am still planning what to do...[73]

Thefts in one of the camps – Mujere – were on a scale that threatened to provoke a wider breakdown of trust, as a rash of stolen nets coincided with a series of break-ins to grocery stores in the camp, and fishermen started to pull their nets from the water and leave for their villages. The 'angry young men' who were the victims took matters into their own hands, circulating among the huts broadcasting their consultation of notorious diviners with powerful medicines threatening terrible ill-fate to the thieves should the nets not be returned. Making theft 'the talk of the camp' was a deliberate strategy to frighten the thieves into returning the stolen goods (as one fisherman explained, 'old men and those who threaten get their nets back').[74] One of those affected wrote:

> Day 1: I never went to the lake because my nets were stolen on 27 August at night...They were the only nets I had. Even now, I don't have even a short piece of net and no money to buy some more...
>
> Day 2: I went to [neighbouring] Senga camp to check for lost nets ... [and to] consult a powerful n'anga who is that side... On our return we found already some of the nets had been returned ... put back in the very place they were stolen. When the thief heard we'd gone to consult that n'anga, obviously he feared ...[75]

But the fact that success was only partial caused others to rush to retrieve their nets. Another Mujere diarist elaborated:

> An urgent meeting was organized to tell all fishermen about the visiting of a herbalist in Sengwa, to sweep the prevailing theft in the camp. We were told that anyone who knows he has a share in the missing of nets and shop-breaking and is kindly asked to forward himself before the charm plays its part....
>
> [next day] I took my nets out of the water because I'm not catching fish and I am worried as thieves are still at large... All the residents of Mujere were anticipating seeing the thieves handing over the nets since they were promised two days as a deadline. We were told that after two days of theft, the victims will be seen whether becoming sick or going naked in public or losing their testicles, but all is in vain. His charm seems not to have any strength, and disbelieving among us is becoming more pronounced... In short that n'anga's failure to catch the thieves has given them power to steal more nets than before, and this will encourage others going to steal as well.[76]

Fishing was also interrupted by disruptions and family obligations in home villages away from the lakeshore, and several of the fishermen spent most of their diaries describing calamities that had little to do with fishing directly, other than taking time away from it. One had to deal with his wife 'going mad', after relatives reported she had deserted her home and was found wandering in the forests; several had to care for sick children or other relatives; one had to return home to attend a funeral, others to treat ill cattle, resolve disputes relating to the movement of livestock, or the aftermath of a fight at a beer party. Yet the difficulties of making an income from fishing, rather than simply fishing for relish, were not only about investing in the social networks that enabled its extraction from the lake and provided the support to offset disasters; they were also about having the networks necessary to sell it.

Fish Trading & the Border Economy

Trading networks link the Binga fishing camps to local rural consumers as well as to the main markets for fish in Zimbabwe's urban centres and across the border in Zambia. Because of the distances involved the dynamics of the Binga fish trade are very different from the better connected, eastern parts of the lake. At the time of the research, the price difference between fresh fish in Binga town and that in Gokwe town was of the order of five times, rising to nearly ten times when compared with prices in some of the remoter fishing camps. In Bulawayo and Harare, the price differences were still greater. Yet fishermen were unable to capitalise on potential profits and could not raise fish prices. They understood the unfavourable terms of trade within Zimbabwe as part of a broader historically created political economy of extraction, through which non-Tonga were the prime beneficiaries of Binga's rich endowment of resources: 'Outsiders are sucking Binga dry – our fish, our goats, our cattle, our money', one fisherman complained.[77] This section aims to shed light on this interpretation of the legal fish trade within Zimbabwe as an expression of Tonga marginality, before turning to the expanding informal trades, such as those across the border, in which state power was also implicated.

The faltering trading networks linking the fishing camps with Zimbabwean towns have received some attention in the literature. Nyikahadzoi, for example, argues that low prices for fish in remote areas such as Binga are produced by an historically created monopsony,[78] such as in those Binga camps where white commercial buyers with refrigeration facilities close to the lake provided the prime market. Yet most of the Binga camps described here depended mainly on small-scale African traders arriving from town by bus (many of them women) or in pick-up trucks who brought mealie-meal, clothes, cabbages and other goods into the camps as well as cash.[79] The actions and preferences of these townsfolk shaped he fishermen's 'effort' and risk in various ways, as well as acting to prevent them putting up or even maintaining 'official' prices.

In all the camps, fishermen bemoaned their inability to increase prices, for which they blamed traders. In Simambo, fishermen said 'Catches are dropping year by year ... but even if the catch was bigger, we couldn't sell it.'[80] 'We just get into the water and then look for traders later,' another fisherman from Malala echoed. 'We can't increase the price ... maybe if there were more traders we could increase the price.'[81] The scarcity of traders meant that fishermen had to dry the fish, which lowered their profits: 'The fish rots before you can get it to the traders – you have to dry. A good fish is a fresh fish, because if you sell fresh, the weight is still OK, but today I sold dry – I was bored, I got nothing, it has no profit. I thought, of course, we're lucky to have this lake, where we can get money, but too many problems...'[82] Fishermen also complained that traders serving Ndebele consumers in Bulawayo only wanted bream and were ignorant of other fish species, such as squeakers, which fishermen argued were tasty but difficult to sell.[83] The preference for bream heightened risks, as it does not dry well (unlike the other two most important species, tiger fish and chessa).

Fishermen competed with each other to cultivate friendships with individual traders, offering porterage, credit and other favours to try to secure them as regulars. 'You need to cultivate your traders', one fisherman explained, 'escort them to the bus stop, carry their bag.'[84] Unlike fishermen further east along the lake shore, Tonga fishermen in Binga do not use their wives as fish traders, which they justified by stereotyping female traders as prostitutes.[85] Nor do they trade themselves in a significant way, which they explain in terms of their lack of contacts in town, describing arguments with kin and friends hosting them, being charged too much rent, buses breaking down en route such that fresh fish rotted, or failing to sell at a reasonable price.[86] But cultivating urban traders required more than relieving the physical effort of the laden trek between road and fishing camp, it also meant taking the risk of giving fish on credit. One described in his diary: 'a bus driver promised me that he wanted 20kg fish, so I promised him. But the bad part of it now, is that when I went footing, I did not find him. All that fish was wasted now and I have to sell at a low price. My friend scolded me, you should not catch fish for someone if they don't pay money ... I did not eat anything that night because of that driver.'[87] Another seemed almost surprised when a trader from Gokwe eventually paid up: 'I was very happy today to see a trader who left Mujere last month while he had 5,000 of my credit [from 150kg of fish]. I was failing to catch him, but fortune came my way today and he gave me my amount. I won't lend like that to traders again...'[88]

Traders undermined fishermen's efforts to control fish prices in a variety of ways, including by making deals with individual fishermen before their return to shore. One diarist described how, 'At the harbour, buyers were impatient, buying by orders before the fish had landed. They could jump into the water... The fishermen were not happy, because they could come out with kgs, but a low price.'[89] In Mujere, traders also organized a 'strike', as Mujere fishermen explained:

> We tried to put the price up to $40, we explained to the traders it was because the price of nets had increased. But then the traders saw they had the advantage. They explained to the bus driver and then they refused to pay and he was waiting there for them, they were united on that and they could see that they had the advantage. They were all from Gokwe, those traders. They'd got the independent people with cars to agree. They were all together. We could see all the fish was going to rot. We just had to give in, they saw they had the upper hand, so you see they wanted to put their own price rather than us who are selling.[90]

Of course, trading is also precarious, and having experienced the roads to some camps (such as Mujere), seen the condition and erratic arrival of the buses, it is hard not to agree with Murphree's 'certain ambivalence concerning these traders' business acumen, wondering whether to admire their persistence or question their sanity'.[91] Urban traders did not factor in the various reasons for depressed catches, such as full moons, when planning their trips. The August batch of diaries from Mujere coincided with a full moon, and the diarists gave vivid accounts of desperate traders watching their ice blocks melt:

> Day 3: People were walking around as if they were tired or somehow else, traders were despairing... A trader called Marinda was crying of his few fish which were rotting due to the lack of ice, no cold power. This was because of the days he had waited but no fish, he'd also tried Sengwa camp, but it was the same problem, full moon, no fish. People were drinking at my shop on credit – I dished it out to them to stop it going sour. One of the traders called Elton was out of his mind because he came thinking his ices were mature enough to last long. A nasty surprise was in store in the morning as there was only hot water in the ice box.

> Day 4: No fish again today… a good number of people were seen roaming around the compound, so stranded, desperate and confused not knowing how they will reach their respective places they have come from. A Mazda driver had come for fish but ended up pirating, taking those traders home with no fish.[92]

The fuel price rises over 2001 weakened these trading networks, and as transport became more expensive, many small-scale traders went out of business, and the camps were disconnected from the main urban markets in Bulawayo, Harare and Gokwe, and to a lesser extent Hwange and Victoria Falls (fish markets in the latter were repeatedly disrupted after 2000 by cholera scares). Increasingly, gillnetters needed a stake in the other, more lucrative commercial trades of the lake that could withstand fuel price rises, such as the *kapenta* business, or moved goods across the border into Zambia. The expanding informal *kapenta* trade involved rig workers selling *kapenta* secretly to artisanal fishermen in the lake and the latter hiding it on islands and selling it on to consumers at undercut prices.[93] On the Zambian side of the border it is estimated that as much as 60 per cent of *kapenta* is sold informally in this way.[94] In Zimbabwe too, the scale of this informal trade appears significant and increasing – indeed, for the gillnet fishers involved, it appeared much more rewarding than gillnet fishing itself.

The way state power was implicated in these new informal trades was particularly apparent in the activities and networks of the Zambians who dominated irregular cross-border activity and were in collusion with the Zimbabwean police. Zambians were blamed for a range of crimes: depleting stocks by using outlawed techniques such as fish 'driving' or small meshes, stealing nets and fish directly, and depressing fish prices and controlling cross-border trades. The fishermen diarists caricatured Zambians as gun-toting thieves, equipped not only with firearms, but also with powerful medicines that gave them the capacity to cause bodily harm and to control the dangers of the lake, including its crocodiles – should they act in conjunction with local n'angas and jealous elders. Zambians' collusion with the police meant they acted with impunity, and fishermen felt impotent against them. 'Zambians are all thieves!' one fisherman asserted. 'If they stay a week, on return to Zambia, they steal; they sleep here at the harbour, armed. They act in liaison with the police. We're afraid.'[95] Another elaborated,

> Zambians poach, they come with small nets catching the young ones, raiding our area. They use the *kutumpula* system [driving fish]… It can take two months for the fish to repopulate. We fear to chase them because they have guns. Zambians are a major problem for us – they sell to our customers at a lower price having stolen our fish. We've put that to the police, but there are scandals in the government. Police are favouring the Zambians. They even buy the fish from the Zambians or they become middlemen for them. Zambians don't know what a scale is! They collude with the police, and then at the end of the day the policeman goes with a heap of money.[96]

A Malala fisherman lamented: 'Zambians slept in the camp this night, and during the night they put their nets on top of mine, then in the morning they also took from my nets, so I got nothing, and the rent of the boat was so high – all I could do was pay the rent for the boat this week.'[97] At Simatelele, fishermen argued that Zambians' poaching was encouraged by a white commercial buyer, to whom they sold at depressed prices.[98] As Zimbabwean fishermen feared to challenge Zambians directly, they described trying to scare them off by pretending to be police by speaking loudly in Shona.[99] 'We can't just approach them, you can't just talk to them and appeal to their humanity, and the police don't care,' one elaborated. 'If I can make such an accusation against a Zambian or I try to chase him, I can be lucky to return with my life and on the next day I'll find my homestead destroyed, or I can find a crocodile has been sent to get me when I next get into the water.'[100]

Although fishermen complained that Zambians dominated the informal trades of the lakeshore and border, they and other Zimbabweans also increasingly participated in these activities, albeit from a position of disadvantage. Zambians were particularly quick to respond to the rampant inflation in Zimbabwe and disparity between official and black market exchange rates, which created a profitable market in Zambia for Zimbabwean manufactured goods (the latter were suddenly half the price of goods across the border), and made the relatively stable Zambian Kwacha instantly desirable. The grocery stores in the camps and villages along the Binga lakeshore had been built to serve local Zimbabwean needs, but in 2001 their shelves were empty as Zambians were crossing the lake and buying up the entire stock. This complaint was not confined to fishing villages: official border posts bustled with Zambian petty traders carting back Zimbabwean goods while newspapers ran stories of Zambian businessmen emptying shops by the lorry-full. War veterans in Victoria Falls and Hwange staged 'protests' expressing popular grievances against shopkeepers and supermarkets said to favour Zambians, and blamed Zambians for forcing up prices and creating scarcities in Zimbabwe.[101]

But Zimbabwean fishermen and others were trying to make the border work for them, despite the risks. To take one example: although for some time Zambian traders had been bringing their nets into Zimbabwean fishing camps (in violation of Zimbabwean mesh restrictions), the sudden increase in the cost of nets in Zimbabwe was forcing people to look to Zambia. Several of the fishermen diarists had tried to go to there themselves: after returning empty-handed when Zambians refused to accept their Zimbabwe dollars, they were working out more complex exchange chains, selling cooking pots to raise funds to travel to Lusaka to buy nets. Other commodities moving across the border on a significant scale included sugar, soft drinks, donkeys and oxen.[102] Fishermen with the boats, contacts and knowledge to cross the lake were obviously well placed to exploit these opportunities, despite fears about conditions in Zambian prisons, among other stereotypes of the country.[103] Increasingly, they were joined by others, including some senior public servants who had been expelled from their offices by war veterans.[104] Any sense of Zimbabwean superiority over Zambians faded quickly as the economic crisis bit, and Zimbabweans were forced into risky informal livelihoods that many had considered beneath them – and in the Zambezi borderlands, had associated primarily with Zambians.

Conclusion

The above is, of course, far from a full account of the real economy of the lake and border, not least as it focuses on legal practices and 'illegal yet licit' activities that were increasingly normalized and could be discussed with outsiders, justified both in terms of the economics of survival and by invoking Tonga traditions of crossing the river and entitlement to the lake. These networks of fishing and trade, and the boundaries of the legal, illegal and 'illegal yet licit' have doubtless been reconfigured many times since 2001, as the crisis in Zimbabwe has deepened. The chapter used Binga fishermen's own accounts of their daily lives to throw further light on how state authority over the lake was being reconfigured, and to explore the effects of this on the material relations of everyday fishing.

Although unregulated economic activities clustered around the resources of the lake and border, these new spaces were not outside state power. Fishermen's embodied conceptions of increased 'effort' were inseparable from calculations of risk, itself a reflection of the ways in which state power ramified through relationships on the lake. State power was thus implicated in fishing in crocodile-infested waters, paddling long distances in unstable craft, investing in unknown strangers as friends, and encounters with thieves and Zambian poachers. The lake's formal regime, though increasingly unenforced, also continued to matter in various ways: its legacies were still apparent in the legal structure of access, in ongoing policy debates about participation and co-management, in the presence of wildlife and in the language of fishing which mimicked and subverted bureaucratic authority. Although the resources of the lake and border provided a degree of insulation against the hyper-inflationary Zimbabwean economy, they did nothing to offset the exposure to violence and to the lack of protection intrinsic in illegal activity. As such, 'being forced' into new, dangerous and uncertain informal livelihoods deepened the sentiment that Lake Kariba was 'unfinished business', and heightened disaffection with the authorities and Zanu(PF) more broadly. As the character of the state changed, its violent and unaccountable nature remained aptly captured in the figure of the crocodile, whose privileged presence in the lake was a constant reminder of state priorities and of Tonga fishers' marginality, made more intimate by the animal's entanglement in the everyday networks of fishing and its ability to invoke older ideas of power associated with witchcraft.

Despite the instability and risks of fishing and cross border-trading, the lake has retained its central place as a symbolic as well as a material resource for Tonga fishermen. Although its exploitation has taken the form of lucrative opportunities for the few, irregular incomes for many, and relish for the desperate, it nonetheless remains significant. Indeed, the water's capacity to renew fish stocks and the border's capacity to create opportunities in periods of economic crisis heightened rather than diminished their importance. As the crisis has deepened since this research, the lake's role as a marker of Tonga identity and grievance is more likely to have grown than faded, and the histories of the dam, the displacement and border will surely continue to be told and retold in new contexts and to new ends. The 'knowledge to cross' has certainly remained as important as ever.

Further upstream, away from the unsettled history of the dam, the deepening economic and political crisis also provoked new debate over relationships with the river, and reconfigured political uses of the landscape. The final chapter of the book will return to the resort at Victoria Falls to explore these processes in what was a major tourist destination and source of both local and national wealth, as the political and economic changes also affected the tourist economy, local livelihoods dependent on it, and claims to the landscape.

Notes

1 The impact of state protection for crocodiles is discussed in more detail in McGregor 2005b.

2 See also Alexander 2006:180.

3 Roitman 2006. See also Roitman 1994. Following McGaffey (1991), I find it useful to distinguish between legal goods traded illegally (in this case household commodities such as sugar, and some aspects of the fish trade) and illegal goods traded illegally (such as drugs or gems), and do not consider the latter.

4 Roitman 2006:248.

5 Roitman 1994:193-4.

6 Ibid.:194, see also 222.

7 Das and Poole 1991:9.

8 Nakayama 2008.

9 Criticisms of these policies in relation to Kariba (discussed below) echo those developed in a broader literature: Geheb and Sarch 2003, Tveten 2002, Gordon 2006, Mosse 2003, Cleaver 2003.

10 Kolding et al. 2003.

11 Ibid.:4.

12 Mesh restrictions were relaxed in the 1960s when research suggested that small meshes did not adversely affect the reproduction of commercially important fish species. Malasha n.d. http://www.fao.org/docrep/006/y5056e/y5056e13.htm (accessed 15 August 2006). On persistent differences in net regulations see Kolding et al. 2003.

13 Kolding et al.(2003) argue this position.

14 The five camps were: Simambo, Malala, Masumu, Kaluluwe and Mujere. As the fishermen were assured anonymity, I have not used individuals' names.

15 The diarists were paid at levels that compared favourably with the income from a week's legal fishing. The diaries were mostly written in Tonga, and were read out to me by a Tonga translator in the author's presence, and followed up with questions. As gillnet fishing is an exclusively male activity, the diarists quoted in the chapter were men (women in the camps are either fish traders or fishermen's wives visiting when there is no agricultural work at home). For a broader discussion of diaries solicited for this study and commissioned diaries more generally, see McGregor 2005c.

16 On these expulsions, see McGregor 2002.

17 On *kapenta*, see Nyikahadzoi 2003. Jackson (1991:3) notes that 'the vast majority of the benefits of white fish production continue to accumulate as "value-added" to two companies that appear to control 85-88% of the frozen white fish trade beyond the lake shore'.

18 Bourdillon et al. 1985, Jackson 1991. The lakeshore on the Zambian side is much more intensively used, not only because people were allowed to settle permanently along it, but also because the slope on the northern banks allows for drawback agriculture. See Scudder n.d.

19 Nyikahadzoi n.d.

20 Duffy 2000:49. The majority of the 170 poachers shot between 1987 and 1993 were Zambians.

21 Scudder n.d.

22 Murphree 1985.

23 Lusumpuko Plan, 1 December 1981. Copy held in Binga Rural District Council.

24 Based on a review of Binga Rural District Council Minutes, 1980-2001. On punishment, see also 'Binga District Report on National Parks Anti-Poaching', March 1985. Fishing/national parks files, Binga DAs office. See also Jackson 1991:2.

25 Kaluluwe compound chairman, 3 April 2001.

26 Interview, Peter Mukuli, Kaluluwe, 3 April 2001.

27 In Mbila camp, where the council had cleared gillnetters in favour of kapenta operators, the gillnetters' chairman felt: 'gillnet fishers are looked down upon by the kapentas. They tell us, no we bought this place from the council, you can go away.' Interview, Mbila 4 April 2001.

28 Jackson 1991, Bourdillon et al.1985.

29 Bourdillon et al.:135-41.

30 Jackson 1991:3.

31 Ibid.:3. See also Nyikahadzoi 1995:7.

32 Kariba Lakeshore Combination Masterplan Preparation Authority, 'Kariba Lakeshore Combination Masterplan' (Final Draft, 1998).

33 Malasha 2002:14-5.

34 Ibid.

35 Interviews, Masumu and Simambo camps, 29 and 28 March 2001.

36 Malasha, ibid.

37 The situation may be different in some of the more formal cooperatives, such as those established by (non-Tonga) former contract workers working in former Irvin and Johnson concession areas, outside Binga district.

38 Jackson 1991, Kolding et al. 2003.

39 Some of the old 'sacred' places of the submerged landscape are still partly visible above the waters, but are

considered powerless, such as a tree on a tiny island once associated with a *malende* shrine close to Mujere. Kaluluwe fishermen described introducing new technologies to their ancestors to avoid disasters on the lake, by making their first trip in new boats in the company of elders. The lakeside in many camps is also now used for baptisms by some Christian churches, and Mujere camp has a separate 'Christian section' in which born-again fishermen live apart from their non-believing colleagues.

40 Diaries, Masumu and Kaluluwe, March 2001.

41 This figure is higher than the monthly averages calculated by Lake Kariba Fisheries Research Institute (LKFRI) for the period 1973-88. Jackson,1991:10.

42 Kaluluwe, March diary, 2001. My understanding of the LKFRI data is that it does not include 'hookers'. Yet there are considerable numbers, including many outside the camps: it is a weekend activity for many families in Binga administrative centre. Workers for various tourist ventures along the lake explained their acceptance of low wages in terms of the advantages of living in workers' compounds close to the lake shore, the possibilities of supplementing them through hooking, and if electricity was provided, the potential for renting fridge space to fishermen, or buying fish themselves for trade.

43 Diary, Masumu, April 2001.

44 In some places, fishermen had erected temporary shelters.

45 Diary, Kaluluwe, March 2001.

46 Wildlife was a particular problem in Mujere and Kaluluwe, as the latter cannot be reached on foot after dark due to elephants from Kavira forest, while the former is located within Sijarira safari area and elephants cause havoc in the camps looking for food, destroying granaries and entering homes. Efforts to spread the benefits from crocodile farming to local communities by paying the council for crocodile eggs did not benefit fishing communities. In 1992, a pilot programme removed 'nuisance crocodiles' from one heavily-fished area, but was not generalized or extended. Games and Moreau, 1977. On hostility to Campfire within the district more generally, see Zambezi Valley Consultants 2000.

47 Interview, Mujere, 7 April 2001.

48 For further detail on the history of crocodile conservation see. McGregor 2005b.

49 Loveridge 1996, Games and Moreau 1997, Child 1987.

50 In other African contexts, the introduction of gillnet fishing had a destructive effect on crocodile populations (Jon Hutton, pers. comm.).

51 Games and Moreau 1997.

52 In the August/September diaries, crocodile damage to nets was much lower, with only 9 out of 25 fishermen recording such damage.

53 Diary, Simambo, March 2001.

54 Diary, Kaluluwe, March 2001.

55 Diary, Kaluluwe camp, March 2001.

56 Diary, Simambo camp, March 2001.

57 Diary, Kaluluwe, March 2001; interviews Kaluluwe 3 April 2001.

58 Diary, Malala, March 2001.

59 Simambo diary, March 2001.

60 Interview, Kaluluwe, April 2001; DNPWLM fisheries officer, Binga, pers. comm.

61 Colson (2000) describes the ambiguity of *bulozi,* as it can be used for either good or evil ends, and argues the prosperous and powerful, particularly old men, are all assumed to have access to it.

62 Ibid.

63 Kaluluwe diaries, March 2001; Interview, Kaluluwe, April 2001.

64 In Binga there is not a significant population of migrant fishermen – unlike the eastern part of the lake where conflicts have erupted in fishing camps between Tonga and Korekore fishermen, who are the descendents of those displaced by the lake, and the non-local fishermen who came to the lakeshores as workers of white fishing concessionaires, particularly Irvin and Johnson.

65 Malala diary and interviews, August 2001.

66 Masumu, March 2001.

67 Interview, chairman, Kaluluwe, 29 March 2001.

68 Interview, Kaluluwe, April 2001.

69 Interview, Masumu, April 2001.

70 Interview, Masumu, April 2001.

71 In most camps, prices were the same; in a few, they had increased by up to 20 per cent.

72 Interview, Kaluluwe, April 2001.

73 Kaluluwe diary, August 2001.

74 Mujere diary, August 2001.

75 Mujere diary, August 2001.

76 Mujere diary, August 2001.

77 Binga, 27 March 2001.

78 Nyikahadzoi 1995:14.
79 After the mid-1990s, traders from Gokwe and Harare became increasingly important, where previously Bulawayo traders had predominated, particularly in the eastern parts of Binga lakeshore. The western parts depended more heavily on selling to Victoria Falls and Hwange in addition to Bulawayo.
80 Emion Mumpande, Simambo, 28 March 2001.
81 Interview, Malala, 2 April 2001.
82 Simambo, diary, March 2001.
83 Traders' demand for bream appears to have affected fishing efforts on the Zimbabwean shores Kolding et al. (2003) note that bream features less prominently in fishermen's catches in the Zambian fisheries.
84 Interview, Masumu, 4 April 2001.
85 Interview, Kaluluwe, 29 March 2001.
86 Interview, Kaluluwe, August 2001.
87 Diary, Malala, April 2001.
88 Diary, Mujere, April 2010.
89 Mujere, August 2001.
90 Interview, Mujere, April 2001.
91 Murphree 1985:41.
92 Mujere, August 2001.
93 For further detail on kapenta fishing see Nyikahadzoi 2003, 2005.
94 Malasha 2003.
95 Interview, Malala, 2 April 2001.
96 Interview, Simambo, 28 March 2001.
97 Malala diary, March 2001.
98 Simatelele, 3 April 2001.
99 Malala diary, March 2001.
100 Malala, interview, 1 April 2001. Similar points were made in the Simambo diaries, March 2001.
101 *Daily News*, 8 August and 21 September 2001; *Financial Times* 8 November 2001.
102 Kaluluwe, August 2001.
103 The fears were about the dangers of being imprisoned without relatives close by to supply food.
104 On these expulsions, see McGregor 2002.

10

Unravelling the Politics of Landscape
A Conclusion

The 'unfinished business' of Lake Kariba and Tonga claims to the river discussed in the last two chapters have begun to bring this book back to where it began. By illuminating the current politics of landscape on the Zambezi, the last two chapters examined the contexts in which stories about crossing, privileged relations with the river or past episodes of intervention are told today, exploring why the idea of being 'river people' has retained its salience. They showed how current claims have been shaped by changing cultures of state power and histories of nation-building, neo-liberal reforms, economic austerity and international validation of multiculturalism and indigeneity.

This final chapter further analyses state and international influences on the politics of local claim-making along the river by returning to Victoria Falls to pick up some lost strands of the narrative before revisiting the main themes of the book. As the waterfall was awarded the status of World Heritage Site in 1989, and is upheld by both Zambian and Zimbabwean governments as their primary national monument and tourist destination, Victoria Falls provides a revealing site for exploring how the power of international tourism, the global infrastructure of the heritage industry and state promoted ideas of nationhood shape local claims.

Although the waterfall and broader Zambezian landscape are central to tourism and state revenue, they are located on a frontier, and are marginal to cultural nationalism in Zimbabwe. As we have seen, local efforts to reclaim the landscape through ethnic mobilization proved divisive, had an uneasy relationship with nationalism, and sat uncomfortably with the ethnic diversity and history of the north-west. Moreover, any momentum for changing the public face of the resort to render it more appropriate for a post-colonial context was also tempered by the importance state officials attached to maintaining flows of tourists and control over the foreign exchange they brought in. The tourist and heritage industries operate through rather than outside state interests and institutions, even in such a commercialized site as the waterfall, and the influence of divergent state nationalisms and cultures of state power is thus no less apparent here than in other sites along the river and other historical periods.

This conjuncture of state and international influences has created both convergences and tensions with local interests. The convergences reflect the similarity between international understandings of heritage and the reified, static notions of culture and tradition promoted through the ethnohistories of the 1950s and 60s.[1] As Neil Parsons has elaborated in the context of Botswana, such movements to recover history were 'basically a cry for the rediscovery of dignity – for the consolations of Heritage', rather than an exercise in critical historical inquiry.[2] Local leaders have seen international interest in local culture as an opportunity not only for 'authenticating contemporary identities by providing them with roots', but, as Fairweather argues, may also consider displays of local culture a source of pride and 'a form of reciprocal exchange'.[3] Such performances can convey more than the content of the heritage on display, and can be used to further demands for rights and development, or to criticize state or government actions. The limited realization of meaningful local stakes in tourist revenue or enhanced rights, however, is testimony to some of the tensions between national and local interests, particularly as the state has moved towards a defensive closure of the public sphere in the context of political and economic crisis. Other conflicts reflect the difficulties inherent in representing local interests through the prism of singular, territorialized notions of ethnic identity.

By exploring the disputes over the waterfall as heritage, the first part of this chapter will take us in a rather different direction from debates over the river as a potent material force, a source of energy or fertility, or a strategic barrier or link. Yet the capacity for particular sites or landscapes to invoke a sense of connection with the past has been an important aspect of the history of conflicting claims to the Zambezi explored in this book. The material and symbolic qualities of landscape are inseparably entangled – the material power of such a large body of water has been an intrinsic part of the inspiration for its appropriation, while fantasies of control would be meaningless in the absence of an underlying materiality. As understandings of the power of the river have changed over time and as the frontier has been shaped by successive episodes of intervention linked to processes of state- and nation-building, however, so the difficulty of unravelling past meanings through today's claims becomes clear, even though legacies, memories and other traces of the past continue to be encapsulated in narratives and bodily performances.

The Politics of Landscape at Victoria Falls

The local actors in the politics of landscape at Victoria Falls between 1980 and 2000–2001 have been diverse. They include chiefs and leaders of local cultural associations – Nambya on the Zimbabwean side of the border, Leya in Zambia – as well as local state officials, politicians, heritage managers, war veterans, small and large tourist businesses and other interest groups. We left Victoria Falls at the end of Chapter 5, after it had been turned into a white playground, devoted to marketing experiences of the 'natural wonder', and celebrating the myths of white Rhodesia and British imperial expansion, particularly through the figure of David Livingstone. Its role as a 'site of memory' in white settler nationhood and the myths of imperial expansion had effaced African histories of settlement, former productive and strategic uses of the river, and the myths and practices associated with its 'places of power'. Particularly in the Zimbabwean resort, a tourist could be forgiven for returning with no understanding of the site's one-time location on the contested margins of powerful pre-colonial African states, or of the politics of living in between that shaped histories of Leya, Nambya,

Dombe and Toka settlement. Tourist itineraries provide little appreciation of how the iron bridge and new geography of connections it created made redundant the old politics of crossing and the related status of the mid-Zambezi ferrymen who commanded the fords, and give no insight into the histories of appropriation that undermined any contribution the river's resources could make to local livelihoods.

After independence, as Zimbabwean officials began to debate 'decolonizing' the public face of the country's most visited site, so they embroiled the politics of the past in the politics of post-colonial nation-building. From the outset, debates among heritage managers and local interest groups were overshadowed by the resort's capacity to generate foreign currency for state and private interests, and by officials representing national interests. The idea of decolonizing the resort by renaming it was rapidly dismissed, on the pragmatic grounds that Victoria Falls was world famous, and obscure local terms might threaten business. Zimbabwean heritage managers joked that Queen Victoria could be accommodated by the post-colonial state without contradiction, as by bringing in tourists, she was now 'paying her taxes'.[4] A similar debate at Zambia's independence had likewise retained the name of the explorer Livingstone, who (unlike Rhodes) was a popular figure among first generation nationalists and retained his status as 'Christian hero and exemplar'. As Holmes elaborates, Livingstone's name has continued to be promoted across the Zambian landscape, with schools, churches and roads named after him, while Zambian school children (unlike their Zimbabwean counterparts) continued to learn of the man 'born in poverty, who dedicated his life to spreading the gospel and ending slavery'.[5]

The process of renaming in Victoria Falls also involved local interest groups, such as the Nambya Cultural Association, albeit in a subordinate role. The association's potential influence was undermined partly by officials' attitudes towards Shona-related 'minorities' in Matabeleland. The Shona-speaking officials who had influence over the north-west after independence regarded Nambya as a distortion of 'proper' Shona and a joke: in debates over renaming Wankie, they overruled the Nambya Cultural Association's preferred 'Whange' with 'Hwange', on the grounds that the name existed in Mashonaland and its pronunciation in the north-west was a 'corruption'.[6] Notwithstanding these problems, many places – streets, schools and the like – have now come to bear names from the Nambya past, drawn from the Association's list of 'Names that Should be Remembered'.

Perhaps more important than these debates over names were the notable continuities in the exclusive focus on the 'natural' heritage of the waterfall, and the broader marketing of the north-west as wilderness and 'elephant country', focussed beyond the Falls on the country's flagship National Park – Hwange – and network of other conserved lands along the river and its hinterland.[7] The suppression of the 'cultural' heritage of the Zimbabwean north-west partly reflected the powers and professional interests of the Department of National Parks and Wildlife Management (DNPWLM). The waterfall itself (as far downstream as the fifth gorge) had been designated the country's primary National Monument since 1937, but management had always been delegated to the DNPWLM. After independence, the DNPWLM jealously guarded its control over Zimbabwe's most lucrative and famous tourist asset, obstructing efforts on the part of the National Museums and Monuments Commission of Zimbabwe (NMMZ) to establish a presence at the resort. The DNPWLM had a similar hold over the marketing of Hwange National Park: although NMMZ was responsible for several national monuments within its borders – the ruins of the stone-walled towns of the pre-colonial Hwange dynasty, Bumbusi and Mtoa – the park authorities paid scant attention to protecting or marketing these relics and blocked archaeological research.[8] This exclusive focus on the natural was further entrenched when the International Union for the Conservation of Nature (IUCN) granted the waterfall World Heritage Status solely on the basis of its 'natural assets', for fulfilling the criteria of being a 'superlative natural feature' and an 'exceptional example of significant ongoing geological processes'.[9]

These continuities in the public face of the resort fostered a dramatic expansion of business. Although the war in Matabeleland curtailed growth for a period in the mid-1980s, particularly after the murder of a group of tourists on the Victoria Falls road, as soon as the Unity Accord restored peace, the resort became a boom town. Numbers of visitors rose exponentially, new hotels and other developments proliferated, and the place became a magnet for national and international investors, as well as local entrepreneurs; it also attracted unemployed youth from all over the country.[10] Despite a growth in domestic tourism, the public face of the town was shaped above all by the type of experience that would appeal to wealthy westerners, mediated by continuity in some business interests, as well as new corporate and other investors. The resort thus continued to market imperial nostalgia, though new 'venture' tourism activities also expanded – white-water rafting and bungee-jumping, targeting a more youthful market. This commercialized, Rider Haggard-inspired fantasy idea of Africa was actively encouraged by state interests – one of most notable of the 1990s developments was a joint venture casino complex, 'The Kingdom', which peddled its nationalist credentials by invoking Great Zimbabwe, but was notably similar to other such commercial development in the region.[11] The wealth generated and flaunted at the resort was in stark contrast to the rest of the country, which was in economic decline by the end of the 1980s, and then in the grip of structural adjustment. It also provided a stark disjuncture with the drought-prone Hwange communal area and an increasingly overcrowded African township, Chinotimba.

The dramatic expansion of tourism stimulated by global recognition created extreme pressures on the land around the waterfall, not only for new hotel and tourist developments, but also for housing the growing workforce. Yet the location of the town within a National Park, the need to prevent expansion from spoiling the natural scenery, and the presence of un-cleared minefields from the liberation war, all provided severe constraints on land for new developments. Some of the new hotel projects controversially went ahead within supposedly conserved 'Rural Zones', which were redefined through 'special planning procedures' amidst intensified calls for a new, legally-binding plan to prevent overdevelopment from spoiling the place.[12] Some state officials felt that the only real brake on new commercial developments had been provided by the landmines – they were the 'single most important deterrent' – and joked that the best way to control new developments was not to remove them.[13] Chinotimba was bursting at the seams, and gained notoriety among Zimbabwe's urban planners, not only for being Zimbabwe's most rapid growing town, but for having some of the most overcrowded and insanitary high density accommodation in the country. Initially designed in 1973 to house 8,000 residents, by the early 1990s Chinotimba was home to an official population of 25,000.[14] Informal, overcrowded rented shacks that sprang up within existing home-yards were nicknamed 'Baghdads', because they evoked images of living conditions in the rubble created by the US bombing of Iraq's capital in 1991.[15]

The long history of marketing the resort primarily for its 'natural' credentials and the associated powers and vested interests of the DNPWLM did not, however, go unchallenged. Over the course of the 1990s, pressures for change came from various directions. By the middle of the decade, the distinction between 'natural' and 'cultural' heritage was increasingly under attack in global heritage discourse,[16] and there was a growing market among international tourists for experiences of the indigenous. As the idea that 'superlative natural scenery' stood 'outside' culture began to be challenged from within the heritage industry, various means were proposed for breaking down the long institutionalised nature/culture dichotomy. Notions of 'cultural landscapes' and 'living, spiritual or intangible heritage' were suggested for achieving this. These shifts were important, particularly in so far as they implied that 'a landscape could be "cultural" as well as "natural" purely through the cultural values that people associated with it' and required no 'material manifestation of its "cultural" features'.[17] As a result of these challenges, and despite an ongoing structural divide within the world heritage system that continued to reify the nature/culture distinction, it began to be possible for state bodies to designate World Heritage Sites for 'mixed' cultural and natural reasons. These international shifts provided the opportunity within Zimbabwe for challenges from cultural heritage managers to DNPWLM's monopoly over the management of (and revenue from) Victoria Falls, as well as providing new openings for local constituencies to demand rights and benefits based on their cultural traditions and ideas of the landscape as sacred.[18]

In this context, Zimbabwe's cultural heritage managers in NMMZ began to argue that the designation of the Victoria Falls as a solely natural landscape had been misplaced, and that it too could have qualified for 'mixed' status, in recognition of the archaeological record, and local religious uses. As James Muringaniza, one of the key actors in this challenge, explained to me:

> The Victoria Falls is listed by the IUCN for its natural features, it only takes into account the gorges, rainforest, animals and waterfall itself. But they don't talk about the shrines, as the waterfall itself is a shrine. It has a cultural force the naturalists didn't see. We are saying that if people want to see the Victoria Falls in its totality, then the cultural and historical aspects should also be brought in, the two are complementary, otherwise it is only one part, a corpse without its spirit or soul, a skeleton without its flesh. [19]

This view was backed up through research on local tradition and religious ideas about the waterfall and river. Muringaniza continued:

> For the locals, the river itself is regarded as life-giving, God given. God is known through the ancestors, whose livelihoods depended on the river, and who they approached by going to the river. So they pay their homage to the spirits in the river. For rainmaking, they take that water onto the land, it symbolizes the rain has now to come to the land. Then at the baobab they complete the performance. The important thing is to summon the spirit to talk to the people. At the Victoria Falls itself, that is seen as the headquarters of their spirits. In the rainforest there was a place to pray, that was kept in a very pristine condition, it was where the ancestors liked to stay.[20]

The research into local values that was a necessary part of this politics began to be conducted in 1995.[21] This was the first time any research on the 'cultural' resources of the site had been undertaken on the Zimbabwean side, and constraints were many. NMMZ was based in Bulawayo, some 500 km away from the waterfall, and officials lacked knowledge of the history of the area and had no prior contact with local leaders in the north-west; landmines and dangerous animal predators did not help access to some important archaeological sites and sacred places. Nonetheless, NMMZ managed to interview Nambya chief Shana, the closest of the Nambya chiefs to the waterfall, and other Nambya elders in Hwange communal area as well as mediums associated with sacred sites close to the waterfall.[22] The resulting report had a narrative of heritage spoilt by development and displacement, which it was important to preserve for 'the Nambya', because 'tradition' (whether 'living' or remembered) 'assured identity and continuity as a people'.[23]

Other opportunities followed for NMMZ to elaborate the cultural significance of the waterfall. When the World Heritage Committee asked NMMZ to report on the 'integrity' and 'authenticity' of the Victoria Falls site, NMMZ argued that 'integrity' was about 'nature', but 'authenticity' related to cultural values.[24] They elaborated: 'Integrity is about the fluctuation of water in the river, its influence on the rainforest and erosive powers …about movement over the years, as the waterfalls shift upstream though erosion; drought can undermine the landscape's integrity, with knock-on effects on vegetation.' Yet 'authenticity is incomplete without the perceptions of the local people who interacted with their ancestors at the gorges. Now they talk only of drought because they cannot perform their rituals. Of course, they have sought out surrogate sites for that, but if there is a death, or if rafts capsize and there is a mishap in the river, the locals say "of course – we can no longer perform our rituals at Victoria Falls".'[25] Though drought affected the integrity of the site, tourism and prohibited access threatened its authenticity.

As global and state heritage managers began to consult local leaders on tradition, so too did other actors. Impact assessments for new hotels and other developments began to require assessments of 'cultural' as well as 'environmental' impact, and projects sometimes requested local Nambya elders to perform cleansing ceremonies to appease ancestors and 'clear the way' for new developments.[26] Moreover, changes in the tourist market increasingly took visitors into the rural areas: white-water rafting, for example, required exit points and camping sites and brought porterage jobs to local communities such as that of Chisuma (closest to one of the main exit points), as well as usage fees and investment in schools and other community projects. The growing popularity of 'cultural tours and itineraries' rather than static displays of traditional villages also had the effect of bringing tourists into the communal areas: Monde village, which was so close to the waterfall that the plumes of mist were still visible, was an obvious choice of destination, and the income from the tour operator allowed the community to replace the school roof. Tours of Monde homes were on a small scale and marketed experiences of 'Africa' rather than a local ethnic identity, perhaps because like other places close to the Falls south of the river, a single ethnic label was problematic (Monde derived its name from a Leya ancestor, but the family now identified as Nambya, and the village itself was home to a wide diversity of different people, many of them retired former workers at the resort, of Ndebele, Lozi, Luvale, Shangaan and other origins).[27] Such schemes were brought within the auspices of Campfire projects managed by the rural district council during the 1990s, causing resentment from both local communities and tour operators over additional council fees, where previously there had been only two parties to the negotiations.[28] Discourses of desecration became more prominent as opportunities for their expression opened up – this was not primarily about stopping developments, but rather about greater local consultation and a more equitable spread of the benefits from tourism.

The rhetoric of desecration elevated by cultural heritage managers and local leaders focussed on several key sites, mostly downstream from the waterfall in parts of the river abutting the communal areas. At Victoria Falls itself, the river's former sacred sites had been long abandoned, as we have seen, partly due to the history of evictions in the colonial period that had seen the flight and expulsion of many Leya to Zambia, and the squeezing of Nambya and Ndebele evictees into the Hwange communal areas. Elders recalled how the vine-covered route down to the sacred place by the boiling point at the foot of the waterfall was turned into a tourist path, complete with steps and a handrail.[29] The last religious ceremony that anyone could remember taking place was in the 1950s, since when the light at the foot of the Falls and sounds of ancestors at the waterside have neither been seen nor heard; stories of the monster that lived in the boiling pot were no longer told (except to tourists). The famous Big Tree (a baobab and former territorial shrine) had also become part of the tourist itinerary in the town: access was difficult for local people from the rural areas, and the bark had been inscribed with travellers' initials.[30] In the gorges below the waterfall at Chamapato, white-water rafters used a former riverine 'place of power' – a pool where water was taken for rainmaking and sacred clay pots were left on the banks. But the site had been ruined, according to local elders, as the rafters used the spot as a campsite and for sun-bathing and were also accused of carrying away some of the clay pots.[31]

In this traditionalist discourse, not everything could be blamed on tourists. Debates were more complex than they appeared in exchanges with state heritage managers and tour operators. In the Hwange communal areas, Ndebele immigrants rather than tourists bore the brunt of blame for desecration, for ignoring local sacred sites and the related authority of Leya and Nambya mediums and chiefs; Christians were also criticised for using pools for baptisms, and abandoning ancestors. Local chiefs and spirit mediums themselves were also the object of criticism, and were vulnerable to accusations of disingenuity and greed, for failing to lead ceremonies with conviction if they had converted to Christianity, or for performing them for money. As we saw in Chapter 7, the last rainmaking pilgrimage to the Bumbusi ruins, which began from the graves of the old north-bank village of Hwange Chirisa below the gorges and proceeded with a river crossing and tour of other graves, had fallen apart in mutual recrimination; the party got lost in the National Park trying to find Bumbusi, the then paramount chief Hwange who led the event was accused of hypocrisy due to his conversion to Methodism, and the escape of the goat that was supposed to be sacrificed was taken as a bad omen.[32] Closer to the waterfall, the spirit medium responsible for maintaining ritual in the gorges – MaMbaita – invoked the spirit of her ancestor Kasoso, whose story of immersion in the Zambezi waters was described in Chapter 2. Yet Kasoso was widely criticized by the elders for abusing her powers, for joining a Luvale fortune teller in Victoria Falls craft village, where 'she started working for cash not the ancestors'. After her death in 1974, a baobab tree sprouted on her grave, which should have become an important ritual site. However, it was cut down by Victoria Falls council workers clearing a strip of land for telephone poles.[33]

Given the multifaceted debate surrounding the decline of traditional ritual practice over previous decades, re-casting and re-validating ritual linked to the landscape as 'cultural heritage' not only gave new meanings to old activities, now fixed in a static and distant past, but also broadened the constituency who could associate with it, including Christians. The rhetoric of 'sacred value' elevated through cultural heritage discourse thus could have the contradictory effect of removing religious value even as it tried to 'preserve' it, infusing sites, rituals and performances with a nostalgic appeal precisely because they were now located in the past. As we have seen, this shift was not new, and was reinforced rather than initiated by the new engagement with state heritage managers and tour operators in the 1990s. Nambya leaders in the north-west had a history of trying to gain cultural recognition from colonial as well as post-colonial states, as we saw in Chapter 7, which was centrally concerned with preserving and codifying cultural practice, claiming the ruins of the pre-colonial towns as ethnic heritage, and rescuing the Nambya language.[34] Nambya history has not featured more prominently in this book, despite the historical importance of the Hwange dynasty in relation to Zambezian politics and the history of authority over the river, because the movement became closely related to the politics of Hwange colliery and the town, and because the central focus of ethnohistories and ideas of cultural heritage promoted in the course of this mobilization focussed not on the river, but on the stone ruins of the old ruling seats of the Hwange dynasty, located within the Hwange National Park.

It is important to revisit the fate of this movement for Nambya recognition here, not only because it began to impinge on claims to the river in the 1990s, but also because it sheds light on the broader problems of reified notions of culture as they are mobilized in movements of this sort, showing the tensions created by trying to squeeze complex histories of fluid, assimilative, nineteenth-century political forms into the prism of modern cultural nationalist movements, invoking an essentialized unit tied to a single mother tongue and particular sites in the landscape. There is nothing intrinsically divisive about such desires to reverse discrimination and deliver development and language reforms, but they do have a fissile and fragmentary potential, as the internal schisms of the movement to promote Nambya culture, language and history demonstrated in the 1990s.

The issue of language was particularly complex and divisive in the Hwange context, not only because many Shona officials regarded Nambya as a Shona dialect rather than a language in its own right, but also because the idea of a single mother tongue was misleading, and two minority languages – Tonga and Nambya – had continued to circulate. Faced with the prospect of Nambya being introduced as the sole minority vernacular and in the light of the Tonga mobilization in the adjacent district of Binga, a group of descendents of the incorporated Dombe people formed a rival cultural association in 1990 – The Dombe Cultural Association – to promote their Dombe identity and the Tonga language. This splinter group focused on the Zambezi and their independent status as 'river people' and made claims to being first, which were linked to a bid for a Dombe chieftaincy.[35] Dombe activists felt that Nambya activism had continued to be linked to notions of hierarchy and status, in which Dombe and Tonga history, language and other cultural forms were looked down upon, denigrated and suppressed. The Dombe group produced rival surveys supporting their demand for Tonga in schools. Faced with disagreement over which of the two minority languages should be used, most schools decided to teach neither, and continued with Ndebele, as before.

This dispute between rival cultural associations provides the context for the recharged significance of stories of Mapeta the ferryman recounted in Chapter 2, whose descendents were the focus of the Dombe chieftaincy bid. Dombe versions of the story of the flight of the Nambya across the Zambezi after the destruction of their state by Ndebele warriors hinged on Mapeta's skill in crossing and his status as an independent indigene of the Deka confluence on the Zambezi.[36] This reversed Nambya versions of the same story, in which Mapeta was a lowly Nambya subordinate, whose status derived solely from Nambya gratitude for assistance in their flight, for which he gained his honorary title and a Nambya bride. In 2000, when these conflicting versions of the same story were being recounted with renewed political charge, everyday interactions and dependencies on the river had long ceased, and Mapeta's old crossing point had been appropriated by the Colliery in the 1950s, as the site where the mine abstracted water from the Zambezi. There were also a range of other developments around the pipeline and pumping station, such as the Colliery Angling club. Access had been further restricted during and after the liberation war, as canoes had been destroyed and the riverbank mined. In Hwange, the Dombe and Nambya along the river had been prohibited from fishing or owning canoes in the first two decades of independence (as this section of the river was deemed an important breeding ground for the lake downstream), as well as from cultivating riverbank gardens. But access was being opened up for the first time during my research. People had justified evading state conservationist restrictions (with encouragement from some development workers) with the argument, 'God gave us this river, it is for us to use by right'.[37]

The dispute between Nambya and Dombe undermined both cultural associations, and administrators threatened to ban them on the grounds they were 'dividing the people'.[38] The issue of language was additionally complex due to the cosmopolitan nature of the two towns of Hwange and Victoria Falls, where the history of migrant labour before independence, and the influx of workers from other parts of Zimbabwe thereafter had led to a diverse mix. In the context of the dispute between the rival cultural associations, the youth of the Colliery town formed their own explicitly non-tribal association 'so as not to be exclusive', and found a common ground for mobilization in the language and idea of development.[39] From the late 1990s, the national opposition party and civic reform movement increasingly absorbed the energies of this younger generation and also incorporated elders, as both urban and rural constituencies in Hwange overwhelmingly showed their disillusion with the ruling party in successive elections from 2000.

The polarization of national politics and the dramatic collapse in tourism adversely affected the opportunities that had begun to open up for promoting local culture and history. By 2000, the resort had only a quarter of the 1.4 million tourists it had hosted the previous year, less than 20 per cent of the hotel rooms were occupied, and Zimbabwe was losing potential new hotel contracts to Zambia.[40] In this context, other local actors and other claims to the river's resources became more prominent, framed within Zanu(PF)'s exclusive brand of nationalism, with its focus on land and race, and the rhetoric of ongoing liberation against the forces of imperialism. As elsewhere, local war veterans were encouraged by the ruling party to take leading roles in this new brand of local politics. At the Victoria Falls, war veterans' protests focussed on whites being prime beneficiaries of the wealth and resources of the resort, and demanded their share of plots for tourist development along the river, accusing local authorities of corruption in the allocation of stands, and of privileging white entrepreneurs. There was some basis for this line of argument in previously articulated grievances over the failures of 'indigenization' in the Victoria Falls context. Planners in the mid-1990s pointed out that this debate had come too late at the resort, 'the "cake" has been shared out already', leaving 'little space for more [development], indigenous or otherwise, to come in'.[41] Aside from achieving a percentage stake in plots along the river, veterans occupied selected white farms and conservancies, as part of the broader politics of land in the country, and also joined with Zanu(PF) youth and others in violent assaults on members of the political opposition around Hwange district and in the town; they intervened in colliery union politics, closed local municipal and rural district council offices and expelled civil servants seen as disloyal.[42] They also briefly and ineffectually re-activated the politics of imperial relics in the resort, and staged protests demanding that the statue of David Livingstone be removed from the Victoria Falls Park. Timed to coincide with the Zanu(PF) congress in 2001, they defaced the plaque on the statue (that proclaimed his status as abolitionist, explorer and bringer of Christian light), claiming it was an 'insult to Zimbabweans'.[43]

This new emphasis on land, race and 'patriotic history' did not totally sideline local ethnic claims.[44] As we saw in Chapter 8, an authoritarian politics of patronage is also compatible with ethnic politics, as patriotic history has depended on an elevated role for chiefs and issues of 'culture' were useful to Zanu(PF) so long as they did not threaten the distribution and control of key resources or disrupt Zanu(PF)s patronage networks. In this context, Nambya poets were invited to recite verses at Great Zimbabwe, and some further concessions were made regarding minority language teaching, albeit at a time when the state education system was in crisis. As elsewhere, chiefs sat on land committees, and local people were among those to gain access to plots within former white land along the line of rail, though long-standing Nambya claims to state land within the National Park were ruled out.[45] It is, of course, still very unclear quite how much of the prime land for tourism and the wildlife conservancies has been claimed by politicians, governors, army and ruling party officials (or given to Malaysian, Gulf and Chinese interests), though persistent rumours and press reports suggest it is substantial.[46]

All of this was fostered by the politics of Zanu(PF) state- and nation-building, despite the internationalism of tourism and heritage funds, and despite the long history of efforts to promote local culture and history on the part of the Nambya Cultural Association. The rise of patriotic history and decline of tourism had eclipsed the momentum that was gaining ground though a collusion of interests on the part of state heritage managers interested in cultural interpretations of the waterfall and local Nambya chiefs and elders. At the time of writing, tourism at Victoria Falls was resolutely still about the waterfall, Livingstone and other white pioneers, game-viewing and rafting. Cultural tours and performances still tended to be marketed as 'African' rather than lending any insight into local history, and state cultural heritage managers still lacked any kind of public space in the resort, and continued to fail in making claims on its revenue.

Across the border, the politics of landscape at the waterfall were shaped by a very different history of nation-building, different cultures of state power, a different post-independence trajectory and a different local politics. As we saw in Chapter 5, in Livingstone there was a tradition of marketing 'cultural heritage' including local African culture and history, that dated back to the 1930s; moreover, responsibility for the national monument of the waterfall was in the hands of the Zambia National Conservation Commission rather than with national parks. When Zambia's economic fortunes began to change in the late 1990s, as global competition for resources intensified, the Kwacha stabilized, and Zimbabwe's political and economic crisis made its more stable northern neighbour newly attractive to international investors, so Livingstone won major new hotel developments, saw an influx of white Zimbabwean and South African entrepreneurs, and new luxury lodges proliferated, alongside a number of small Zambian-run businesses. As Zimbabwean war veterans were staging protests against David Livingstone's continued presence, on the other side of the river preparations were underway for celebrating 100 years of the layout of the town, and 150 years of Livingstone's discovery of the falls, and state officials were busy reviving old imperial traditions such as the Zambezi regatta, and planning renovations to some of the infrastructure, such as the bridge (with Chinese funds), the Livingstone Museum (with EU funds) and the main road. In this context, a distinctively Zambian politics of place began to unfold, as the stakes in new tourist wealth became higher, and controversy over who benefited escalated, given the profound disjuncture between luxury developments and a town with a decaying industrial base, high unemployment, non-functioning public service infrastructure and a population with the highest HIV/AIDs rates in the country.[47]

The politics of landscape in Livingstone require brief discussion here, to follow through the story of the Leya evicted from Zimbabwe, because the landscape of the waterfall was being more elaborately and more loudly (if also controversially) claimed by local leaders, underlining the disjuncture of the state border, the effects of which have constituted one storyline in this book.

The most prominent of local ethnic claims in Livingstone were those made by chief Mukuni on the part of his Leya people. As we saw in earlier chapters, Mukuni's Leya were, in the late nineteenth century, only one of many groups with historical connections to the waterfall, associated with particular islands in the river, and who commanded its crossing points. Mukuni's voice was louder than that of other Leya or Toka chiefs, as he was closest to the Falls, and was also an educated man, a politician and entrepreneur, who sat on the Zambian National Tourist Board. Mukuni had developed the most successful of 'village cultural tours', on a much larger and more organized scale than anything on the Zimbabwean side. The tours involved semi-permanent teams of local guides working to greet the truckloads of visitors disembarking at the chief's palace who, for a fee of several US dollars, were shown around homes and taken to a curio stall; the tourists were also invited to watch seasonal and other rituals performed in the village, which had been re-invigorated by these new audiences and deployed headdresses and other costumes brought back from Mukuni's visits around the African continent. The plans for the village were ambitious, and included a prospective grand river crossing ceremony (modelled on the Lozi Kuomboka), a village museum, with feasibility studies for the various projects being conducted by Dutch tourism management students.[48] Mukuni had also re-instated 'Bedyango' – the 'priestess' or female counterpart to his own male chieftainship – who was based permanently in the village palace. She represented indigeneity and connections to the landscape of the waterfall through descent from the wife of the apical Mukuni ancestor who had migrated to the waterfall and married there. The position was cast as rectifying gender distortions created in the colonial period when an exclusively male chiefly line and authority was elaborated. When I met Bedyango, in early 2000, shortly after her accession, the young woman was still learning the names of the sacred places in the waterfall associated with the 'mists of the Mukuni ancestors', the line of chiefs, and a version of history that established not only intimacy with the landscape, but also past territorial authority on a scale belied by the historical circumstances of the small riverine chiefs in the old politics of the mid-Zambezi frontier described in Chapters 2 and 3.

In response to the Zimbabwean war veterans, Mukuni advocated moving the controversial statue of Livingstone across the Zambezi to Zambian soil where it would be welcome. He claimed that the man was part and parcel of Zambian heritage, the heritage of Livingstone town and of his own Leya people, and thus had a legitimate, even acclaimed, presence in the landscape. 'The Zambians have a good deal of affection for Livingstone, unlike the Zimbabweans', he argued. 'We have changed a great many of our colonial place names since independence, but we have kept the name of Livingstone out of a deep respect. For Zimbabwe the statue merely represents tourism and money. We would like the statue, but we would prefer not to fight over it.'[49] Visitors to Mukuni village were also shown a tree under which the explorer is said to have sat and talked with Mukuni's forebears and been granted Leya guides and porters. Even the place name 'Livingstone' was deemed appropriate – chiefly ritual had involved a mythical stone, the swallowing of which ensured continuity between generations, such that, in Mukuni's interpretation, 'living stone' was apposite as it represented the tradition of the Mukuni chiefly line.[50] For Mukuni, tourism was an opportunity not to be missed – in a place beset by repeated drought, where rainfed agriculture was never going to be reliable and stable jobs proved fickle, 'tourists are easier to reap than crops'.[51]

Mukuni's energetic embrace of tourism, and his success in impressing his own version of Leya history on the public face of the town, did not, however, go undisputed. The prominent marketing of Mukuni and Leya heritage was also linked to controversial efforts to assert authority over and above other Tonga, Toka and Leya chiefs in Zambia's Southern Province, and in urban politics too, had been matched with a campaign to elevate Leya over Lozi, including by controversially renaming the central 'Barotseland Park' as Mukuni park (this in a context where the original name derived from the elevated status of the Lozi in the early colonial period, and where urban politics had a history of bifurcating into Lozi/Tonga factions).[52] Mukuni's claims to privileged links with the waterfall were equally problematic for state heritage managers, who saw such exclusivity as undermining their own efforts to create a more inclusive and cosmopolitan sense of local history that incorporated the broad range of local ethnic groups and chiefs in the rural areas, as well as the mixed population of the town.[53] Nor was Mukuni's embrace of Livingstone uncontroversial: some managers in the Livingstone Museum resented international visitors' desire to see exhibits relating to the man above all else, and complained that EU funds for the Museum's refurbishment, which were pegged to a new 'Livingstone gallery', undermined their own efforts to create exhibits meaningful for domestic tourists. The new gallery was 'distorting' local and national history, they argued, and they felt that Livingstone the man had enough space devoted to him in the general gallery where he was integrated within a broader Zambian national historical narrative.[54]

This brief discussion of local claim-making at Victoria Falls and Livingstone, revolving around the politics of ethnic and national heritage and access to revenue from tourism, is important as one context in which the stories focussing on the waterfall recounted in the opening chapter of the book are told today. The focus at the resort on tourist revenue and public marketing of heritage has, however, taken us away from the materiality of the Zambezi waters themselves. International interests in local culture and heritage did not provide an opportunity for disputing land or water rights per se. Rather they provided an opportunity for local entrepreneurs to market a saleable version of their past and culture, and to use the idea of being a 'river people' with an authentic, spiritual relationship with the river as a resource in the politics of recognition, and as a lever for developmental funds. Yet conflicts over the water as a resource will undoubtedly continue to shape claim-making in the future, especially as water scarcity has continued to be a critical feature of rural life in the mid-Zambezi valley and the prospect of 'water wars' in the region promises to bring further state and non-state actors into the politics of landscape and Zambezian water rights.

At the time of this research, various further interventions were being mooted. The proposed Batoka dam for the gorges below the Victoria Falls had only just been shelved after a decade of feasibility studies and consultations, while pegging was underway for a dam on the Gwai/Shangani, heralding future displacements for people from parts of Hwange, Binga and Lupane. Meanwhile, the 15-year-long NGO campaign to implement the Matabeleland Zambezi Water Project (and related Trust) to supply Zambezi water by pipeline to Bulawayo continued and at times it seemed close to initiation. New proposals framing the project as a broader 'developmental corridor' along the pipeline, for example, stimulated a renewed speculative land grab on the part of ministers and generals along the pipeline's course.[55] At the same time, Botswana, Namibia and South Africa continue to eye Zambezian waters: commentators see their future abstraction of water as a certainty within a 20-year time frame, given the lack of alternative sources of water to support rapid urbanization and modern urban lifestyles, and given the enthusiasm for ambitious high modernist schemes in both countries.[56] The new Zambezi Watercourse Commission established under SADC in July 2004 as a forum for eight states to negotiate water rights is only beginning to find its feet, and its dynamics are so far unclear.[57] Yet the constraints on effective water partitioning will surely be huge when dealing with so many states and interests, and given the wide disparities in the power and bargaining weight of the parties. The story of the river as a border and resource told in this book has focussed on a particular stretch of the river and the two states of Zimbabwe and Zambia, but upstream and downstream connections have always been crucial, and will become prominent in the Commission's future debates.

Future interventions will surely rejuvenate the local claims and grievances described in this book, whether or not new interest groups advocate genuine local consultation and participation. The campaign over the Bulawayo pipeline, for example, has for Tonga leaders simply highlighted their own marginalization, as 50 years of campaigns have yet to solve water scarcity among those displaced by the dam and living in its immediate vicinity. Moreover, the precedent for genuine local collaboration set by other transboundary state projects has not been encouraging. We glimpsed the beginnings of such a dynamic in the new transnational planning bodies responsible for Kariba, discussed in Chapter 9, which had not transformed property rights fundamentally and had taken local representatives away from the practicalities of everyday control over the lake, making the new management structures increasingly irrelevant to the developing real economy of the lake.

Rather than speculating further over new developments and interests, however, in the final section of the book I will instead look back over the 150 years of claim-making we have considered, as each time authority over the Zambezi has been reconfigured and new interventions planned, so contemporary actors have reassessed versions of the politics of the past and entangled them within new claims.

Conclusion

Looking back over the chapters of the book, we can see that the idea of being a 'river people', and the historical fixity and locality it proclaims, has co-existed with histories of migration and mobility, profound transformation of the landscape and networks of political relationships that have extended far beyond the valley itself. Although control over crossing granted a degree of independence, and the sacred 'places of power' in the landscape created the valley as home for small decentralized groups, state claims to the landscape have long shaped local politics: relations with these more powerful state others has been a persistent theme. From the nineteenth-century frontier politics of Kololo, Lozi and Ndebele states, through the colonial and post-colonial politics of the Zambezi borderlands, the strategic role of the river has consistently embroiled local riverine politics in broader processes of state-making. The book has endeavoured to highlight the role of these translocal relationships in the history of claims and related appropriation of the river's energy, resources and surrounding landscape. My argument has been less that the politics of the Zambezi frontier can shed light on the shifting character of these more powerful states (though at times it has done so), than that the history of the making of the border landscape cannot be understood outside the broader history of state- and nation-building.

The material transformations, reconfigured access, and shifting strategic and symbolic role of the river associated with this process has been driven by a modernist narrative of 'conquest over nature', which Blackbourn has shown to be powerful for a long time in European contexts. This metaphor has been a potent driving force in the three episodes of European intervention along the Zambezi and their legacies discussed here – from Livingstone's explorations to the engineering works that re-shaped the river's strategic role and re-channelled its energies. The metaphor of conquering nature provided a powerful means of justifying colonial occupation and white settlement, and was central in upholding racial hierarchies. The technological projects along the river that made such conquest visible and self-evident were thus transformed into symbolic 'sites of memory' for white Rhodesia and effaced older African claims and meanings in the process. While the history of African nationalism in the borderlands and post-colonial minority mobilization can be understood in part as a history of resistance to these interventions, the evident power of technology to transform, rendered tangible through the steel and concrete of monumental bridges and dams, has also left its legacies in an ardent desire for modernity and development among those who have borne its costs.

I have tried to avoid the 'optimistic' over-arching narrative that can accompany this metaphor, just as I have tried to avoid its pessimistic counterpart. Rather I have tried to represent the viewpoints of the different actors at different historical junctures. However, the tone of the book, and the views of the current cast of actors it discusses, have inevitably been shaped by recent scientific assessments of the value of the interventions involved. Thus, although the weight of Lake Kariba is such that it has caused an increase in earthquakes and has posed a risk to the dam's safety, so far the dam has stood the test of time. It has not suffered siltation on a scale to pose a threat, and has held up during exceptional floods, partly due to the hasty modifications made during the course of its construction, when it was nearly breached by unprecedented flows. Fifty years after its completion, the World Commission on Dams evaluated Kariba as successful in its own narrow technical terms – as an efficient generator of cheap hydroelectric power.[58] The productivity of the fishery might have been over-estimated but was still an important secondary benefit, along with tourism; and the social costs of displacement are now acknowledged not just as planning errors, but as injustices that deserve recompense. Adverse ecological effects downstream, which have not been considered here, are now incorporated into dam planning. In short, international opinion regards the resettled Tonga as 'the main losers', against 'those who gained' – the 'millions of electricity consumers, the copper mines and other industries ... National Parks ... fishermen and workers in the fish industry, and those who find employment in tourism'.[59] Aside from legitimating claims to reparations, this has no doubt also encouraged state and non-state actors in the broader water-scarce region to continue to advocate large as well as small engineering works as the way ahead, and to support water-demanding urban lifestyles and further industrial development. Even Thayer Scudder, a long-standing critic of the social and environmental effects of Kariba and other large dams, sees these structures as 'a necessary development option for providing water and energy resources'.[60]

If one of the consequences of the history of intervention has been a convergence of opinion over the desirability of development (especially on the part of those who bear the costs and see few of the benefits), another has been the spread of modern cultural nationalist ideas about identity and culture, the self and the past, and notions of citizenship and rights. The ethnic frame of local claim-making along the river today partly reflects the legacies of colonial processes of subjectification – which elevated tribe and race as prime social cleavages and the basis of subjecthood. Though part of the story told here has been of resistance to colonial rule and related appropriation of the landscape to the myths of white Rhodesian and 'natural wilderness', mobilization around minority identities involved collaboration between local intellectuals, missionaries and colonial officers. The modern, territorialized notions of ethnic identity they have fostered include a reified notion of culture and bounded spatial units that replicate European ideas of national identity. These collective identities have had a potent landscape dimension, and are seen to inhere in particular 'sites of memory'. The narrative told in this book has counterposed settler claims to nationhood in the region, with the emergence of ethnicized claims to the same landscapes, rooted in different readings of symbolic sites along the river, and framed by processes of African state- and nation-building.

For those claiming the status of 'river people' today, notions of ethnic heritage and related ethnohistories have involved recasting former 'places of power' in the landscape associated with ancestors as memorial sites to a fast disappearing culture and as relics of the past. This shift has allowed such sites to be celebrated by modern, Christianised, educated, sometimes urban, elites, as well as facilitating their commodification. All such 'sites of memory', however, do not cluster along the river, as there is nothing deterministic about relations between landscape and identity: those who lived closely with the river in the late nineteenth century did not all regard themselves as river people, and have not all invoked the riverine landscape as a central focus. For the Tonga displaced by the dam, I have argued that it is the displacement itself rather than the old, now submerged places of power along the river that has emerged as the foundational memorial 'site' in modern public identity.

The ethnic frame for claim-making, particularly where it has involved 'minorities', has not generally worked against nationalism; for the architects of ethnic heritage in these movements (who were also African nationalists) there was no contradiction between securing inclusion and equal rights within the nation and recognition as a tribe. An ethnicized frame for understanding national belonging has, moreover, been validated in a post-colonial context by globalized discourses of multiculturalism, and, at times, through post-colonial state-building and civil society reform movements mobilizing around rights and the constitution. The movements for cultural recognition that have featured in this book are not inherently backward-looking and divisive, and their developmentalism is as striking as their cultural assertion. I have tried to show that the sense of grievance that fuels them does not have an intrinsic politics that can be evaluated as either positive or negative, but rather is available either for incorporation into a schismatic politics of patronage, or for a forward-looking inclusive cosmopolitan politics of rights, citizenship and rule of law.

Though the idea of being a 'river people' is not always deployed at the level of conscious manipulation implied by strategic essentialism, this book has focussed on the politics of such movements. We have seen how the idea of a privileged relationship with the river has been used as a resource in different times and contexts, how older discourses have become entangled and sedimented in current rhetorics, encapsulated in some ways, even as they have changed over time. Throughout this consideration of the historical politics of landscape on the Zambezi, I decided to use Nora's contested notion of 'sites of memory', because the term has a utility in drawing attention to the ways in which notions of identity and belonging have been reproduced through the power and nostalgic popular appeal of particular sites, 'where memory crystallizes and secretes itself'.[61] By historicizing the construction of these sites at different periods of the past, and by considering the contested narratives and practices of each historical layer, I hope, however, to have avoided implying any 'fundamental opposition' between memory and history, or any romance of community and tradition that would cast embodied ritual as 'authentic', primordial, timeless or unchallenged.[62] Rather, to use Raphael Samuel's metaphor, I hope to have 'unravelled' both the changing content and political charge of such ethno-nationalist constructions.[63]

The traditional sacred places in the mid-Zambezian landscape no less than later 'sites of memory' were intensely political and their powers were upheld in the context of disputed narratives and not simply through unthinking emotion and embodied practice. Even as the landscape worked to naturalize authority and ritual at such sites, and created a sense of belonging and community, there was an intrinsic politics involved, related to who came first, to the remembering and forgetting of old hierarchies or slave status, which was implicit in who led ceremonies, in what language, and with what repertoire of song and dance. Moreover, in the late nineteenth century, authority along the river was shaped by the realities and insecurities of life on a contested frontier, the remembered violence of which has at various times worked to undermine nostalgia for the pre-colonial past. The river – or rather, sites along it – has been incorporated so incompletely into such very different and disputed political projects, that historical contestation undermines any romance of a singular united past. The settler nation-building projects that appropriated the river at different times were contested among settlers themselves quite aside from provoking opposition from African nationalists and local interests. By virtue of their location on a frontier, these sites have often been marginal to post-colonial nation-building, and have been appropriated by local, marginalized groups whose recognition has been incompletely achieved, and whose grievances remain unsettled.

I hope that the lengthy genealogy of the river's role as frontier has become clear, reflected not only in its designation as 'natural' colonial state border, but also in its role as barrier for pre-colonial African states. As such, the colonial view of the river as a space of disorder and violence, occupied by people who were 'other', was reframed rather than invented, and was made credible by drawing on internal African frontier discourses and related political and moral hierarchies. As notions of modernity and backwardness were mapped onto earlier ideas of frontier violence, each episode of intervention along the river was legitimated partly through juxtaposition of the developmental force of colonial expansion with its counterpoint in the primitive stagnation and wilderness of the frontier. The river's border status has been repeatedly recharged, not only by each successive technological intervention, but also by episodes of war or economic decline on one or the other side. Each of these changes has provoked a new round of frontier discourses and has caused a shift in border identities. We left the border, most recently, at a time when Zimbabweans were implicating themselves in the risky and violent cross-border economies previously seen as inferior 'Zambian' spaces. Although these new livelihoods demanded a blurring of the boundaries between legal and illegal, and the regimes of authority over resources and the border became notably 'illegible', these unregulated networks were enmeshed with bureaucratic controls, and state agents were implicated within them. They threw light on how the Zimbabwean state was being reconfigured in an era of economic decline, heightened repression, and authoritarian discourses of 'patriotic history'; even in remote peripheries along the Zambezi, state power continued as a formidable presence that affected everyday life and relationships. As the struggle to survive and make a living deepened in Zimbabwe, the materiality of the water, and the border itself, became increasingly important, as the stocks of fish that provided the basis for livelihoods or simply relish were replenished annually by the rains, and economic discrepancies created by the border provided a resource that could create value at a time when money was depreciating, the state was not remunerating workers beyond its repressive arms, and insecurity was rife.

Amidst this long history of mobility, of social and political change and radical physical transformation of the mid-Zambezian frontier, the landscape itself has proved duplicitous in its ability to conjure a sense of permanence and continuity and to mask conflict. Yet the diametrically opposed interests that have invoked a connection with the river at different times, and the contrasting ends to which such ideas have been used, have exposed this deception for what it is. The politics of landscape on the Zambezi has been shaped by the processes of post-colonial state- and nation-building, and the persistent marginality of much of the valley. Yet the idea of privileged relations with the river stretching back through time has not lost its salience. Rather the idea of being 'river people' has continued to act as a means of speaking back from the margins, asserting rights to resources, inclusion and development. Knowledge of the skills to cross remain as useful as at any time in the past 150 years.

Notes

1 On the convergence between 'history' and 'heritage' see Lowenthal 1998, Parsons 2006.
2 Parsons ibid.
3 Fairweather 2006.
4 Discussion, NMMZ staff, Bulawayo, 17 July 2000.
5 Holmes 1993:xiv. On Zimbabwean post independence school texts' treatment of colonialism and whites, see Barnes 2004, 2007.
6 The history of the Nambya Cultural Association is discussed in greater length in McGregor 2005a.
7 The growth of 'venture tourism' around white-water rafting in the gorges downstream from the waterfall reinforced this emphasis on natural scenery. On growth of the venture tourist sector, see IUCN 1996:55-56.

8 McGregor 2005a.
9 IUCN 1996:20 based on a joint bid from Zimbabwe's DNPWLM and the Zambian National Heritage Conservation Commission.
10 The number of international visitors rose 4 fold in the early 1990s. IUCN ibid.:1.
11 Pikirayi 2006.
12 IUCN 1996:59.
13 Kumirai et al. 2000:10; J. Muringaniza, pers. comm., Bulawayo, 18 July 2000.
14 IUCN 1996: 44.
15 Residents' grievances were many, and focussed on the lack of housing and sky-high rents, irregular and inadequate water supply, lack of schools, overloaded sewerage system and burst sewer pipes, and by the public health issues created by a huge and growing mountain of rubbish that greeted those entering the township. IUCN 1996:60-61. Interviews, Victoria Falls residents, 20-21 July 2000.
16 In 1992, UNESCO adopted the idea of 'cultural landscapes', three categories of which were defined and incorporated into its Operational Guidelines. Fontein 2006:190-191.
17 Fontein 2000:45.
18 In 2003, the Matopos Hills in Zimbabwe was re-designated a 'cultural landscape', 20 years after being recognized purely for its 'natural' credentials. Ranger 1999:228.
19 Interview, J. Muringaniza, Natural History Museum, Bulawayo, 17 July 2000; see also Kumirai et al. 2000.
20 Interview, J. Muringaniza, Bulawayo, 17 July 2000.
21 Muringaniza 1995.
22 Ibid.
23 Ibid.
24 Kumirai et al. 2000.
25 Interview J. Muringaniza, 17 July 2000; Kumirai et al. ibid.
26 Group interview, Magomba village, Chisuma, 23 March 2000. Locals told of two nearby projects – Matupula camp and Gorges Lodge – initially developed with little consultation. When chalets were destroyed in a fire, it was said the ancestors were offended, and those responsible for the site approached local communities, who brewed beer and danced to intercede for success for the camp. Magomba is of Leya descent, but now identifies as Nambya.
27 Interview, Timothy Shakani Kwaju, Monde, 22 March 2000.
28 Hasler 1993. Group interview, Magomba family, Chisuma, 23 March 2000.
29 Interview, Maxon Musaka Ndlovu, Chidobe, 27 March 2000.
30 Interviews, Maxon Musaka Ndlovu Chidobe, 27 March 2000; group interview, Magomba family home, Chisuma, 23 March 2000.
31 Muringaniza 2000.
32 McGregor 2005a.
33 Interview Maxon Musaka Ndlovu, Chidobe, 27 March 2000; Esinath Mambaita Kasoso, Milonga, 11 April 2000.
34 Detailed in McGregor 2005a.
35 Interview, Mathias C. Munzabwa, Hwange, 3 April 2000, and others in the Dombe Cultural Association.
36 Interview, Mankonga Mapeta, Simangani, 24 March 2000; Mathias C. Munzabwa, Hwange, 3 April 2000.
37 Interviews, view, Josiah Siamwenda Chuma, Deka drum, 21 March 2000; Mankonga Mapeta, Simangani, 24 March 2000.
38 This charge was made both by the District Administrator and Ministry of Education officials. Nambya Cultural Association members, pers. comm.
39 Interview, Hwange Youth Development Association, Hwange 15 August 2001.
40 'Sun sets on Zimbabwe tourism', by Alex Machipisa', http://www.bbc.co.uk, 14 March 2001 (accessed July 2006).
41 IUCN 1996:8.
42 For further detail, see McGregor 2002:9-37.
43 *Zimbabwe Standard*, 28 April 2001.
44 On patriotic history, see Ranger 2004.
45 In the early days of the occupations, chief Shana sat on the district fast track land force, but felt 'we should have been more involved, but you can't act, everything is political and already done.' Interview, 17 March 2001.
46 N. Ncube, 'Chefs in fresh land grab orgy in Matabeleland North', *Financial Gazette*, 25 August 2005
47 IUCN 1996.
48 Interview chief Mukuni, 12 and 14 September 2001; interview Bedyango, 12 September 2001.
49 *Daily Telegraph*, 1 August 2004.
50 Interview, chief Mukuni 12 and 14 September 2001; address to heritage conference participants, 5 July 2004; interview Bedyango 12 July 2001.

51 Interview, chief Mukuni 14 September 2001; address to heritage conference, 5 July 2004.
52 See discussion in *The Livingstonian* 5 (1999) and 12 (2000).
53 Mufuzi 2004.
54 Ibid.
55 On the history of the MZWT, see Gunby et al. 2001.
56 L. Swatuk, presentation to the conference, 'The Power of Water', Edinburgh, April 2007.
57 L. Munjoma n.d.,IRIN, 14 July 2004, 'Commission to Manage Zambezi Waters', available from http://www.irinnews.org/report.aspx?reportid=50649 (accessed 27 June 2007).
58 WCD (2000:v) calculates that electricity prices fell 30% over the period 1961-77, in a context where other costs rose by 75%. See also the discussion in Scudder 2005:188-210.
59 WCD, ibid.:xi.
60 Scudder 2005:293.
61 Nora 1989:7.
62 Ibid.:8.
63 Samuel 1998.

Sources & Bibliography

Archival Sources

Unless otherwise stated, archival references are from the National Archives in Zimbabwe, or from the British National Archives at Kew (codes DO/CO).

NATIONAL ARCHIVES OF ZIMBABWE, HARARE (NAZ)

A3 18/28	CNC Taylor, A Short History of the Tribes of the Province of Matabeleland', Bulawayo,11 January 1904.
A11/2/17/1-2	Victoria Falls Misc and Correspondence
B 13/5-7	ANC Hemans to Supt. Natives, 26 May 1916
D 3/8/1	Criminal cases, Sebungwe 1903-13
D 3/8/2	Criminal cases, Sebungwe 1914-18
D 3/8/3	Criminal cases, Sebungwe 1919-26
D 3/37/1-8	Criminal cases, Wankie, 1891-1920
LO 5/6/8	William Reid to administrator, Bulawayo, 23 February 1897
MS 936	Vetoka Correspondence
NB 3/1/6	Correspondence, 1906
NB 6/1/2	Sebungwe Annual report, 1899
NB 6/1/3-4	District Annual Reports, 1900-03, 1907
NB 6/1/5-7	District Annual Reports, 1904-7
NB 6/1/8-10	District Annual Reports, 1908-9
NB 6/4/1	Monthly reports, Sebungwe: August 1897; September 1897; October 1897.
NB 6/4/4	Monthly reports, Sebungwe: August-December 1899, January-May 1900
NB 6/4/4	Monthly report, Wankie: February 1903
NB 6/4/5-6	Monthly reports, 1903-5
NB 6/4/7	Monthly reports, Sebungwe: June-December 1906
NB 6/6/1	CNC, Annual report for Matabeleland Province, 1898
NB 6/5/2/2	Val Gielgud, Report on Sebungwe district, 8 January 1898
NB 6/5/2/5	ANC Green, Report on patrol to Zambezi valley, 14 December 1897
NBH 1/1/1	Annual Reports, Wankie, 1903, 1912
NGB 2/7/1	Correspondence, Sebungwe District
NGB 2/2/5	Corresponence, Sebungwe, Native Marriage Ordinance
NGB 2/3/1	Correspondence, Sebungwe
NGB 3/1/1	Annual Report, Sebungwe, 1912, 1914
N3/23 1-3	Native Marriage, Garidzela, 1905-6

N9/1/16-26	Annual Reports, 1913-1924
S 246/277-8	Victoria Falls Power Concession
S 482/134/48-9	Victoria Falls Power Concession
S 142/13/21	Victoria Falls – press cuttings of opening of bridge
S 707	Wankie Correspondence 1911-1923
S 235/501-518	Annual Reports, Sebungwe, Wankie 1923-1946
S 273	Court Records, Gokwe 1924-6,1928
S 1563	Annual Reports 1937-1947
S 1194/1614/1/1-7	Victoria Falls Reserve 1936-42
S 1194/1614/2/1	Victoria Falls Correspondence
S 1563	Annual Reports, 1936, Wankie District
S 2806/1970	Sebungwe Land 1940-1956
S 2403/2681	District Annual Report Wankie, 1952
S 2827/2/2/3 vol 2/3	District Annual Report, Wankie, Gokwe 1955
S 2827/2/2/4 vol 1/3	District Annual Reports, Wankie, Gokwe 1956
S2827/2/2/5 vol 2/3	District Annual Reports, Wankie, Binga, Gokwe 1957
S 2827/2/2/6 vol 2/3	District Annual Reports, Wankie, Binga, Gokwe 1958
S 2827/2/2/7 vol 1/2	Disrict Annual Reports, Wankie, Binga, Gokwe 1959
S 2827/2/2/8 vol 1/3/5	District Annual Reports, Wankie, Binga, Gokwe1961
S 2929/5/1-2	Binga – Delineation Report
S 2929/5/7	Wankie – Delineation Report
S 2929/7/3	Gokwe – Delineation Report
S 924/9114/1-4	Victoria Falls Native Village Scheme
ORAL FL 1	Sir Patrick Fletcher
ORAL CA 5	Sir John Caldicott
ORAL ST 6	Albert Rusbridge Stumbles
ORAL 227	Richard John Powell
ORAL 241	Rupert Meredith Davies
ORAL QU 2	H.J. Quinton

LIVINGSTONE MUSEUM, LIVINGSTONE, ZAMBIA

'Short History of the Baleya People of Kalomo District', unpublished manuscript, 1957

G 5/2/1 Duncan Watt, 'Thesis on the Town of Livingstone', ms.

G 102 Government messenger Sekwaswa, 'History of Livingstone District', unpublished ms, n.d., ca. 1936

BULAWAYO REFERENCE LIBRARY

Kariba Box *Kariba: Opening by her Majesty Queen Elizabeth the Queen Mother Tuesday 17 May* Federal Power Board 1961.

Royal Occasion 1960: The Kariba Project, S. Rhodesian Government Printer, n.d.

Rhodesia's Lake Kariba, Rhodesia National Tourist Board, ca. 1968

Rhodesia Calls, Mardon Printers Ltd, ca.1966

Kariba: The First Ten Years, Central African Power Corporation, 1971

Victoria Falls Box *The Victoria Falls of Southern Rhodesia*, Government Printer, 1936

A Guide Book to the Victoria Falls, Bulawayo, 1960

BULAWAYO RECORDS OFFICE

Records for Wankie District, 1966-1979. Now transferred and catalogued by the National Archives of Zimbabwe.

Box 5/6/9R	Wankie Local Board, 1974-78
Box 4/6/5R	LAN 21 1971-77

Box 1/7/5F Wankie Annual Reports, 1969, 1972
Box 5/8/8/F Running File 1975, 1977; Civil Defence, Victoria Falls

NATIONAL ARCHIVES, KEW, LONDON, UK

DO 35/4602 Kariba - Correspondence
DO 35/4604 Kariba – Fortnightly summary of news
DO 35/4605 Kariba – Correspondence
DO 35/4606 Kariba – Corresponence, incl. Mr Hunt's diary, 1957
DO 35/4607 Kariba - Correspondence
CO 1015/1529 Strike by workers at Kariba dam, 1959
CO 1015/958 Kariba, Nyasaland chiefs' visit, 1956
CO 1015/946 Kariba, Letter, Sir A. Benson to W. Gorell Barnes 1 April 1955
CO 1015/948 Kariba, Minute by W. G. Wilson, 1 May 1956
Federal Newsletter 1955-1959

RHODES HOUSE OXFORD

Report of the Land Commission 1946 (Northern Rhodesia)
South Africa Handbook No. 6 [n.d. ca. 1903]
David Livingstone Unveiling Ceremony, *The Address and Programme*, August 1934

THOMAS COOK ARCHIVES, PETERBOROUGH

Traveller's Gazette

DISTRICT ADMINISTRATOR'S OFFICE & RURAL DISTRICT COUNCIL OFFICE, BINGA

Binga District Council Minutes/Rural District Council Minutes
Drought Committee File
Binga District Team Minutes
Binga District Lusumpuko Plan, December 1981
Social Development File
Binga Development Association (BIDA) File
Fishing/National Parks File
Binga Chiefs File

DISTRICT ADMINISTRATOR'S OFFICE & RURAL DISTRICT COUNCIL OFFICE, HWANGE

Hwange Chiefs File
Hwange District Council Minutes/Rural District Council Minutes
Hwange District Team Minutes

OTHER

Nambya Cultural Association Papers

Bibliography

PUBLISHED SOURCES

Alexander, J. 2006. *The Unsettled Land: Land and Politics in Zimbabwe*, Oxford, James Currey.
Alexander, J. and J. McGregor, 1997. 'Modernity and Ethnicity in a Frontier Society: Understanding Difference in Northwest Zimbabwe', *Journal of Southern African Studies*, 23, 2, pp. 187-203.
— 2004. 'War Stories: Guerrilla Narratives of Zimbabwe's Liberation War', *History Workshop Journal*, 57, pp. 79-100.
Alexander, J., J. McGregor and T. Ranger, 2000. *Violence and Memory: One Hundred Years in the 'Dark Forests' of Matabeleland, Zimbabwe*, Oxford, James Currey.
Alexander, J. and T. Ranger 1998. 'Competition and Religious Integration in Northwest Zimbabwe', *Journal of Religion in Africa*, 28, 1, pp. 3-31.
Allan, W. and M. Gluckman (eds), 1968. *Land Holding and Land Usage Among the Plateau Tonga of Mazabuka District: A Reconaissance Survey*, Manchester, Manchester University Press.
Armstrong, L. and Neville Jones, 1936. 'The Antiquity of Man in Rhodesia as Demonstrated by Stone Implements of the Ancient Zambezi Gravels, South of Victoria Falls', *Journal of the Royal Anthropological Institute*, 66, pp. 331-48.
Arnot, F.S. 1969. *Garenganze: Or 7 Years Pioneer Mission Work in Central Africa*, London, Frank Cass.
Asiwaju, A.I. (ed.) 1984. *Partitioned Africans: Ethnic Relations Across Africa's International Boundaries 1884-1994*, London, Hurst.
Baines, T. 1864. *Explorations in South West Africa*, London, Longman, Green, Roberts & Green.
Baldwin, W.C. 1863. *African Hunting and Adventure from Natal to the Zambezi*, London, Richard Bentley.
Balneaves, E. 1971. *Elephant Valley*. London, John Gifford.
Bamgbose, A., R. Fardon and G. Furniss (eds), 1994. *African Languages, Development and the State in Africa*, London, Routledge.
Bamgbose, A. 2000. *Language and Exclusion: The Consequences of Language Policies in Africa*, Hamburg, Lit Verlag.
Barnes, T. 2004. 'Reconciliation, Ethnicity and School History in Zimbabwe', in B. Raftopoulos and T. Savage (eds), *Zimbabwe: Injustice and Political Reconciliation*, Cape Town, Institute for Justice and Reconciliation, pp. 140-59.
— 2007. 'History has to Play its Role: Constructions of Race, Reconciliation and Secondary School Historiography in Zimbabwe, 1980-2002', *JSAS*, 33, 3.
Barrell, J. 2000. 'Death on the Nile: Fantasy and the Literature of Tourism', in C. Hall (ed.), *Culture of Empire: A Reader*, Manchester, Manchester University Press.
Beach, D. 1980. *The Shona and Zimbabwe 900-1850*, Gweru, Mambo Press.
Beinart, W. 1999. 'The Renaturing of African Animals: Film and Literature in the 1950s and 1960s', in P. Slack (ed.), *Environments and Historical Change*, Oxford, Oxford University Press.
Beinart, W. 2000. 'African History and Environmental History', *African Affairs*, 99, pp. 269-302.
Beinart, W. and J. McGregor (eds), 2003. *Social History and African Environments*, Oxford, James Currey.
Bender, B. (ed.) 1993. *Landscape: Politics and Perspectives*, Oxford, Berg.
Berger, J. 1977. *Ways of Seeing*, London, Penguin.
Berman, B. 1998. 'Ethnicity, Patronage and the African State: The Politics of Uncivil Nationalism', *African Affairs*, 97, 388, pp. 305-41.
Berman, B., D. Eyoh and W. Kymlicka (eds), 2004. *Ethnicity and Democracy in Africa*, Oxford, James Currey.
Berry, S. 1993. *No Condition is Permanent: The Social Dynamics of Agrarian Change in Sub Saharan Africa*, Madison WI, Wisconsin University Press.
Blackbourn, D. 2006. *The Conquest of Nature: Water, Landscape and the Making of Modern Germany*,

London, Jonathan Cape.
Blok, A. 2000. 'The Enigma of Senseless Violence', in G. Aijmer and J. Abbink (eds), *Meanings of Violence: A Cross Cultural Perspective*, Oxford, Berg.
Bourdillon, M.F.C., A. Cheater and M.W. Murphree, 1985. *Studies on Fishing on Lake Kariba*, Gweru, Mambo Press.
Bridges, R. C. 1994. 'Maps of East Africa in the Nineteenth Century', in J. Stone (ed.), *Maps and Africa*, Aberdeen University, Aberdeen African Studies Group.
Brown, R. 1966. 'Aspects of the Scramble for Matabeleland' in E. Stokes and R. Brown (eds), *The Zambezian Past: Studies in Central African History*, Manchester, Manchester University Press, pp. 63-93.
Cannadine, D. 2002. *Ornamentalism: How the British Saw their Empire*, London, Penguin.
Caplan, G. 1970. *The Elites of Barotseland 1878-1969*, London, Hurst.
Carruthers, J. and M. Arnold, 1995. *The Life and Works of Thomas Baines*, South Africa, Fernwood Press.
Castree, N. 2003. 'Environmental Issues: Relational Ontologies and Hybrid Politics', *Progress in Human Geography*, 27, 2, pp. 203-11.
Chapman, J. 1868. *Travels in the Interior of South Africa 1849-1863 Vols 1 and 2*, London, Bell and Daldy.
Child. B. 1987. 'The Management of Crocodiles in Zimbabwe', in G. J. W. Webb, S.C. Manolis, P.J. Whitehead (eds), *Wildlife Management:Crocodiles and Alligators*, Australia, Surrey Beatty and Sons.
Christie, A. 1924. *The Man in the Brown Suit*, London, Bodley Head.
Clark, J.D. 1952. 'The Native Tribes' in Clark, J.D. (ed.) *The Victoria Falls*, Livingstone, Government Printer.
Clark, P. 1936. *Autobiography of an Old Drifter: The Life of Percy Clark*, London, Harrap
Clay, G. 1968. *Your Friend Lewanika: Litunga of Barotseland 1842-1916*, London, Chatto and Windus.
Cleaver, F. 2003. 'Reinventing Institutions: Bricolage and the Social Embeddedness of Natural Resource Management', in T. A. Benjaminsen and C. Lund (eds), *Securing Land Rights in Africa*. Frank Cass, London.
Clements, F. 1959. *Kariba: The Struggle with the River God*, London, Methuen.
Coillard, F. 1897. *On the Threshold of Central Africa: A Record of Twenty Years Pioneering Among the Barotsi of the Upper Zambezi*, London, Hodder and Stoughton.
Colson, E. 1950. 'A Note on Tonga and Ndebele' *Northern Rhodesia Journal*, 1, 2, pp. 35-42.
— 1960. *The Social Organization of the Gwembe Tonga*, Manchester, Manchester University Press.
— 1962. *The Plateau Tonga of Northern Rhodesia. Social and Religious Studies*, Manchester, Manchester University Press.
— 1969. 'Spirit Possession Among the Tonga of Zambia', in J. Beattie and J. Middleton (eds), *Spirit Mediumship and Society in Africa*, London, Routledge.
— 1970. 'The Assimilation of Aliens Among the Zambian Tonga', in R. Cohen and J. Middleton (eds), *From Tribe to Nation in Africa: Studies in Incorporation Processes*. San Francisco, Chandler, pp. 35-53.
— 1971. 'Heroism, Martyrdom and Courage: An Essay on Tonga Ethics' in T.O. Biedelman (ed.), *The Translation of Culture: Essays to Edward Evans Prichard*, London, Tavistock, pp. 19-35.
— 1975. *The Social Consequences of Resettlement. The Impact of the Kariba Resettlement upon the Gwembe Tonga*, Manchester, Manchester University Press.
— 1977. 'A Continuing Dialogue: Prophets and Local Shrines Among the Tonga of Zambia' in Werbner, Richard (ed.), *Regional Cults*, London, Academic Press, pp. 119-39.
— 1996. 'The Bantu Botatwe: Changing Political Definitions in Southern Zambia', in D. Parkin, L. Caplan and H. Fisher (eds), *The Politics of Cultural Performance*, Oxford, Berghahn, pp. 61-80.
— 1997. 'Places of Power and Shrines of the Land', in *Paideuma: Mitteilungen zur Kulturkunde*, 43, pp. 47-59.
— 2000. 'The Father as Witch', *Africa*, 70, 3, pp. 333-58.
— 2004. 'Leza into God – God into Leza', in B. Carmody (ed.), *Religion and Education in Zambia*, Ndola, Mission Press.
— 2006. *Tonga Religious Life in the Twentieth Century*, Lusaka, Bookworld Publishers.

Comaroff, J. and J. Comaroff, 1997. *Of Revelation and Revolution: The Dialectics of Modernity on a South African Frontier Volume 2*, Chicago, Chicago University Press.
Conyers, D. and F. Cumanzala, 2004. 'Community Empowerment and Democracy in Zimbabwe: A Case Study from Binga District', *Social Policy and Administration*, 38, 4, pp. 383-398.
Cosgrove, D. 1998. *Social Formation and Symbolic Landscape*, University of Wisconsin Press [first ed. London, Croom Helm, 1984].
Coupland, R. 1928. *Kirk on the Zambezi. A Chapter of African History*, Oxford, Clarendon Press.
Cowan, J. M-B. Dembour and R. Wilson, 2003. *Culture and Rights: AnthropologicalPerspectives*, Cambridge, Cambridge University Press.
Crewel, J. 1994. *A History of the Victoria Falls Hotel: Ninety Glorious Years*, Zimbabwe.
Croxton, A.H. 1973. *Railways of Rhodesia. The Story of the Beira, Mashonaland and Rhodesia Railways*, Newton Abbott, David and Charles.
Cunnison, I. 1951. 'History on the Luapula', *Rhodes Livingstone Papers*, 21, Oxford University Press.
Curriculum Development Unit, 1990. *A Report on the Survey of the Teaching/Learning of Minority Languages in Zimbabwe*, April 1990, Harare, Ministry of Education and Culture.
Das, V. and D. Poole, 1991. 'State and Its Margins: Comparative Ethnographies', in V. Das and D. Poole (eds), *Anthropology in the Margins of the State*, Oxford, James Currey.
Debenham, F. 1955. *The Way to Ilala: David Livingstone's Pilgrimage*, London.
Donham, D. and W. James (eds), 2002. *The Southern Marches of Imperial Ethiopia*, Oxford, James Currey.
Driver, F. 2001. *Geography Militant: Cultures of Exploration and Empire*, Oxford, Blackwell.
Dubow, S. 2000. 'A Commonwealth of Science: the British Association in South Africa, 1905 and 1929', in S. Dubow (ed.), *Science and Society in Southern Africa*, Manchester, Manchester University Press, pp. 66-99.
Duffy, R. 2000. *Killing for Conservation: Wildlife Policy in Zimbabwe*, Oxford, James Currey.
Duncan J. and D. Gregory (eds), 1999. *Writes of Passage: Reading Travel Writing*, London, Routledge.
Dzingirai, V. 2003. 'Campfire is not for Ndebele Migrants: The Impact of Excluding Outsiders from Campfire in the Zambezi Valley', *Journal of Southern African Studies*, 29, 2, pp. 445-59.
Dzingirai V. and M. Bourdillon, 1998. 'Religious Ritual and Environmental Control in the Zambezi Valley: The Case of Binga', Harare, University of Zimbabwe, Centre for Applied Social Studies, Working Paper.
Edney, M. *Mapping an Empire: The Geographical Construction of British India, 1765–1843*, University of Chicago Press.
Englund, H. and F. Nyamanjoh (eds), 2004. *Rights and the Politics of Recognition in Africa*, London, Zed Books.
Fagan, B. 1967. *Iron Age Cultures in Zambia, Vol 1*, London, Chatto and Windus.
Fairhead, J. and M. Leach, 1987. *Misreading the African Landscape*, Cambridge, Cambridge University Press.
Fairweather, I. 2006. 'Heritage, Identity and Youth in Postcolonial Namibia', *JSAS*, 32, 4, pp. 719-36.
Flint, L. 2006. 'Contradictions and Challenges in Representing the Past: The Kuomboka Festival of Western Namibia', *JSAS*, 32, 4, pp. 701-18.
Fontein, J. 2000. 'UNESCO, Heritage and Africa: An Anthropological Critique of World Heritage', Occasional Paper 80, Centre of African Studies, University of Edinburgh
— 2006. *The Silence of Great Zimbabwe: Contested Landscapes and the Power of Heritage*, London, University College London Press.
Ford, J. 1971. *The Role of Trypanosomiases in African Ecology*, Oxford, Clarendon Press.
Foskett, R. (ed.), 1965. *The Zambezi Journal and letters of John Kirk 1858-1863*, London, Oliver and Boyd.
Games, I. and J. Moreau. 1997. 'The Feeding Ecology of Two Nile Crocodile Populations in the Zambezi Valley', in J. Moreau (ed.), *Advances in the Ecology of Lake Kariba*, Harare, Zimbabwe, University of Zimbabwe Publications.
Gann, L.H. 1965. *A History of Southern Rhodesia: Early Days to 1934*, London, Chatto and Windus.
— 1954. *A History of Northern Rhodesia: Early Days to 1953*, London, Chatto and Windus.
— 1958. *The Birth of a Plural Society. The Development of Northern Rhodesia Under the BSAC 1894-*

1914, Manchester, Manchester University Press.
Geheb, K. and M-T. Sarch, 2003. *Africa's Inland Fisheries: The Management Challenge*, Kampala, Uganda, Fountain Publishers.
Gelfand, M. 1957. *Livingstone the Doctor: His Life and Travels. A Study in Medical History*, Oxford, Basil Blackwell.
— (ed.) 1968. *GuBulawayo and Beyond: Letters and Journals of the Early Jesuit Missionaries to Zambezia 1879-1887*, London, Chapman.
Gibbons, A. Sgt. Hill 1904. *Africa from South to North Through Marotseland*, London, Bodley Head
Giles-Vernick, T. 2001. 'Lives, Histories and Sites of Recollection' in White, L., S. F. Miescher and D. W. Cohen (eds), *African Words, African Voices: Critical Practices in Oral History*, Bloomington, Indiana University Press.
Gillies, C. 1999. *Kariba at the Millenium*, Bulawayo, Zimbabwe.
Girard, R. 1977. *Violence and the Sacred*, translated by Patrick Gregory, Baltimore, Johns Hopkins Press.
Gluckman, M. 1951. 'The Lozi of Barotseland in Nortwestern Rhodesia' in E. Colson and M. Gluckman (eds), *Seven Tribes of British Central Africa*, London, Oxford University Press.
— 1941. *Economy of the Central Barotse Plain*, Rhodes Livingstone Institute, Paper No. 7.
Godlewska A. and E. Smith (eds), 1994. *Geography and Empire*, Manchester, Manchester University Press.
Gordon, D. 2006. *Nachituti's Gift: Economy, Society and Environment in Central Africa*, University of Wisconsin Press.
Gray, R. 1960. *The Two Nations: Aspects of the Development of Race Relations in the Rhodesias and Nyasaland*, Oxford, Oxford University Press.
Guelke, L. and J. Guelke, 2004. 'Imperial Eyes on South Africa: Reassessing Travel Narratives', *Journal of Historical Geography*, 31, 1, pp. 11-31.
Gunby, D., R. Mpande and A. Thomas, 2001. 'The Campaign for Water from the Zambezi for Bulawayo', in A. Thomas, S. Carr and D. Humphreys (eds), *Environmental Policies and NGO Influence: Land Degradation and Sustainable Resource Management in Sub-Saharan Africa*, London, Routledge, pp. 72-91.
Haarden, B. 1996. *A River Lost: The Life and Death of the Columbia*, New York, Norton.
Haarthoorn, A.M. 1970. *The Flying Syringe: Ten Years of Imobilizing Wildlife in Africa*, London, Cox and Wyman Ltd.
Hachipola, S.J. 1998. *A Survey of the Minority Languages of Zimbabwe*, Harare, University of Zimbabwe Press.
Hamilton, C. 1998. *Terrific Majesty: The Powers of Shaka Zulu and the Limits of Historical Invention*, Cambridge, MA, Harvard University Press.
Hanna, A.J. 1965. *The Story of the Rhodesias and Nyasaland*, London, Faber and Faber.
Harding, C. 1905. *In Remotest Barotseland: Being an Account of a Journey of Over 800 Miles Through the Wildest and Remotest Part of Lewanika's Empire*, London, Hurst and Blackett Ltd.
Harraway, D. 1989. *Primate Visions: Gender, Race and Nature in the world of Modern Science*, London, Verso.
Hasler, D. 1993. 'Brokering the Community Basis of Tourism Potential in the Hwange Communal Lands', Harare, Centre for Applied Social Sciences.
Hayes, M.E. 1977. 'The Nambiya People of Wange', *Native Affairs Department Annual*, 11, 4.
Hazelwood, A. 1967. 'The Economics of Federation and Dissolution in Central Africa' in A. Hazelwood (ed.), *African Integration and Disintegration*, Oxford, Oxford University Press.
Helly, D. 1987. *Livingstone's Legacy: Horace Waller and Victorian Myth-Making*, Athens OH, Ohio University Press.
Hemans, H.N. 1912. 'History of the Abenanzwa Tribe', *Proc. of the Rhodesia Scientific Association*, vol 12, pp. 85-112.
— 1935 *Log of a Native Commissioner*, London, Witherby.
Henson, G.L. 1973. 'History and Legend of the Vanambiya', *Native Affairs Departmental Annual*, 59, pp. 32-4.
Hiller, V. W. 1949. 'The Concession Journey of Charles Dunell Rudd', in C.E. Fripp and V.W. Hiller (eds), *Gold and the Gospel in Mashonaland 1888*, London, Chatto and Windus.

Hirsch, E. and M. O'Hanlon, 1995. *The Anthropology of Landscape*, Oxford, Oxford University Press.

Hofmeyer, I. 1994. "*We Spend our Years as a Tale that is Told": Oral Historical Narrative in a South African Chiefdom*, London, James Currey.

Hole, H. M. 1905. 'Notes on the Batonga and Bstshukulumbwi Tribes', *Proc. of the Rhodesia Scientific Association*, v, part II pp. 62-7.

— 1995. *The Passing of the Black Kings*, Africana Book Society, Bulawayo, Reprint [first ed., 1932].

Holleman, J.F. 1969. *Chief, Council and Commissioner: Some Problems of Government in Rhodesia*, Oxford University Press.

Holmes, T. (ed.), 1990. *David Livingstone, Letters and Documents, 1841-72*, London, James Currey.

— 1993. *Journey to Livingstone: Exploration of an Imperial Myth*, Edinburgh, Canongate Press.

Holub, E. 1881. *Seven Years in South Africa: Travels, Researches and Hunting Adventures Between the Diamond Fields and the Zambezi, 1872-1879, 2 Volumes*, London, Sampson Low, Marston, Searle and Rivington.

Hopgood, C.R. 1940. *Tonga Grammar*, London.

Howarth, D. 1961. *The Shadow of the Dam*, London, Collins.

Hughes, D. 2006. 'Whites and Water: How Euro-Africans Made Nature at Kariba Dam', *Journal of Southern African Studies*, 32, 4, pp. 823-39.

— 2006. *From Enslavement to Environmentalism: Politics on a Southern African Frontier*, Washington University Press.

IUCN, 1996. *Strategic Environmental Assessment of Developments Around Victoria Falls*, Zimbabwe, IUCN.

Isaacman, A. and B. Isaacman, 2004. *Slavery and Beyond: The Making of Men and Chikunda Ethnic Identities in the Unstable World of South-Central Africa, 1750-1920*, Portsmouth, NH, Heinemann.

Jackson, J. 1991. 'The Artisanal Fishery of Lake Kariba (Eastern Basin). A Sociological Input into Lakeshore Planning and Fisheries Management', Harare, Centre for Applied Social Studies.

Jalla, A.D. 1928. *Sur les Rives du Zambeze: Notes Ethnographiques*, Paris, Societé des Missions Evangeliques.

Jardine, N. and E. Spary (eds), 1996. *Cultures of Natural History*, Cambridge, Cambridge University Press.

Jaspan, M.A., 1953. *The Ila-Tonga Peoples of Northwest Rhodesia*, Edinburgh, International African Institute.

Jeal, T. 1973. *Livingstone*, London, Penguin.

Johnson, D. H., 1981. 'The fighting Nuer: Primary Sources and the Origins of a Stereotype', *Africa* 51, 1, pp. 509-27.

Jones, A.M. and H. Carter, 1967. 'The Style of a Tonga Historical Narrative', *African Language Studies*, 8, pp. 93-126.

Kagoro, B. 2004. 'Constitutional Reform as Social Movement', in B. Raftopoulos and T. Savage (eds), *Zimbabwe: Injustice and Political Reconciliation*, Cape Town, Institute for Justice and Reconciliation, pp. 236-256.

Kimbrough, R. (ed.), 1983. *J. Conrad, the Heart of Darkness. A authoritative Text: Backgrounds and Sources Criticism*, London and New York, W.W. Norton, Third edition.

Klein, K.L., 2000. 'On the Emergence of Memory in Historical Discourse', *Representations*, 69, pp. 127-50.

Knight, J. (ed.), 2000. *Natural Enemies: People-Wildlife Conflict in Anthropological Perspective*, London, Routledge.

Kolding, J., B. Musando and N. Songore, 2003. 'Inshore Fisheries and Fish Population Changes in Lake Kariba' in E. Jul Larsen, J. Kolding, R. Ovcra, J. R. Nielsen, P. Van Zwieten (eds), *Management, Co-Management or No Management? Major Dilemmas in Southern African Freshwater Fisheries*, FAO Fisheries Technical Papers 426/2, FAO, Rome.

Kopytoff, I. (ed.), 1989. *The African Frontier: The Reproduction of Traditional African Societies*, Bloomington, Indiana University Press.

Kreike, E. 2004. *Re-Creating Eden: Land Use, Environment and Society in Southern Angola and Northern Namibia*, Portsmouth NH, Heinemann.

Kuper, A. 2005. *The Reinvention of Primitive Society. Transformations of a Myth*, London, Routledge.
Lagus, C. 1959. *Operation Noah*, London, William Kimber.
Lamplugh, G.W. 1908. 'The Gorge and Basin of the Zambezi Below the Victoria Falls, Rhodesia', *Geographical Journal*, 31, pp. 133-52.
— 1907. 'The Geology of the Zambezi Basin Around the Batoka Gorge (Rhodesia)', in *Quarterly Journal of the Geological Society*, 73, pp. 62-79.
— 1906, 'Notes on the Occurrence of Stone Implements in the Valley of the Zambezi and Around the Victoria Falls', *Journal of the Royal Anthropological Institute*, 36, pp. 159-60.
Lancaster, C. 1974. 'Ethnic Identity, History and "Tribe" in the Middle Zambezi Valley,' *American Ethnologist*, 1, pp. 707-30.
Lancaster, C. and K. Vickery (eds), 2007. *The Tonga-Speaking Peoples of Zambia and Zimbabwe: Essays in Honour of Elizabeth Colson*, Maryland, University Press of America.
Larby, P.M. (ed.), 1987. *Maps and Mapping of Africa*, London, Scolma.
Legg, S. 2005. 'Contesting and Surviving Memory: Space, Nation and Nostalgia in Les Lieux de Memoire', *Society and Space*, 23, pp. 481-504.
Leopold, M. 2004. *Inside West Nile: Violence, History, Representation on an African Frontier*, Oxford, James Currey.
Livingstone, D. 1857. *Missionary Travels and Researches in South Africa*, London, Ward, Lock and Co.
Livingstone, D. and C. Livingstone, 1865. *Narrative of an Expedition to the Zambesi and its Tributaries*, London, Charles Murray.
Loveridge, J. P. 1996. *A Review of Crocodile Management in Zimbabwe*, Dept. of Biological Sciences, University of Zimbabwe, Harare, Zimbabwe.
Lowenthal, D. 1998. *The Heritage Crusade and the Spoils of History*, Cambridge, Cambridge University Press.
Luck, R. A. 1902. *Visit to Lewanika*, London, Simpkins, Marshall, Hamilton Kent and Co. Ltd.
Luig, U. 1999. 'Constructing Local Worlds: Spirit Possession in the Gwembe Valley, Zambia', in H. Behrend and U. Luig (eds), *Spirit Possession: Modernity and Power in Africa*, Oxford, James Currey.
Luig, U. and A. van Oppen, 1997. 'Landscape in Africa: Process and Vision. An Introductory Essay', in *Paideuma: Mitteilungen zur Kulturkunde*, 43, pp. 7-45.
Lunn, J. 1997. *Capital and Labour on the Rhodesian Railways System 1888-1947*, London, Macmillan.
Machena, C. and H. Moinuddin, 1993. *Management of the Kale Kariba Inshore Fisheries (Zimbabwe): A Proposal*. Zambia-Zimbabwe SADC, Fisheries Project (Lake Kariba) Report 19.
Mackenzie, F. 2000. 'Contested Ground: Colonial Narratives and the Kenyan Environment 1920-1945', *Journal of Southern African Studies*, 26, 4, pp. 698-714.
Mainga, J. 1973. *Bulozi Under the Luyana Kings: Political Evolution and State Formation in Precolonial Zambia*, London, Longmans.
Malasha, I. n.d. 'Colonial and Postcolonial Fisheries Regulations: The Cases of Zambia and Zimbabwe' FAO, http://www.fao.org/docrep/006/y5056e/y5056e13.htm [accessed 15 August 2006].
Malasha, I. 2003. 'The Outcome of a Co-Managerial Arrangement in an Inland Fishery: The Case of Lake Kariba (Zimbabwe)', in K. Geheb and M-T. Sarch, *Africa's Inland Fisheries*, pp. 89-106.
— 2002. 'Fishing in a Bathtub: A Comprehension of the Conflicts in the Lake Kariba Inshore Fishery (Zimbabwe)', Harare, Centre for Applied Social Studies.
Mann, K. and R. Roberts (eds), 1991. *Law in Colonial Africa*, Portsmouth NH, Heinemann.
Mansfield, C. 1911. *Via Rhodesia. A Journey Through Southern Rhodesia*, London, Stanley and Paul and Co.
Matless, D. 1998 *Landscape and Englishness*, London, Reaktion Books.
Matthews, T. 1981. 'Portuguese, Chikunda and Peoples of the Gwembe Valley: The Impact of the "Lower Zambezi Complex" on Southern Zambia', *Journal of African History*, 22, 1, pp. 23-41.
Maxwell, D. 1999. *Christians and Chiefs in Zimbabwe: A Social History of the Hwesa People c. 1870-1990s*, Edinburgh, International African Institute.
McClintock, A. 1995. *Imperial Leather: Race, Gender and Sexuality in the Colonial Conquest*, London,

Routledge.
McGaffey, J. 1991. *The Real Economy of Zaire*, Philadelphia, University of Pennsylvania Press.
McGregor, J. 2002. 'The Politics of Disruption: War Veterans and the Local State in Zimbabwe', *African Affairs*, 101, 402, pp. 9-37.
—2003. 'Living with the River: Landscape and Memory in the Zambezi Valley, Zimbabwe' in Beinart, W. and J. McGregor (eds), *Social History and African Environments*, Oxford, James Currey.
— 2003. 'The Victoria Falls 1900-1940: Landscape, Tourism and the Geographical Imagination', *Journal of Southern African Studies*, 29, 3, pp. 639-56.
— 2005. 'Landscape, Politics and the Historical Geography of Southern Africa', *Journal of Historical Geography* 31, pp. 205-19.
— 2005. 'The Social Life of Ruins: Sites of Memory and the Politics of a Zimbabwean Periphery', *Journal of Historical Geography*, 31, pp. 316-37.
— 2005. 'Crocodile Crimes: People versus Wildlife and the Politics of Postcolonial Conservation in Lake Kariba, Zimbabwe', *Geoforum*, 36, 3, pp. 353-69.
McGregor, J. and L. Schumaker, 2006. 'Heritage in Southern Africa: Imagining and Marketing Public Culture and History', *Journal of Southern African Studies*, 32, 4, pp. 649-66.
McKittrick, M. 2006. 'The Wealth of these Nations: Rain, Rulers and Religion on the Cuvelai Floodplain', in T. Tvedt and T. Oestigaard (eds), *A History of Water. Volume III: The World of Water*, London, IB Tauris.
Merrington, P. 2002. 'Cape to Cairo: Africa in Masonic Fantasy', in G. Harper (ed.), *Comedy, Fantasy and Colonialism*, London, Continuum, pp. 140-57
Miller, D.P. and P.H. Reill (eds), 1996. *Visions of Empire: Voyages, Botany and Representations of Nature*, Cambridge University Press.
Mitchell, D. 1996. *The Lie of the Land: Migrant Workers and the California Landscape*, Minneapolis, University of Minnesota Press.
Mitchell, D. 2003. 'California Living, California Dying: Dead Labor and the Political Economy of Landscape', in K. Anderson, S. Pile, and N. Thrift (eds), *Handbook of Cultural Geography*, London , Sage, pp. 233-48.
Mohr, E. 1876. *To the Victoria Falls of the Zambezi*, London, Sampson Low, Marston, Searle and Rivington.
Molyneux, A.J.C. 1905. 'The Physical History of the Victoria Falls' *Geographical Journal*, 25, 1, pp. 40-55.
Monk, W. (ed.), 1858. *Dr Livingstone's Cambridge Lectures*, Cambridge, Deighton Bell.
Moore, D. 2005. *Suffering for Territory: Race, Place and Power in Zimbabwe*, Durham NC, Duke University Press.
Moreau, J. 1950. 'Bakule Menyo', *Native Affairs Departmental Annual.*
Mosse, D. 2003. *The Rule of Water: Statecraft, Ecology and Collective Action*, New Delhi, Oxford University Press.
Mubitana, K. 1990. 'The Traditional History and Ethnography', in Phillipson, J. (ed.), *Mosi-oa-Tunya: A Handbook to the Victoria Falls*, Harare, Longman Zimbabwe.
Mulford, D. 1967. *Zambia: The Politics of Independence, 1957-1964*, Oxford, Oxford University Press.
Mumpande, I. 2000. *A Report on the Education Related Problems in Binga District*, Zimbabwe, Binga, Catholic Commission on Justice and Peace.
Munjeri, D. 1991 'Refocusing of Reorientation: The Exhibit or the Populace? Zimbabwe on the Threshold', in I. Karp and S. Lavine (eds), *The Poetics and Politics of Museum Display*, Washington, Smithsonian Institute Press.
Munjoma, L. n.d. 'Zambezi Watercourse Commission sets Transboundary Perspective', *The Zambezi*, vol 6, 1. SARDC.
Muntemba, M. 1970. 'The Political and Ritual Sovereignty Among the Leya of Zambia', in *Zambia Museum Journal*, 1, pp. 28-39.
Murphree, M. 1985. 'The Binga Inshore fishery' in M.F.C Bourdillon, A.P. Cheater and M.W. Murphree *Studies of Fishing on Lake Kariba*, Gweru, Mambo Press.
Murphy, P. (ed.) 2005. *British Documents on the End of Empire. Central Africa. Part I Closer Association 1945-1958. Part II Crisis and Dissolution 1959-1965*, London HMSO.

Muzondidya, J. 2005. 'Zimbabwe for Zimbabweans. Invisible Subject Minorities and the Quest for Justice and Reconciliation in Post-Colonial Zimbabwe', in B. Raftopoulos and T. Savage (eds), *Zimbabwe: Injustice and Political Reconciliation*, Cape Town, Institute for Justice and Reconciliation, pp. 213-23.

Nakayama, S. 2008. '"City Lights Emblaze Village Fishing Grounds": The Re-Imaginings of Waterscape by Lake Malawi Fishers', *JSAS*, 34, 4.

Nora, P. 1989. 'Between Memory and History: *Les Lieux de Memoires*', *Representations*, 26, pp. 7-25.

— (ed.), 2002. *Rethinking France: Les Lieux de Memoire. Vol 1 The State*, Chicago, Chicago University Press.

Nora, P., and L. Kritzman (eds), 1996. *Realms of Memory: The Construction of the French Past. Vols 1-3*, New York, Columbia University Press.

Nugent, P. and Asiwaju, A.I. 1996. *African Boundaries: Barriers, Conduits and Opportunities*, New York and London, Francis Pinter.

Nugent, P. 2003. *Smugglers, Secessionists and Loyal Citizens on the Ghala-Togo Frontier*, Oxford, James Currey.

Nyikahadzoi, K. 2003. 'Contesting Inequalities in Access Rights to Lake Kariba's *Kapenta* Fisheres: An Anaysis of the Politics of Natural Resource Management', in Geheb and Sarch (eds), *Africa's Inland Fisheries*, pp. 74-86.

Nyikahadzoi, K. 1995. 'Lake Kariba: Inshore Fishery Management: Experiences, Problems and Opportunities', Harare, Centre for Applied Social Studies.

— n.d. 'Multi-user conflicts in Lake Kariba', http://www.fao.org/fi/alcom/an21muc.htm [accessed 15 August 2006].

Oates, F. 1881. *Matabeleland and the Victoria Falls. A Naturalist's Wanderings in the Interior of South Africa*, London, Kegan Paul and Co.

Palley, C. 1966. *The Constitutional History and Law of Southern Rhodesia 1888-1965 With Special Reference to Imperial Control*, Oxford, Clarendon.

Palmer, R. and N. Parsons (eds), 1977. *The Roots of Rural Poverty in Central/Southern Africa*, London, Heinemann.

Parsons, N. 2006. 'Unravelling History and Cultural History in Botswana', *Journal of Southern African Studies*, 32, 4, pp. 667-83.

Peel, J. 1992. 'The Colonization of Consciousness', *Journal of African History*, 33, pp. 328-9.

— 1995. 'For Who Hath Despised the Day of Small Things? Missionary Narratives and Historical Anthropology', *Comparative Studies in Society and History*, 37, 3, pp. 581-607.

Phillips, R. 1997. *Mapping Men and Empire: A Geography of Adventure*, London, Routledge.

Phillipson, D.W. 1974. 'Iron Age History and Archaeology in Zambia', *Journal of African History*, 15, 1, pp.1-25.

— (ed.) 1990. *Mosi-oa-Tunya: A Handbook to the Victoria Falls*, Harare, Longman Zimbabwe.

— 1990 'The Early History of Livingstone Town' in D. W. Phillipson (ed.), *Mosi-oa-Tunya: A Handbook to the Victoria Falls*, Harare, Longman Zimbabwe.

Phimister, I. 1994. *Wangi Kolia: Coal, Capital and Labour in Colonial Zimbabwe 1894-1954*, Johannesburg, Witwatersrand University Press.

Pikirayi, I. 2006. 'The Kingdom, the Power and Forever More: Zimbabwe Culture in Contemporary Art and Architecture', *Journal of Southern African Studies*, 32, 4, pp. 755-70.

Posselt, F.W.T. 1935. *Fact and Fiction: A Short Account of the Natives of Southern Rhodesia* Bulawayo, Rhodesian Printing and Publishing Company.

Power, M. 2003. 'Geographers and the Tropics' in *Rethinking Development Geography*, London, Routledge.

Pratt, M.L. 1992. *Imperial Eyes: Travel Writing and Transculturation*, London, Routledge.

Raftopoulos, B. 1997. 'The Labour Movement in Zimbabwe 1945-65', in B. Raftopoulos and I. Phimister (eds), *Keep on Knocking: A History of the Labour Movement in Zimbabwe 1900-97*, Harare, Baobab Books.

— 2006. 'The Zimbabwean Crisis and the Challenges for the Left', *Journal of Southern African Studies*, 32, 2, pp. 203-19.

Raftopoulos, B. and I. Phimister, (eds) 1997. *Keep on Knocking: A History of the Labour Movement in Zimbabwe 1900-97*, Harare, Baobab Books.

Raftopoulos B. and T. Savage (eds), 2004. *Zimbabwe: Injustice and Political Reconciliation*, Cape Town, Institute for Justice and Reconciliation.
Ranger, T. O. 1960. *Crisis in Southern Rhodesia*, London, Fabian Commonwealth Bureau.
— 1981. 'Making Northern Rhodesia Imperial: Variations on a Royal Theme 1924-1938', *African Affairs*, pp. 349-73.
— 1994. 'African Identities: Ethnicity, Nationality and History. The Case of Matabeleland 1893-1993', in J. Heidrich (ed.) *Changing Identities: The Transformation of Asian and African Identities Under Colonialism*, Berlin, Centre for Modern Oriental Studies.
— 1999. *Voices from the Rocks: Nature, Culture and History in the Matopos Hills, Zimbabwe*, Oxford, James Currey.
— 2003. 'Women and Environment in African Religion', in W. Beinart and J. McGregor (eds), *Social History and African Environments*, Oxford, James Currey, pp. 72-86.
— 2004. 'Historiography, Patriotic History and the History of the Nation', *Journal of Southern African Studies*, 30, 2, pp. 215-34.
Reynolds, P. 1991. *Dance Civet Cat: Child Labour in the Zambezi Valley*, London, Zed Books.
Reynolds, R. and C. Crawford Cousins, 1993. *Lwaano Lwanyika: Tonga Book of the Earth*, London, Panos [1st ed., Save the Children Fund, Harare 1991].
Rich-Dorman, S. 2005. 'Democrats and Donors: Studying Democratization in Africa' in T. Kelsall and J. Igoe (eds) *Donors, NGOs and the Liberal Agenda in Africa*, Durham NC, North Carolina Press, pp. 33-59.
Roberts, A.D. 1968. 'The Nineteenth Century in Zambia', in Ranger, T. (ed.), *Aspects of Central African History*, London, Heinemann.
Robins, E. and Legge, R. 1959. *Animal Dunkirk: The Story of Lake Kariba and "Operation Noah" the Greatest Animal Rescue Since the Ark*, London, Herbert Jenkins.
Roitman, J. 2006. 'The Ethics of Illegality in the Chad Basin', in J. Comaroff and J. Comaroff (eds), *Law and Disorder in the Postcolony*, University of Chicago Press.
— 1994. 'Productivity in the Margins: The Reconstitution of State Power in the Chad Basin', in V. Das and D. Poole (eds), *Anthropology in the Margins of the State*, Oxford, James Currey.
Rotberg, T. 1966. *The Rise of Nationalism in Central Africa: The Making of Malawi and Zambia 1873-1964*, Cambridge Massachusetts, Harvard University Press.
Rotberg, R. 1988. *The Founder. Cecil Rhodes and the Pursuit of Power*, Oxford University Press.
Said, E. 1993. *Culture and Imperialism*, London, Vintage.
Samuel, R. 1998. *Island Stories: Unravelling Britain. Theatres of Memory Volume II*, London, Verso.
Schama, S. 1995. *Landscape and Memory*, London, Fontana.
Schapera, I. (ed.), 1959. *David Livingstone: Family Letters, 1841-1856*, London, Chatto and Windus.
— 1960. *Livingstone's Private Journals 1851-53*, London, Chatto and Windus.
Schapera, I. (ed.), 1963. *Livingstone's African Journal 1853-6*, Vols 1 and 2, London, Chatto and Windus.
Schumaker, L, 2001. *Africanizing Anthropology: Fieldwork, Networks and the Making of Cultural Knowledge in Central Africa*, Durham, Duke University Press.
Scudder, T. 1962. *The Ecology of the Gwembe Tonga*, Manchester, Manchester University Press.
— 1965. 'The Kariba Dam; Manmade Lakes and Resource Development in Africa', *Bulletin of Atomic Scientists*, December 1965, pp. 6-11.
— 2005. *The Future of Large Dams: Dealing With Social, Environmental, Institutional and Political Costs*, London, Earthscan.
Scudder, T. and E. Colson, 2002. 'Long Term Research in Gwembe Valley, Zambia', in R.V. Kemper and A. P. Royce (eds), *Chronicling Cultures: Longterm Field Research in Anthropology*, New York and London, Altamira Press.
Selous, F.C. 1911. *A Hunter's Wanderings in Africa* [first ed. 1881], London, Macmillan.
— 1893. *Travel and Adventure in South East Africa*, London, Fowland Ward and Co.
Simmons, J. 1955. *Livingstone and Africa*, English Universities Press.
Smith, E.W. 1956. 'Sebetwane and the Makololo', *African Studies*, 15, 2, pp. 49-74.
Smith, E.W. and A.M. Dale. 1920. *The Ila Speaking Peoples of Northern Rhodesia*, London, Macmillan.
Solway, J. 2004. 'Reaching the Limits of Universal Citizenship: 'Minority' Struggles in Botswana' in

Berman et al. *Ethnicity and Democracy*, pp. 129-47.
Stewart, P. and Andrew Strathern (eds), 2003. 'Introduction', *Landscape, Memory and History: Anthropological Perspectives*, London, Pluto.
Stirke, D.W. 1922. *Eight Years Among the Barotse*, London, John Bale Sons and Danielsson Ltd.
Stokes, E. 1966. 'Barotseland: The Survival of an African State', in E. Stokes and R. Brown, *The Zambezian Past: Studies in Central African History*, Manchester, Manchester University Press.
Stoler, L.A. 1989. 'Rethinking Colonial Categories: European Communities and the Boundaries of Rule', *Comparative Studies in Society and History*, 31, pp. 134-61.
Stone, J. 1995. *A Short History of the Cartography of Africa*, Lampeter, Mellen Press.
Symausoonde, J. 1947. *Naakoyo Wwamba Cano Cakwe (Old Naakoyo Tells Her Story)*, Ndola, African Literature Committee of Northern Rhodesia.
Sykes, F. 1898. *With Plumer in Matabeleland*, London, Constable.
—1905. *Official Guide to the Victoria Falls*, Bulawayo, Argus Co.
Tabler, E. (ed.), 1963. *Trade and Travel in Early Barotseland: The Diaries of George Westbeech, 1885-6 and Captain Norman Macleod, 1875-76*, London, Chatto and Windus.
Tabler, E. (ed.), 1967. *To the Victoria Falls via Matabeleland. The Diary of Major Henry Stabb, 1875*, Cape Town, Struik.
Taylor, S. 1897. *The Mighty Nimrod: A Life of Frederick Courteney Selous*, London, Collins.
Teede J. and F. Teede, 1991. *Zambezi: River of the Gods*, London, Andre Deutsch.
Terorde, P. 1881. *Vom Cap Zum Sambezi*, Freiburg im Briesgau.
Thomas, N. 1994. *Colonialism's Culture: Anthropology, Travel and Government*, London, Polity Press.
Thomas, T. M. 1873. *Eleven Years in Central Africa*, London, John Snow.
Thorpe, C. 1951. *Limpopo to Zambezi: Sixty Years of Methodism in Southern Rhodesia*, London, The Cargate Press.
Tilley, H. 2003. 'African Environments and Environmental Sciences: The African Research Survey, Ecological Paradigms and British Colonial Development 1920-1940', in W. Beinart and J. McGregor (eds), *Social History and African Environments*, pp. 109-30.
Torrend, J. 1891. *A Comparative Grammar of the South-African Bantu Languages*, London, Kegan Paul, Trench, Trubner.
Tremmel, M. 1994. *The People of the Great River: The Tonga Hoped the Water Would Follow Them*, Gweru, Mambo Press.
Turner, V. 1951. *The Luyana Peoples of Barotseland*, Ethnographic Survey of Africa, London, Oxford University Press.
Tvedt, T. and E. Jakobsson. 2006. 'Introduction: Water History is World History', in T. Tvedt and E. Jakobsson (eds.) *A History of Water: Water Control and River Biographies*, London, I.B. Tauris.
Tveten, I. 2002. '"If You Don't Fish, You Are Not a Caprivian": Freshwater Fisheries in Caprivi, Namibia', *Journal of Southern African Studies*, 28, 2, pp. 421-41.
Van Binsbergen, W.M. 1992. *Tears of Rain: Ethnicity and History in Central Western Zambia*, London and New York, Kegan Paul International.
Van Onselen, C. 1976. *Chibaro African Mine Labour in Southern Rhodesia, 1900-1933*, London, Pluto Press.
—1974. 'The 1912 Wankie Colliery Strike', *Journal of African History*, 15, 2, pp. 275-89.
Van Oppen, A. and U. Luig. 1997. 'The Making of African Landscapes', *Paideuma: Mitteilungen zur Kulturkunde* 43, pp. 7-46.
Van Sittert, L. 2005. 'The Nature of Power: Cape Environmental History, the History of Ideas and Neoliberal Historiography', *Journal of African History*, 46, 1, pp 127-34.
Vickery, K. 1986. *Black and White in Southern Zambia. The Tonga Plateau Economy and British Imperialism 1890-1939*, New York, Greenwood Press.
Wallis, J.P.R. (ed.), 1941. *Thomas Baines of Kings Lynn. Explorer and Artist 1821*-1875, London, Jonathan Cape.
—(ed.) 1953. *The Barotseland Journal of James Stevenson-Hamilton 1898-1899*, London, Chatto and Windus.
—1956. *The Zambezi Expedition of David Livingstone II Vols*, London, Chatto and Windus.

Weinrich, A.K.H. 1977. *The Tonga People on the Southern Shore of Lake Kariba*, Gweru, Mambo Press.
Wellington, J.H. 1955. *Southern Africa: A Geographical Study*, Cambridge, Cambridge University Press.
Werbner, R. 1991. *Tears of the Dead: A Social Biography of an African Family*, Edinburgh, Edinburgh University Press.
Werbner, R. 2004. *Reasonable Radicals and Citizenship in Botswana: The Public Anthropology of Kalanga Elites*, Bloomington IN, Indiana University Press.
West, M. 2002. *The Rise of an African Middle Class: Colonial Zimbabwe 1898-1965*, Bloomington, Indiana University Press.
Whatmore, S. 2002. *Hybrid Geographies: Natures, Cultures, Spaces*, London, Sage.
White, L. 2001. 'True Stories: Narrative, Event, History and Blood in the Lake Victoria Basin', in L. White, S. Miescher and D. Cohen, (eds), 2001, *African Words, African Voices: Critical Practices in Oral History*, Bloomington and Indianopolis, Indiana University Press.
White, L. S. Miescher and D. Cohen (eds), 2001. *African Words, African Voices: Critical Practices in Oral History*, Bloomington and Indianopolis, Indiana University Press.
Wills, A.J. 1967. *An Introduction to the History of Central Africa*, London, Oxford University Press.
Wilson, T. and H. Donnan (eds), 1998. *Border Identities: Nation and State at International Frontiers* Cambridge, Cambridge University Press.
Wolmer, W. 2006. *From Wilderness Vision to Land Invasions: Conservation and Development in Zimbabwe's South East Lowveld*, Oxford, James Currey.
Wood, J.R.T. 1983. *The Welensky Papers: A History of the Federation of Rhodesia and Nyasaland*, Durban, Graham Publishing.
Woods, J. and S. Manning, 1960. *A Guide Book to the Victoria Falls*, Bulawayo.
Woods, M. 1997. 'Researching Rural Conflicts: Hunting, Local Politics and Actor-Networks', *Journal of Rural Studies*, 14, 3, pp. 321-40.
Worby, E. 1994. 'Maps, Names and Ethnic Games: The Epistemology and Iconography of Colonial Power in Northwestern Zimbabwe', *Journal of Southern African Studies*, 20, 3, pp. 371-92.
Worby, E. 1998. 'Inscribing the State at the "Edge of Beyond"; Danger and Development in Northwestern Zimbabwe', *Political and Legal Anthropology Review*, 21, 2, pp. 55-70.
World Commission on Dams Secretariat [WCD], 2000. *Kariba Dam Case Study: Zimbabwe and Zimbabwe*, final draft September, Cape Town, World Commission on Dams Secretariat.
Worster, D. 1985. *Rivers of Empire: Water, Aridity and the Growth of the American West*, New York, Hill and Wang.
Youngs, T. 1994. *Travellers in Africa: British Travelogues, 1850-1900*, Manchester University Press.
Zambezi Mission Record Editor, 1902. 'History of the Zambezi Mission', *Zambezi Mission Record*, 2, 17.

UNPUBLISHED SOURCES

Beccerril, F.A. 1988. A Brief History of Hwange Diocese, unpublished ms.
Beinart W. with K. Brown and D. Gilfoyle, 2005. 'Experts and Expertise in Colonial Africa Reassessed: Colonial Environmental Science and the Interpenetration of Knowledge', seminar paper, presented at Emory University.
Bolding, A. 2004. *In Hot Water: A Study on Sociotechnical Intervention Models and Practices of Water Use in Small Holder Agriculture, Nyanyadzi Catchment*, Zimbabwe, PhD thesis, Wageningen.
Chikumbi, D. C. 1998. 'A Specialist Study on Archaeology and History Issues', in *The Environmental Impact Assessment on the Proposed Sun International Hotel Development at Victoria Falls, Zambia*. Report available from National Heritage Conservation Commission, Livingstone.
Cobbing, J. 1976. 'The Ndebele Under the Khumalos 1820-1896', DPhil, University of London.
Dahl, J. 1997. 'A Cry for Water: Perceptions of Development in Binga District, Zimbabwe', PhD thesis, Dept of Geography, University of Goteborg, Series B, no. 92.
Glyn, Capt. R.G., 1863. *Journal of His Expedition from Durban to Victoria Falls on the Zambezi*. Copy of manuscript, held at Zimbabwe National Archives.
Holst, R. 1983. 'Continuity and Change: The Dynamics of Social and Productive Relations in

Hwange, Zimbabwe, 1870-1960', MPhil, University of Manchester.
Husbands, W. 1994. 'Nature, Society and the Origins of Tourism at the Victoria Falls', unpublished paper.
Jalla, A.D. 1920. *Litaba za Sicaba sa Malozi* (London), translated transcript by S.D. Jones, 'History, Traditions and Legends of the Barotse Nation', [n.d.] available from Royal Commonwealth Society Library, London.
Kariba Lakeshore Combination Masterplan Preparation Authority, 1998. 'Kariba Lakeshore Combination Masterpoan', final draft.
Kumirai, A., S. J. Muringaniza and D. Munyikwa, 'Victoria Falls/Mosi-oa-Tunya World Heritage Site (Zambia and Zimbabwe): Issues and Values', paper presented at UNESCO meeting on Authenticity/Integrity in an African Context, 29-30 May 2000, Masvingo, Zimbabwe.
Manyena S.B., 2003. 'Missing the Tonga. The Impact of the Land Reform Programme on Binga District', unpublished report.
Matthews, T. 1976. 'The Historical Traditions of the Peoples of the Gwembe Valley, Middle Zambezi', unpublished PhD thesis, University of London.
Mavhunga, C. 2003. '"Sold Down the River"? Forced Resettlement and Landscape Transformation: Lessons from Kariba Dam, 1950-1963', unpublished paper.
McGregor, J. 2005. 'Notes on the Use of Diaries', unpublished paper.
Mufuzi, F. 2004. 'The Dilemma Faced by African National Museums in the Quest to Mount Public Displays that Seek to Promote National Unity and Identity: The Case of the Livingstone Museum', paper presented to the conference 'Heritage in Southern and Eastern Africa: Imagining and Marketing Public Culture and History, Livingstone, July 2004.
Mukuni, Chief (Siloka II), 1957. 'Short History of the Baleya People of Kalomo District', unpublished manuscript, held at Livingstone Museum.
Muringaniza, J. 1995. 'Report on the Cultural Resources on the Zimbabwean Side', paper prepared for IUCN Strategic Environmental Assessment of Developments Around the Victoria Falls Area.
Nyikahadzoi, K. 2005. 'Inequalities and Rule Conformance in the Management of the Kapenta Fishery at Lake Kariba, Zimbabwe', unpublished paper.
Nambya Cultural Association, n.d [ca. 1975]. 'The BaNambya Tribe', unpublished ms.
Ncube, G. 1986. 'A Study of the Tonga People of Sebungwe District (Binga District) under BSAC rule (1896-1923) and Their Response to Colonization', BA Honours Dissertation, UZ.
— 1994. 'A History of Northwest Zimbabwe: Comparative Change in Three Worlds', MPhil thesis, University of Zimbabwe.
Nkomo, N. n.d. 'The Hammar and the Anvil: The Autobiography of Nicholas Nkomo', unpublished ms.
Ranger, T.O., n.d., 'No Missionary, No Exchange, No Story? Narrative in Southern Africa', unpublished paper.
— 1997. 'New Approaches to African Landscape', unpublished paper.
Scudder, T. n.d. 'The rise and fall of the Kariba Lake Fisheries: The Tonga Response (1959-1970)', unpublished paper, available from Binga District Library.
Steele, M. 1972. 'The Foundation of a "Native Policy": Southern Rhodesia, 1923-1933', DPhil thesis, Simon Fraser University.
Thornton, S. 1978. 'Municipal Employment in Bulawayo, 1895-1935', unpublished seminar paper, University of York.
Warndoff, T. and A. Gonga, 1994. *Nutritional and Health Status of Young Children and Household Level Food Security During a Severe Drought in Binga District*; Binga District Food and Nutrition Committee During April/May 1992, Harare, Save the Children Fund.
Watt, D. n.d. 'Thesis on the Town of Livingstone', thesis ms, Livingstone Museum [cover/title page missing].
Zambezi Valley Consultants, 2000. 'Report on a Study on Land Use and Wildlife Management in Binga District', report for USAID/WWF MAPS Project, Harare.

Index

Lightning Source UK Ltd.
Milton Keynes UK
UKHW021550280422
402157UK00003B/199